Plant Phosphorus Nutrition

This book is an up-to-date reference on phosphorus nutrition in plants. Phosphorus has no substitute in food production, and the use of phosphate (Pi) fertilisers has increased crop yields to feed billions of people. This book covers phosphorus metabolism and phosphorus sensing molecular mechanisms and signalling in plants. It covers functions of phosphorus and crosstalk with other nutrients. It discusses how plants sense Pi deficiency and coordinate the responses via signalling pathways and networks for the regulation of Pi-deficiency responses.

FEATURES

- Discusses the latest developments in phosphate management in plants
- Provides insights on emerging topics for sustainable approaches to managing phosphate shortage
- Throws light on the resilience of plants to phosphate deficiency
- Provides extensive updates that serve as primary points for further research
- Explains molecular and physiological mechanisms of phosphate transport

This book compiles the latest research from experts in the field. It is useful for advanced graduates and researchers in plant sciences and agriculture.

Plant Phosphorus Nutrition

Edited by
Hatem Rouached
INRAe, Montpellier, France and Michigan State University, USA

Santosh B. Satbhai
Indian Institute of Science Education and Research (IISER), Mohali, India

CRC Press is an imprint of the
Taylor & Francis Group, an **informa** business

First edition published 2024
by CRC Press
2385 NW Executive Center Drive, Suite 320, Boca Raton FL 33431

and by CRC Press
4 Park Square, Milton Park, Abingdon, Oxon, OX14 4RN

CRC Press is an imprint of Taylor & Francis Group, LLC

© 2024 selection and editorial matter, Hatem Rouached and Santosh B. Satbhai; individual chapters, the contributors

Reasonable efforts have been made to publish reliable data and information, but the author and publisher cannot assume responsibility for the validity of all materials or the consequences of their use. The authors and publishers have attempted to trace the copyright holders of all material reproduced in this publication and apologize to copyright holders if permission to publish in this form has not been obtained. If any copyright material has not been acknowledged please write and let us know so we may rectify in any future reprint.

Except as permitted under U.S. Copyright Law, no part of this book may be reprinted, reproduced, transmitted, or utilized in any form by any electronic, mechanical, or other means, now known or hereafter invented, including photocopying, microfilming, and recording, or in any information storage or retrieval system, without written permission from the publishers.

For permission to photocopy or use material electronically from this work, access www.copyright.com or contact the Copyright Clearance Center, Inc. (CCC), 222 Rosewood Drive, Danvers, MA 01923, 978-750-8400. For works that are not available on CCC please contact mpkbookspermissions@tandf.co.uk

Trademark notice: Product or corporate names may be trademarks or registered trademarks and are used only for identification and explanation without intent to infringe.

ISBN: 978-1-032-51652-3 (hbk)
ISBN: 978-1-032-57595-7 (pbk)
ISBN: 978-1-003-44007-9 (ebk)

DOI: 10.1201/9781003440079

Typeset in Times
by Deanta Global Publishing Services, Chennai, India

Contents

Foreword vii
Editors ix
Contributors xi

Chapter 1   Phosphorus Management for Agriculture and the Environment 1

Debolina Chakraborty and Rishi Prasad

Chapter 2   Molecular Mechanisms Underlying Phosphate Sensing and Signalling in Plants..... 18

Viswanathan Satheesh, Abdul Wahab, Muthusamy Ramakrishnan, and Mingguang Lei

Chapter 3   Phosphate Homeostasis and Root Development in Crop Plants 30

Dhriti Singh and Santosh B. Satbhai

Chapter 4   Crosstalk between Phosphate and Other Nutrients 48

Xianqing Jia, Long Wang, and Keke Yi

Chapter 5   Hormonal Control of Phosphate Uptake and Assimilation... 59

Shreya Gupta and Amar Pal Singh

Chapter 6   The Pivotal Role of Phosphate in Shaping Beneficial Plant-Microbe Interactions.... 83

Arianna Capparotto and Marco Giovannetti

Chapter 7   Phosphorus and Plant Immunity 98

Anurag Kashyap, Swagata Saikia, Shenaz Sultana Ahmed, and Munmi Sarma

Chapter 8   Biotechnological Approaches for Improving Phosphate Uptake and Assimilation in Plants 110

Rumi Rumi, Kanika Maurya, Mandavi Pandey, Pawandeep S. Kohli, Poonam Panchal, Alok K. Sinha, and Jitender Giri

Chapter 9 Analysis and Comparison of Alphafold-Structure Predictions between Pi-Uptake Transporters Recovering Phosphate in Natural Environments 129

Nussaume Laurent, Desnos Thierry, Jinsheng Zhu, David Pascale, Kumiko Miwa, and Kanno Satomi

Chapter 10 A General Perspective of Phosphorus Research in Plants 151

Hatem Rouached

Index 155

Foreword

Nitrogen and phosphorus (P) are the main elements limiting plant growth in both agricultural and natural ecosystems. Plants essentially acquire P from the soil as inorganic phosphate (Pi; PO_4^{3-}). Although P is an abundant element in soils, the concentration of soluble Pi found in soil solution, the only form that plant roots can acquire, is kept at very low concentrations because Pi is strongly sorbed onto soil particles and forms insoluble complexes with iron, aluminium, and calcium. This is why the use of fertilisers rich in readily soluble Pi is of primordial importance in agriculture, allowing farmers in many regions of the world to optimise crop yield. However, in contrast to nitrate and ammonium used in fertilisers, which are derived from the chemical conversion of atmospheric nitrogen, Pi is mined from P-rich deposits and thus represents a finite resource. Considering the strategic importance of P in agriculture, it is striking that P mining is essentially concentrated in very few regions of the world, with Northern Africa (Morocco, Tunisia, Algeria) and China representing more than 65% of the global production. Recent geopolitical shockwaves on energy and food prices resulting from the war in Ukraine have reminded us of the problematic outcome of the high concentration of essential goods in a few countries. Because Pi is also a limiting factor for the growth of aquatic plants, increased input of Pi into lakes and rivers from agricultural fertiliser runoff contributes to massive algal blooms, resulting in the eutrophication of large bodies of water. Management of Pi resources by plants and farmers is thus essential for both the health of the planet and its people.

This book provides a timely review of a variety of topics essential for our understanding of fundamental aspects of Pi nutrition in plants and its practical impacts on agriculture and our environment. Sustainable use of P resources, both in the context of fertiliser use and its environmental impact on water systems, will necessitate the adoption of optimal P management strategies (Chapter 1). It will also depend on the breeding of new crops that will be able to maintain maximal yield under a reduced input of P fertilisers. This represents an important shift since traditional breeding goals have typically aimed at maximising yield under a nutrient-rich environment. In this area, the development of novel breeding strategies involving genomics and gene-editing technologies should allow breeders to develop more P-resilient crops (Chapter 8). Such resiliency is likely to be mediated, in part, through the development of a robust root system in crops capable of maximising the extraction of this vital nutrient from the soil. This in turn will depend on gaining a better understanding of the ways the plant root system adapts in response to Pi availability (Chapter 3) as well as how transporters can acquire Pi via a deeper knowledge of their structure-function parameters (Chapter 9).

Plant adaptation to Pi deficiency relies on an elaborate network of proteins involved in sensing the Pi supply and modifying plant development and metabolism to sustain growth and survival under such stress. The last decade has seen important progress in our understanding of the proteins and metabolites involved in mediating Pi sensing and signalling, including the discovery of the interaction between inositol pyrophosphate, SPX proteins, and key transcription factors participating in this network (Chapter 2). In nature, plants are typically subjected to multiple concomitant stressors. There has thus been a recent shift towards studying how plants adapt to Pi deficiency in combination with other stressors. On the abiotic side, recent work has revealed the presence of intricate crosstalk between pathways involved in adaptation to distinct nutrient deficiencies or excess, such as nitrogen, iron, and phosphate (Chapter 4). Very exciting new fundamental knowledge on how Pi nutrition impacts the adaptation of plants to biotic stress, such as plant pathogens and herbivory, has been gained, showing that the plant Pi status impacts numerous pathways involved in plant immunity as well as influencing the plant-associated microbiome (Chapters 6 and 7). The multidimensional repercussions of the plant Pi status on its development, metabolism, and the activity of pathways impacting biotic and abiotic stress responses implicate the modulation of several phytohormones that have complex antagonistic and synergistic interactions (Chapter 5).

The topics assembled in this book will be valuable to both foundational researchers as well as breeders and biotechnologists to gain an overall view of the progress accomplished in the last decade on plant adaptation to Pi deficiency and provide leads as to how this knowledge can be used to develop future crops capable of combing high yield with reduced fertiliser input.

Yves Poirier
Department of Plant Molecular Biology
University of Lausanne
Lausanne, Switzerland

Editors

Hatem Rouached gained scientific and managerial skills during his PhD (University of Montpellier II, France, 2002–2005) and postdoc (University of Lausanne, Switzerland, 2005–2009), and as Senior Scientist in Crop Design (BASF Company, Ghent, Belgium, 2010–2012). In 2010, Hatem obtained the academic diploma for the ability to supervise PhD students (HDR University of Bourgogne, France). During his PhD and postdoctoral period, he acquired expertise in the areas of molecular and physiological aspects of plant nutrition. In particular, he studied the molecular mechanisms controlling the sulfate and phosphate transport and signalling in *Arabidopsis* through the study of the SULTR and PHO1 gene family. In 2012, he was recruited to INRA, Montpellier, France, as a permanent researcher. Since then, he developed an original research programme in the Biochemistry and Molecular Biology of Plants (B&PMP) Research Unit, aiming to decode the genetic and molecular basis of the interaction between the homeostasis of macro- and micro-nutrients, particularly phosphates and metals in plants. Hatem is currently furthering this research topic at Michigan State University, at the Plant Resilience Institute, East Lansing, Michigan, United States. He is also active in review and editorial services, and is an editor for *The Plant Journal*, *Critical Reviews in Biotechnology*, *Scientific Reports*, *Frontiers in Plant Science*, *PLOS One*, and the *International Journal of Molecular Science*. In addition, Hatem has organised several special issues in the plant mineral nutrition field.

Santosh B. Satbhai is currently working as an assistant professor in the Department of Biological Sciences, Indian Institute of Science Education and Research (IISER), Mohali, India. He earned his doctoral degree from Nagoya University, Japan. He has published papers in several reputed international journals such as *Nature Genetics*, *Nature Communications*, *Nature*, *DNA Research*, *PLOS Genetics*, *Journal of Experimental Botany*, *Annals of Botany*, and more. Additionally, he has also published book chapters in the "Methods in Molecular Biology" series (Springer). His findings (published in *Nature Communications*) were covered in many online news media websites. How plant growth and development are influenced and regulated by environmental stresses is a major research question in his laboratory. He was awarded the prestigious Ramalingaswami Re-entry Fellowship by the Department of Biotechnology (DBT), Ministry of Science and Technology, Government of India. Dr. Satbhai is an active member of several professional scientific societies, such as the American Society of Plant Biologists (ASPB), the Plantae Community, and the Asia Association of Plant Scientists (AAPS).

Contributors

Shenaz Sultana Ahmed
Department of Plant Pathology
Assam Agricultural University
Assam, India

Arianna Capparotto
Department of Biology
University of Padua
Padua, Italy

Debolina Chakraborty
Department of Crop, Soil, and Environmental
 Sciences
Auburn University
Auburn, Alabama, United States

Marco Giovannetti
Department of Biology
University of Padua
Padua, Italy
Department of Life Sciences and Systems
 Biology
University of Turin
Turin, Italy

Jitender Giri
National Institute of Plant Genome Research
New Delhi, India

Shreya Gupta
National Institute of Plant Genome Research
New Delhi, India

Xianqing Jia
State Key Laboratory of Efficient Utilization
 of Arid and Semi-arid Arable Land in
 Northern China
Institute of Agricultural Resources and
 Regional Planning
Chinese Academy of Agricultural Sciences
Beijing, China

Anurag Kashyap
Department of Plant Pathology
Assam Agricultural University
Assam, India

Pawandeep S. Kohli
National Institute of Plant Genome Research
New Delhi, India

Nussaume Laurent
CEA, CNRS,
Aix Marseille University, UMR7265, BIAM
St Paul lez Durance, France

Mingguang Lei
Shanghai Center for Plant Stress Biology
CAS Center for Excellence in Molecular Plant
 Sciences
Chinese Academy of Sciences
Shanghai, China

Kanika Maurya
National Institute of Plant Genome Research
New Delhi, India

Kumiko Miwa
Institute for Advanced Research
and
Graduate School of Science
Nagoya University
Nagoya, Japan

Poonam Panchal
National Institute of Plant Genome Research
New Delhi, India

Mandavi Pandey
National Institute of Plant Genome Research
New Delhi, India

David Pascale
CEA, CNRS, Aix-Marseille University,
 UMR7265, BIAM, St Paul lez Durance,
 France

Rishi Prasad
Department of Crop, Soil, and Environmental
 Sciences
and
Department of Animal Science
Auburn University
Auburn, Alabama, United States

Muthusamy Ramakrishnan
Co-Innovation Center for Sustainable Forestry in Southern China
and
Bamboo Research Institute
Nanjing Forestry University
Nanjing, China

Hatem Rouached
Plant Resilience Institute
and
Department of Plant, Soil, and Microbial Sciences
Michigan State University
East Lansing, Michigan, United States

Rumi Rumi
National Institute of Plant Genome Research
New Delhi, India

Swagata Saikia
Department of Plant Pathology
Assam Agricultural University
Assam, India

Munmi Sarma
Department of Chemical Sciences
Tezpur University
Assam, India

Santosh B. Satbhai
Indian Institute of Science Education and Research
Mohali, Punjab, India

Viswanathan Satheesh
Shanghai Center for Plant Stress Biology
CAS Center for Excellence in Molecular Plant Sciences
Chinese Academy of Sciences
Shanghai, China
and
Genome Informatics Facility
Office of Biotechnology
Iowa State University
Ames, Iowa, United States

Kanno Satomi
Institute for Advanced Research
Nagoya University
Nagoya, Japan

Amar Pal Singh
National Institute of Plant Genome Research
New Delhi, India

Dhriti Singh
Indian Institute of Science Education and Research
Mohali, Punjab, India

Alok K. Sinha
National Institute of Plant Genome Research
New Delhi, India

Desnos Thierry
CEA, CNRS, Aix-Marseille University, UMR7265, BIAM, St Paul lez Durance, France

Abdul Wahab
Shanghai Center for Plant Stress Biology
CAS Center for Excellence in Molecular Plant Sciences
Chinese Academy of Sciences
Shanghai, China

Long Wang
State Key Laboratory of Efficient Utilization of Arid and Semi-arid Arable Land in Northern China
Institute of Agricultural Resources and Regional Planning
Chinese Academy of Agricultural Sciences
Beijing, China

Keke Yi
State Key Laboratory of Efficient Utilization of Arid and Semi-arid Arable Land in Northern China
Institute of Agricultural Resources and Regional Planning
Chinese Academy of Agricultural Sciences
Beijing, China

Jinsheng Zhu
CEA, CNRS, BIAM, UMR7265
Aix Marseille University
Saint-Paul-lez-Durance, France

1 Phosphorus Management for Agriculture and the Environment

Debolina Chakraborty and Rishi Prasad

1.1 SOIL PHOSPHORUS: OVERVIEW

Phosphorus (P) is an essential nutrient for plant growth and is the second most limiting macronutrient after nitrogen (N). Up to 0.5% of P is present in plant dry weight as an integral component of several plant structures, and it helps in plant processes such as photosynthesis, respiration, energy storage, and transfer (Vance et al., 2003). Phosphorus deficiency can reduce plant growth and can potentially limit crop yield; however, excess P in soil is a potential threat to water quality. Sources of P to fresh water can be categorised as a) point sources and b) nonpoint sources. Point sources, such as pipes that drain into rivers and streams from industry or sewage treatment, are much easier to identify compared to the nonpoint sources, such as surface and subsurface loss of P from agricultural lands, which is usually considered the primary nonpoint source (USEPA, 2013). Phosphorus loss from agricultural soils has been recognised as one of the primary causes of deterioration of surface water quality in the United States (USEPA, 2000; Sharpley et al., 2013). Excess P from agricultural lands can enter freshwater bodies through a wide range of hydrologic processes such as erosion, interflow, overland flow, matrix flow, or preferential flow, and can increase the fertility status of natural water (eutrophication), causing excessive algal growth. The eutrophication process can eventually result in the death of fish, reduce the recreational value of the water body, and render water unfit for human consumption. Purification of water to remove algal toxins is exceptionally costly. An estimated $2.2 billion is spent annually on cleaning up the direct effects of eutrophication in the United States (Dodds et al., 2009).

Phosphorus exported during agricultural runoff is a combination of dissolved P and particulate P (derived from eroded soil particles). Dissolved P is predominantly phosphate ions (dissolved reactive P, DRP) along with P sorbed to colloidal particles, organic P compounds, and non-reactive mineral forms, including polyphosphates and phosphonates (Reid et al., 2018). Most of the P in soil is in the particulate form, which led to the misconception that controlling soil erosion would effectively control agriculture's contribution to P loss. In reality, a significant portion of P losses can be in the dissolved form (Baker et al., 2014; Joosse and Baker, 2011). Since DRP is readily bioavailable, it primarily drives the algal blooms. Phosphorus concentrations that cause eutrophication can range from 0.01 to 0.03 ppm (Sharpley et al., 1996; USEPA, 1986). For effective soil P management and to prevent negative impact on the water quality, there is a need to understand soil P forms, transformation, and their cycling in soil.

Phosphorus in the soil exists in organic and inorganic forms. In most soils, 50 to 75% of the P is inorganic. Organic P compounds range from stable compounds that have become part of the soil organic matter to readily available undecomposed plant residues and microbes. Predominant organic P forms include inositol phosphates, phospholipids, nucleic acids, and their derivatives (Stevenson, 1982). Inorganic P compounds range from labile (soluble) P to moderately labile calcium

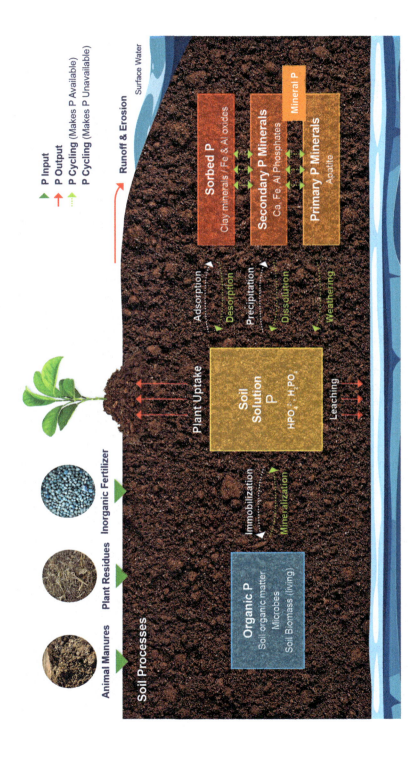

FIGURE 1.1 The soil phosphorus (P) cycle (adapted from Prasad and Chakraborty, 2019).

(Ca) phosphates and very stable iron (Fe) and aluminium (Al) oxides. Plants can absorb P only in the form of orthophosphate that is dissolved in the soil solution. The quantity of P in the soil solution at a given time is < 1% of the total quantity of P in the soil (Pierzynski, 1991). Hence, to meet plant P demands, P in the soil solution must be replenished several times over the life cycle of the plant.

Once external P enters the soil through an organic (such as manure, biosolids, dead plants, animal debris) or inorganic source (such as chemical fertilisers), the P cycles between several soil P pools through chemical and biochemical processes (sorption–desorption, precipitation–dissolution, weathering, and mineralisation–immobilisation). The added P can exist in the soil as phosphate (HPO_4^{2-} or $H_2PO_4^-$) within soil solution, organic P, sorbed P, or mineral P. The P cycle (Figure 1.1) shows the pathways through which P can be added and removed from the soil, along with the P forms and the different P transformation processes.

Soil contains minerals (primary and secondary) that can break down over time by **weathering**, and P is released in the soil solution for plant uptake. **Mineralisation** is another process that releases inorganic P forms (HPO_4^{2-} or $H_2PO_4^-$) through microbial conversion of organic P. In contrast, **immobilisation** is the biological conversion of inorganic P to organic P in the soil. As the microbes die, this microbial P will become available to the plant over time. Inorganic P is also released from soil organic matter or during the decomposition of plant residues. Mineralisation and immobilisation of P are affected by temperature, moisture, aeration, and pH. **Sorption** refers to the process in which P binds to the soil particles. Since phosphate is negatively charged, it is attracted and binds to positively charged minerals such as Fe and Al oxides and hydroxides. **Desorption** is usually a slow process and refers to the release of sorbed P into the soil solution. Sorption and desorption reactions equilibrate within the soil solution. The sorption process involves two steps: a) adsorption and accumulation of P on the surfaces of solid particles and soil constituents, and b) absorption and diffusion of P into solid and soil matrix (Corey, 1981; Sposito, 1986). **Precipitation** is a slow process involving a permanent change into metal phosphates. Phosphorus can become unavailable if plant-available inorganic P reacts with dissolved Fe, Al, or Ca to form phosphate minerals. These metal phosphates can release P in soil solution upon dissolution; however, the process is very slow. Phosphorus availability is driven by soil pH. Optimum soil pH between 6 and 7 will result in maximum P availability. At low pH, P tends to be sorbed with Fe and Al compounds in the soil, whereas, at high pH, P can precipitate with Ca.

The processes contributing to P removal from the soil are a) plant uptake, b) runoff and erosion, and c) leaching. Plants can take a considerable amount of P throughout their life cycle, although the concentration of phosphate at a given time in the soil solution is low. For example, total P uptake for grain crops depends on yield and can range from 4.4 lbs acre^{-1} P in flax grain (15 bsh acre^{-1}) to 23 lbs acre^{-1} in corn grain (150 bsh acre^{-1}) (Pierzynski and Logan, 1993). Surface runoff is the primary pathway for P loss from agricultural lands to water bodies. Erosion moves the smaller soil particles that typically contain the highest P and occurs in conjunction with runoff water that carries these soil particles. Leaching is the loss of soluble P as water percolates down the soil profile. The risk of P loss through leaching is considered less compared to surface runoff, as P is held very tightly by the soil particles. However, P loss through leaching can be prominent in sandy soils or regions with tile-drainage via preferential pathways such as root channels, worm burrows, or cracks, and in soils with a history of P loading.

1.2 HOW DOES PHOSPHORUS ACCUMULATE IN SOIL?

Soils have a finite capacity to retain P. With repeated P additions, soils become saturated with P, resulting in an increased P desorption and greater soluble P concentration during runoff events. Phosphorus accumulates in soil when P input via fertiliser or manure exceeds P removal by crops. The problem is more severe in regions where livestock industries are intensified. Rapid growth and intensification of crop and animal farming have created a regional and local imbalance in P

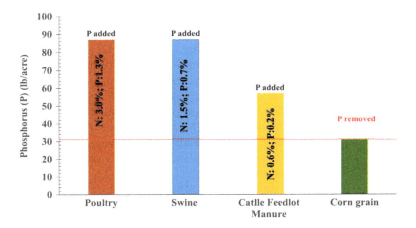

FIGURE 1.2 Phosphorus (P) surplus in soil when manure (from poultry, swine, and cattle) is applied based on nitrogen (N) needs (200 lb acre^{-1}) for optimal corn yield (180 bu acre^{-1}).

inputs and outputs (Sharpley and Beegle, 2001). Intensive livestock production areas or confined animal feeding operations generate enormous amounts of manures (solid or liquid), which are land-applied as the nutrient source for crop production. Manures are applied at rates designed to meet crop N requirements since long-distance transportation of manure is not cost-effective. This often causes excess P application compared to the agronomic P requirements resulting from the mismatch between manure nutrient content and crop nutrient needs. For example, poultry litter (PL) typically has a fertiliser (N-P$_2$O$_5$-K$_2$O) grade of 3-3-2 (Watts et al., 2019). Poultry litter, if repeatedly used year after year, can cause P to build up very quickly. For example, bermudagrass hay harvested at 6 US ton acre^{-1} has a nitrogen requirement that is 300 lbs N acre^{-1}. To meet the N needs of hay, 5 US tons of PL (3-3-2) would be required. Bermudagrass P uptake rate is only 5 lb P per US ton. The mismatch between P removal (30 lbs acre^{-1}) and input (131 lbs acre^{-1}) causes a net balance of 100 lbs acre^{-1} of P, assuming the P runoff or leaching losses are less than 1 lb acre^{-1} yr^{-1}. This practice of applying 5 US tons PL every year will cause the soil P to build up above the environmental threshold. Another example of P imbalance caused due to use of different manure types is presented in Figure 1.2. On average, 30% of P input to farming systems is output in crops and animal produce.

Research conducted in areas dominated by high-density poultry operations in the US, such as the Shenandoah Valley of Virginia (Pease et al., 1998), the Sand Mountain region of Alabama (Kingery et al., 1994), and Eastern Oklahoma (Sharpley et al., 1993), have shown long-term manure application resulted in significant accumulation of soil P. Globally, long-term repeated applications of P have led to an agronomic surplus of 11.5 Tg P y^{-1}, most of which accumulates in agricultural soils as legacy P (Zhang et al., 2020). Pavinato et al. (2020) reported long-term external P application (49 years) through fertiliser- and manure-caused accumulation of 33.4 Tg of legacy P in Brazilian agricultural soils. This accumulated "legacy" P can serve as a potential P source and increase the risk of P loss via surface and subsurface pathways (Jarvie et al., 2013, King et al., 2017; Penn et al., 2014; Sharpley et al., 2003). A large amount of legacy P is one of the primary reasons for the failure of conservation practices in improving water quality during the last two decades (Sharpley et al., 2013).

1.3 PHOSPHORUS MANAGEMENT

An essential tool in phosphorus management is a soil test. The goal of external P application is to achieve maximum crop yield. Maintaining P concentration in the soil solution in an optimum range for plant growth (>0.2 mg L^{-1}) while preventing excess P loss in surface water (>0.03 mg L^{-1}) is a challenge. Soil P testing methods were developed primarily to group the soils into various fertility ratings or indexes (Beegle and Durst, 2014; Kamprath and Watson, 1980; Peck and Soltanpour, 1990). Soil test P (STP) does not provide the total concentration of P in the soil or the actual soil available P; instead, it is used as an index to assess the agronomic response to external P application rates.

1.4 ASSESSMENT OF SOIL PHOSPHORUS FOR AGRONOMIC USE

The principle for any agronomic soil test extractant is to extract a portion of P that will be potentially available to the plant. However, the amount of P extracted varies between different extractants depending on the composition and strength of the chemical extractant used. The extracting solutions for the soil test can be dilute acids (such as Bray I, Bray II, Mehlich 1, Mehlich 3, and Morgan), buffered alkaline extractants (such as Olsen and AB-DTPA), or non-conventional extractants (simple DI water or iron oxide strip). A list of the commonly used soil test extractants used for agronomic purposes is listed in Table 1.1. The extraction mechanism is a combination of dissolution, desorption, and sometimes chelation of the various P forms present in the soil (Maguire et al., 2005). Before selecting a specific extractant, extensive field calibration and verification are required. Depending on the strength of the statistical relationship between the soil test value and the plant response, a soil test extractant is selected (Corey, 1987). Based on the response of extractable nutrients and crop yield, the soil nutrient level can be categorised as below optimum, optimum, and above optimum. If STP values are optimum (also referred to as the critical value for yield), the likelihood of increased crop yield with additional P application is minimal (Figure 1.3); instead, the accumulated P could be of environmental concern. Greater STP concentration leads to higher dissolved P concentrations in surface runoff.

Recommendations for P fertiliser are based on either fertilising the soil (build-maintain), fertilising the crop (sufficiency), or a combination of both of these, termed hybrid. In fertilising the soil approach, P fertiliser is recommended for optimum crop production and maintaining the STP at an optimum level by replacing the projected P removed by the crop (Beegle, 2005). Maintaining STP value in the optimum range throughout later years is a long-term investment with expected returns in the future. The sufficiency approach is designed to supply enough P based on the probability of a

TABLE 1.1
Common Soil Test Phosphorus Extractants Used for Agronomic Purposes

Extractant	Composition	References
AB-DTPA[#]	1 M NH_4HCO_3 + 0.005 M DTPA (pH 7.5)	Soltanpour and Schwab (1977)
Bray 1	0.03 M NH_4F + 0.025 M HCl	Bray and Kurtz (1945)
Bray 2	0.03 M NH_4F + 0.1 M HCl	Bray and Kurtz (1945)
Lancaster	First step: 0.05 M HCl; Second step: 0.037 N NH_4F + 0.03 N $AlCl_3 6H_2O$ + 1.58 N CH_3COOH + 0.125 N $CH_2(COOH)_2$ + 0.187 N $CH_2CHOH(COOH)_2$ (pH 4)	Cox (2001)
Mehlich 1	0.05 M HCl + 0.0125 M H_2SO_4	Mehlich (1953)
Mehlich 3	0.015 M NH_4F + 0.2 M CH_3COOH + 0.25 M NH_4NO_3 + 0.013 M HNO_3 +0.001 M EDTA	Mehlich (1984)
Morgan	0.52 N CH_3COOH + 0.72 N $NaC_2H_3O_2$ (pH 4.8)	Morgan (1941)
Modified Morgan	0.62 N NH_4OH + 1.25 N CH_3COOH (pH 4.8)	McIntosh (1969)
Olsen	0.5 M $NaHCO_3$ (pH 8.5)	Olsen et al. (1954)

(Modified from Beegle, 2005)
[#]Ammonium bicarbonate-diethylenetriaminepentaacetic acid

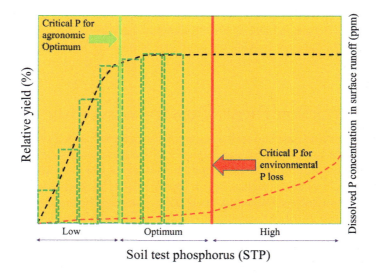

FIGURE 1.3 Soil test phosphorus and its relation to agronomic and environmental critical P concentration. Greater STP concentration is associated with higher dissolved P concentrations in surface runoff (adapted from Sharpley et al., 2003).

profitable crop response. Rather than a single approach, producers often utilise a hybrid system for profitable crop production.

Most of the P fertilisation approaches are also driven by fertiliser prices. Increasing fertiliser prices resulted in a shift in utilising P from organic (manures) or other alternative P sources (such as struvite). There remains a gap in understanding the contribution of these alternative P sources to crop production. With the increasing use of these alternative P sources in the future, more research will be needed to understand whether the current soil test methods would remain valid for P fertiliser recommendation and environmental P loss risk assessment.

1.5 ASSESSMENT OF ENVIRONMENTAL PHOSPHORUS LOSS

For effective assessment of P loss risk, factors associated with P transport (surface runoff, soil erosion, and subsurface loss pathways) and site management (soil P, P source, P application rate, method, and timing) should be evaluated (Heathwaite et al., 2005; Sharpley et al., 2001, 2006) (Table 1.2). Several studies in the past (Abboud et al., 2018; Fischer et al., 2018; Vadas et al., 2009) have aimed to develop parameters that can be used for environmental P loss risk assessment. The P Index is used in the United States, which accounts for both the source and transport factors associated with P loss (Heathwaite et al., 2005). However, there has been an increased concern over the accuracy of the P Index and its effectiveness in improving water quality (Drewry et al., 2011; Sharpley et al., 2012; Nelson and Shober, 2012). Alternatively, soil test indicators (soil test P; degree of P saturation; P saturation ratio; soil P storage capacity) are also used as a tool (stand-alone or used together) for environmental P loss risk assessment. The limitation of these soil test indicators, though, is they have methodological issues due to variation in the soil solution ratio (Koopmans et al., 2002) and fail to account for the P transport factor (Sharpley et al., 2012). However, despite these limitations, these methods are more practical and easier to use than the complex P Index.

There are several practices and approaches that can be applied to manage P and improve environmental P sustainability. For simplicity, we have presented information on the indicators that can be used to identify the P loss risk from nonpoint sources as well as the important best management practices (BMPs) that can be implemented simultaneously to reduce the risk.

TABLE 1.2
Factors Influencing Phosphorus (P) Loss

Site Factors	Description
1. Soil P measured using STP[#]	Greater STP is correlated with higher P loss in surface runoff.
2. Application rate	Greater application rates are correlated with higher P loss.
3. Application method	P loss is less with subsurface injection compared with surface broadcast with no incorporation.
4. Application timing	High P loss occurs if P is applied during spring ice melt or before heavy rains.
5. P source	P in some fertilisers (MAP)[#] and manure (poultry litter) are more soluble and susceptible to P loss.
6. Grazing animals	P loss increases if animals are grazed and have unlimited access to water or fed in an area close to water bodies.
7. Irrigation management	Improper irrigation at high rates can increase P loss by increasing surface runoff and erosion.
Transport factors	
1. Soil texture/hydrological group	Sandy textures promote subsurface loss, whereas clay soils have a greater surface loss.
2. Field slope	The greater the field slope, the higher the surface runoff.
3. Soil erosion rate	Greater erosion rates have higher total P loss.
4. P application distance and connectivity to water	The closer the water body is to an agricultural field where P is applied, the shorter the time for P transport to the water body.
5. Vegetative buffer width	The greater the width of the vegetative buffer, the lower the P loss.
6. Sensitive or outstanding water body proximity	The closer the sensitive or outstanding water body is to a P source, the greater the point of impact.

(Modified from Sharpley et al., 2001, 2006)
[#] STP: Soil test P; MAP: monoammonium phosphate

1.5.1 Soil Test Indicators for Environmental Phosphorus Loss Assessment

1.5.1.1 Soil Test Phosphorus

Soil test P was originally developed to measure P availability and P fertiliser recommendations for agronomic purposes. In the last two decades, STP has also been used as a management tool by environmental specialists and policymakers in the US for P loss risk assessment to minimise water quality problems from nonpoint sources (Smith et al., 2016; Leytem et al., 2017). Several studies in the past (Daniel et al., 1994; Pote et al., 1996, 1999) have shown a strong relationship between STP concentration and P loss in runoff water, with greater STP concentration leading to higher P concentration in the runoff. Maguire and Sims (2002) showed a strong relationship between DRP in leachate with conventional (Mehlich 1, Mehlich 3) and non-conventional (iron oxide P, water) STP methods. Previous studies have shown that the strength of the relationship between STP and P in runoff ranges from $r^2 = 0.99$ to 0.41 (Table 1.3). The relationship between STP and P concentration in runoff water is soil specific. Hence, it is not feasible to assign universally acceptable STP values above which P loss risk increases rapidly.

One of the standard and relatively simple methods that have been used for environmental P management is to use the STP "threshold" approach, above which P applications are limited or not permitted. The concept of environmental STP threshold is based on the fact that once soil reaches a critical STP level, further P addition will be detrimental to the environment compared to its agronomic benefit (Daniel et al., 1998). Research has shown that in soil tests above the STP threshold, referred to as the "change point", P loss during runoff events increases rapidly with an increase in

TABLE 1.3
Relationship Between Soil Test Phosphorus (STP, Mg Kg⁻¹) and Dissolved Reactive Phosphorus (DRP, Mg L⁻¹) from Previous Studies

Soil Order	STP Method	Coefficient of Determination (r^2)	Reference
Ultisols	Mehlich 3	0.82–0.87	Pote et al. (1999)
Ultisols	Olsen	0.75–0.87	Pote et al. (1999)
Ultisols	Morgan	0.82–0.89	Pote et al. (1999)
Entisols, Inceptisols, Ultisols	Mehlich 3[#]	0.62	Sims et al. (2002)
Entisols, Inceptisols, Ultisols	Mehlich 3[##]	0.72	Sims et al. (2002)
Alfisols	Mehlich 3	0.41–0.94	Torbert et al. (2002)
Alfisols, Entisols, Ultisols	Mehlich 3	0.89	Dari et al. (2018)
Chernozemic	Modified Kelowna extraction	0.93–0.99	Little et al. (2006)
Alfisols, Molisols	Bray 1	0.65–0.66	Andraski and Bundy (2003)

[#] Relationship between Mehlich 3 P and DRP in runoff.
[##] Relationship between Mehlich 3 P and DRP in leachate.

STP (McDowell and Sharpley, 2001). A split-line model is used to determine the STP threshold, above which P loss increases disproportionately with an increase in STP values.

Heckrath et al. (1995) reported that DRP concentration in tile drain flow was low when the Olsen P concentration was < 57 mg kg⁻¹, and the DRP concentration increased rapidly above this value. The critical STP threshold above which P loss increases significantly can vary with soils (Hesketh and Brookes, 2000). The goal of developing a sustainable agriculture system is to have an optimum STP concentration sufficient for crop production and should be below the environmental threshold (Bai et al., 2013).

Although STP is currently used for agronomic and environmental P management, it is not a suitable indicator for environmental P loss risk. Soil test P fails to account for the P retentive capacity of the soil and can lead to ambiguous conclusions (Hooda et al., 2000). The relationship between STP and P in runoff water can differ significantly between soils since P released into the solution depends on the P retention and release characteristics (Cox and Hendricks, 2000; Sharpley, 1995). Sharpley (1995) reported DRP concentration in runoff from clay and silt loam soils with the same Mehlich 3 P concentration (200 mg kg⁻¹) of 0.28 and 1.36 mg L⁻¹, respectively.

1.5.1.2 Degree of Phosphorus Saturation

Degree of P saturation (DPS) was introduced as a tool to quantify the environmental P loss risk for non-calcareous sandy soils in areas with intensive livestock production in the Netherlands (Breeuwsma and Schoumans, 1987; Breeuwsma et al., 1995). Soil DPS is based on the saturation of P sorbing sites for soil and determines P release (intensity factor) along with the soil P level (capacity factor) (Breeuwsma and Silva, 1992; Sharpley et al., 2020).

The DPS is calculated using acid ammonium oxalate (DPS_{Ox}) as the ratio of extractable P to the sum of Fe and Al (Eq. 1)

$$DPS_{Ox} (\%) = \text{Oxalate extractable P}/\alpha(\text{Oxalate extractable [Fe+Al]}) \times 100 \qquad (1)$$

where oxalate extractable P, Fe, and Al in Eq. (1) are expressed in mmol kg⁻¹.

In the Netherlands, a DPS of 25% has been established as the critical value above which the potential for environmental P loss risk increases (Breeuwsma et al., 1995). Empirical factor α represents the proportion of oxalate extractable Fe and Al available for P sorption. The factor α varies with oxalate extractable Fe and Al among soils and within a soil profile and is determined by the slope of the equation relating the Langmuir sorption maximum (derived from traditional batch isotherms) to the molar concentration of oxalate extractable Fe and Al (Schoumans, 2013). Pautler and Sims (2000) identified an α of 0.68 for Delaware soils, while Nair and Graetz (2002) used an α of 0.55 for Florida sandy soils.

Oxalate extractant was used originally for the DPS calculations, as oxalate extracts most of the reactive Al and Fe present in the soil and proportionally represents the P retention capacity of the soil (Kleinman et al., 2003). However, the instability of the oxalate solution under standard laboratory conditions makes it challenging to adopt the method on a routine basis in most soil testing laboratories (Sharpley et al., 2020; Kleinman and Sharpley, 2002; Schoumans, 2009). Additionally, the determination of α involves traditional batch isotherm, which is laborious, time-consuming, and not routinely used. In order to use DPS as a soil test indicator, there is a need to assess α across a range of soil conditions. Challenges associated with the DPS determination make it difficult to be widely adopted for environmental P loss risk assessment.

1.5.1.3 Phosphorus Saturation Ratio

Phosphorus saturation ratio (PSR) was introduced as a modification to the DPS concept in various parts of the US to quantify environmental P loss risk and is determined by using acid ammonium oxalate or commonly used soil test extractant (Mehlich 1 or Mehlich 3) (Chakraborty et al., 2021; Dari et al., 2018; Nair, 2014). Soil PSR is the molar ratio of extractable P to the sum of extractable Fe and Al without using the corrective factor α. The PSR values are soil specific. Two soils can have the same PSR value; however, the P loss risk may not be the same for both soils as it depends on the P retentive capacity of the soil.

Previous studies identified a PSR change point or threshold value above which the P loss risk in leachate or runoff water increases rapidly (Casson et al., 2006; Nair et al., 2004). The amount of P lost from the soil after P application from organic or inorganic sources will depend on the solubility of the material and the sorption capacity of the soil and can be best determined by extraction with water (water-soluble P, WSP). Threshold PSR is obtained using the relationship between WSP and PSR for a particular group of soils (Figure 1.4). The threshold PSR is unique for a soil type and the extractant used. The threshold PSR values used for different soil across the world are summarised in Table 1.4.

The PSR concept was developed for sandy surface soils; however, research has shown its broader applicability to subsurface and wetland soils (Chakraborty et al., 2011; Mukherjee et al., 2009). The threshold PSR for Florida sandy soils using oxalate, Mehlich 1, and Mehlich 3 extractants were 0.10, 0.10, and 0.08, respectively (Nair et al., 2004). Dari et al. (2018) obtained a threshold PSR of 0.1 for a wide range of acidic soils in the US using Mehlich 3 extractant. Despite the differences in methodology in obtaining a threshold PSR, the PSR value for most surface soils ranges from 0.10 to 0.15.

1.5.1.4 Soil Phosphorus Storage Capacity

Soil P storage capacity (SPSC) provides a direct estimate of the amount of P that can be added to a certain volume or mass of soil before exceeding a threshold soil equilibrium concentration, i.e., before the soil becomes an environmental concern (Nair and Harris, 2004). The SPSC can effectively differentiate the P loss risk of unimpacted soils with low P retention capacity from impacted soil with high P retention capacity. Soil P storage capacity can also be used to determine the P loss risk from previous loadings (legacy P) without needing prior P application history.

Soil P storage capacity depends on a threshold PSR as determined for a specific range of soils. The SPSC is calculated using the following generalised equation:

$$\text{SPSC (mg kg}^{-1}\text{)} = (\text{Threshold PSR} - \text{Soil PSR}) * [\text{extractable Fe} + \text{Al}] * 31. \qquad (2)$$

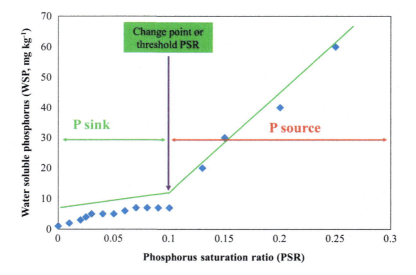

FIGURE 1.4 Relationship between water-soluble phosphorus (WSP) and phosphorus saturation ratio (PSR) (modified from Chakraborty and Prasad, 2021).

TABLE 1.4
Threshold Phosphorus Saturation Ratio (PSR)# for World Soils

Location	Soil Order	Soil Extractant Used	Phosphorus Runoff Indicator	Threshold PSR	Reference
UK	Alfisol, Entisol, Inceptisol, Spodosol	Oxalate	Desorbed P (0.01 M $CaCl_2$)	0.10	Hooda et al., 2000
Delaware, US	Entisols, Inceptisols, Ultisols	Mehlich 3	Dissolved runoff P	0.14	Sims et al., 2002
			Column leachate P	0.21	
Florida, US (uplands)	Entisols, Ultisols	Oxalate	Water-soluble P	0.10	Nair et al., 2004
		Mehlich 1	Water-soluble P	0.10	
		Mehlich 3	Water-soluble P	0.08	
Florida, US (wetlands)	Spodosols	Mehlich 1	Water-soluble P	0.10	Nair et al., 2015
Wide range of acidic soils within the US and Puerto Rico	Alfisols, Entisols, Inceptisols, Spodosols, Ultisols	Mehlich 3	Water-soluble P	0.10	Dari et al., 2018
Alabama	Ultisols	Oxalate	Water-soluble P	0.11	Chakraborty et al., 2021
		Mehlich 1	Water-soluble P	0.06	
		Mehlich 3	Water-soluble P	0.05	

(Modified from Nair, 2014 and Sharpley et al., 2020)

* Represents soils collected from Arkansas, Florida, Georgia, Maryland, and Pennsylvania.
# PSR is the molar ratio of extractable P to the sum of extractable Fe and Al, determined using oxalate or common soil test extractant (Mehlich 1, Mehlich 3). The phosphorus runoff indicator used in the determination of threshold PSR can vary. Past studies have used desorbed P; dissolved runoff P from rainfall simulation experiments; column leachate; and water-soluble P by extracting the soil with DI water.

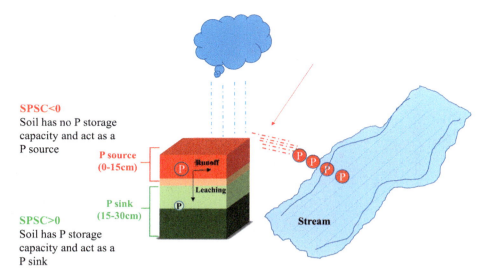

FIGURE 1.5 Illustration of soil phosphorus storage capacity (SPSC) used as a screening tool to understand the P loss risk from a soil profile (30 cm depth) where surface 0–15 cm soil acts as a P source and the subsurface 15–30 cm soil acts as a P sink (adapted from Chakraborty and Prasad, 2021).

P, Fe, and Al can be determined in oxalate or soil test extractant (Mehlich 1 or Mehlich 3) and expressed in moles. Since soil can have only one P storage capacity, SPSC calculated using ammonium oxalate is reliable and accurate (McKeague and Day, 1966). However, SPSC can also be calculated using common extractants (Mehlich 1 or Mehlich 3) via calibration with oxalate extraction (Nair and Harris, 2014; Chrysostome et al., 2007; Nair et al., 2010; Chakraborty et al., 2021).

The SPSC can be either > 0 (positive) or < 0 (negative). When SPSC is positive, the soil has additional capacity to retain P (P sink). In contrast, when SPSC is negative, the soil cannot hold additional P (P source) (Figure 1.5). Several studies have shown that WSP is minimal when SPSC is positive, whereas WSP increases linearly when SPSC is negative (Figure 1.6). Dari et al. (2018) validated SPSC using field water quality data where DRP increased in runoff water with an increase in negative SPSC values. Information on the storage capacity of an entire soil profile will greatly benefit nutrient management since the amount of P that can be applied under specific loading conditions can be determined. Since SPSC is additive when expressed on a volumetric basis, a single value can be obtained for a given soil profile to any desired soil depth. This can help identify locations within a watershed subject to P loss.

1.6 BEST MANAGEMENT PRACTICES TO REDUCE PHOSPHORUS LOADING TO WATER BODIES

Best management practices are effective, practical conservation practices developed for a particular region and can mitigate or prevent water pollution. Most BMPs are designed and have been proven beneficial to retard the movement of P into water bodies. Adoption of BMPs should go hand in hand with crop production and not be implemented as a remedial measure once the field becomes a P source of higher P loss risk category. Depending on the soil type, climate, P loading history, and other management factors, a stand-alone or combination of BMPs can be selected. Many of the practices are already used by producers for environmental protection and are economically beneficial. The BMP can be categorised into either controlling the P source or controlling the P transport (Table 1.5). Once the BMP is implemented, it is important to maintain and monitor its efficacy.

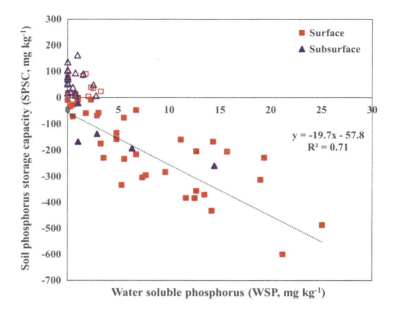

FIGURE 1.6 Relationship between soil phosphorus storage capacity (SPSC) and water-soluble phosphorus (WSP) of surface and subsurface soils (adapted from Chakraborty et al., 2021).

TABLE 1.5
Best Management Practices (BMPs) to Reduce Phosphorus (P) Loadings to Water Bodies

BMPs for Managing P Loss at the Source

1. Apply P inputs based on agronomic STP[#] recommendations and balance P input with P output once the soil test reaches a critical level.
2. Deep band P or inject P in soils where the risk of subsurface P loss is less.
3. Do not apply P in the autumn or spring thaw period.
4. If STP is higher than critical values, focus on P mining through crop removal.
5. Calibrate spreaders before P application.
6. Avoid grazing animals close to a stream. Fencing keeps animals away from the waterways.
7. If manure is used, test manure and determine the appropriate application rate.
8. Do not apply P fertiliser or manure before heavy rainfall events.
9. Avoid over-irrigation to prevent erosion and runoff.
10. P application should be precise to avoid high P zones in the field.
11. Move manure off the farm/county/state if the P imbalance is high.
12. Find alternate uses of manure, such as converting it into a by-product that can be used for a purpose other than fertilising.

BMPs for Managing P Loss in the Transport Phase

1. Maintain a grassed waterway to trap eroded P.
2. Observe setback distances for manure applications near water bodies.
3. Adopt conservation practices such as cover crops to reduce erosion and runoff.
4. Adopt no-till or minimum tillage if the farm has a high risk of sediment loss.
5. Adopt riparian zones to trap eroded P.
6. Stream bank stabilisation to minimise erosion.
7. Residue management to prevent washing during the fallow rainfall period.
8. Wetlands to trap eroded P.
9. Install filter strips to trap eroded P.

(Adapted from Sharpley et al., 2006 and Singh et al., 2020)
[#] STP: Soil test P

1.7 SUMMARY

Long-term repeated P application in excess of crop removal leads to farm P imbalance, promoting soil P accumulation. The accumulated soil P can be potentially lost during runoff events causing water quality degradation. Soil test P is the starting point to evaluate the P status for agronomic purposes. Once the STP approaches the critical P concentration, the yield plateaus and no additional yield gains are observed. Any additional P application beyond the critical P concentration increases the environmental risk of P loss. One caveat of STP is that it does not effectively assess the environmental P loss risk since it fails to account for the P retentive capacity of the soil. This limitation of STP is overcome by using the threshold PSR, which can be used as a screening tool to differentiate between a P source (soil PSR > threshold PSR) and a P sink (soil PSR < threshold PSR). Soil P storage capacity based on the threshold PSR can be effectively used to quantify the amount of P that can be safely added to the soil without any environmental concern. The PSR/SPSC can be determined easily from any routine soil testing laboratory. Adoption of BMPs can have a significant effect on reducing P loading from nonpoint sources. However, the efficacy of BMPs must be monitored to ensure their functionality. Adopting soil test indicators and implementation and regular monitoring of BMPs will benefit environmental P management.

REFERENCES

Abboud, F.Y., N. Favaretto, A.C.V. Motta, G. Barth, and G.D. Goularte. 2018. Phosphorus mobility and degree of saturation in oxisol under no-tillage after long-term dairy liquid manure application. *Soil Till. Res.* 177: 45–53. https://doi.org/10.1016/j.still.2017.11.014.

Andraski, T.W., and L.G. Bundy. 2003. Relationships between phosphorus levels in soil and in runoff from corn production systems. *J. Environ. Qual.* 32(1): 310–316.

Bai, Z., H. Li, X. Yang, B. Zhou, X. Shi, B. Wang, D. Li, J. Shen, Q. Chen, W. Qin, O. Oenema, and F. Zhang. 2013. The critical soil P levels for crop yield, soil fertility and environmental safety in different soil types. *Plant Soil* 372(1–2): 27–37.

Baker, D.B., R. Confesor, D.E. Ewing, L.T. Johnson, J.W. Kramer, and B.J. Merryfield. 2014. Phosphorus loading to Lake Erie from the Maumee, Sandusky and Cuyahoga rivers: The importance of bioavailability. *J. Great Lakes Res.* 40(3): 502–517. https://doi.org/10.1016/j.jglr.2014.05.001.

Beegle, D. 2005. Assessing soil phosphorus for crop production by soil testing. In: *Phosphorus: Agriculture and the Environment*, ed. J.T. Sims and A.N. Sharpley, 123–143. Agron. Monogr. 46. CSSA, ASA, and SSSA, Madison, WI.

Beegle, D.B., and P.T. Durst. 2014. Managing phosphorus for crop production. *Penn State Ext* UC055: 1–6. https://extension.psu.edu/managing-phosphorus-for-crop-production.

Bray, R.H., and L.T. Kurtz. 1945. Determination of total, organic, and available forms of phosphorus in soils. *Soil Sci.* 59(1): 39–45.

Breeuwsma, A., and O.F. Schoumans. 1987. Forecasting phosphate leaching on a regional scale. In *Proceedings of the International Conference on Vulnerability of Soil and Groundwater to Pollutants*, ed. W. van Duijvenboode and H.G. van Waegeningh, 973–981. TNO Committee on Hydrological Research, The Hague Proceedings of the and Information No. 38.

Breeuwsma, A., and S. Silva. 1992. Phosphorus fertilization and environmental effects in the Netherlands and the Po region (Italy). In *Agriculture Research Department Report 57*. Winand Staring Centre for Integrated Land, Soil, and Water Research, Wageningen, The Netherlands.

Breeuwsma, A., J.G.A. Reijerink, and O.F. Schoumans. 1995. Impact of manure on accumulation and leaching of phosphate in areas of intensive livestock farming. In *Animal Waste and the Land-Water Interface*, ed. K. Steele, 239–249. New York: Lewis Publishers.

Casson, J.P., D.R. Bennett, S.C. Nolan, B.M. Olson, and G.R. Ontkean. 2006. Degree of phosphorus saturation thresholds in manure-amended soils of Alberta. *J. Environ. Qual.* 35(6): 2212–2221. https://doi.org/10.2134/jeq2006.0085.

Chakraborty, D., and R. Prasad. 2021. Phosphorus management: Use of soil phosphorus storage capacity. ANR- 2739. https://www.aces.edu/blog/topics/crop-production/phosphorus-management-use-of-soil-phosphorus-storage-capacity/.

Chakraborty, D., R. Prasad, A. Bhatta, and H.A. Torbert. 2021. Understanding the environmental impact of phosphorus in acidic soils receiving repeated poultry litter applications. *Sci. Total Environ.* 779: 146267. https://doi.org/10.1016/j.scitotenv.2021.146267.

Chakraborty, D., V.D. Nair, M. Chrysostome, and W.G. Harris. 2011. Soil phosphorus storage capacity in manure-impacted Alaquods: Implications for water table management. *Agric. Ecosyst. Environ.* 142(3–4): 167–175.

Chrysostome, M., V.D. Nair, W.G. Harris, and R.D. Rhue. 2007. Laboratory validation of soil phosphorus storage capacity predictions for use in risk assessment. *Soil Sci. Soc. Am. J.* 71(5): 1564–1569. https://doi.org/10.2136/sssaj2006.0094.

Corey, R.B. 1987. Soil test procedures: Correlation. In *Soil Testing: Sampling, Correlation, Calibration, and Interpretation*, ed. J.R. Brown, 15–22 SSSA Spec. Publ. 21. SSSA, Madison, WI.

Corey, R.C. 1981. Adsorption vs. precipitation. In *Adsorption of Inorganics at Solid-Liquid Interfaces*, ed. M.A. Anderson and A.J. Rubin, 161–182. Ann Arbor Sci, Ann Arbor. MI.

Cox, F.R., and S.E. Hendricks. 2000. Soil test phosphorus and clay content effects on runoff water quality. *J. Environ. Qual.* 29(5): 1582–1586.

Cox, M.S. 2001. The Lancaster soil test method as an alternative to the Mehlich 3 soil test method. *Soil Sci.* 166(7): 484–489. https://doi.org/10.1097/00010694-200107000-00006.

Daniel, T.C., A.N. Sharpley, and J.L. Lemunyon. 1998. Agricultural phosphorus and eutrophication: A symposium overview. *J. Environ. Qual.* 27(2): 251–257.

Daniel, T.C., A.N. Sharpley, D.R. Edwards, R. Wedepohl, and J.L. Lemunyon. 1994. Minimizing surface water eutrophication from agriculture by phosphorus management. *J. Soil Water Conserv.* 49: 30–38.

Dari, B., V.D. Nair, A.N. Sharpley, P. Kleinman, D. Franklin, and W.G. Harris. 2018. Consistency of the threshold phosphorus saturation ratio across a wide geographic range of acid soils. *Agrosyst. Geosci. Environ.* 1(1): 1–8. https://doi.org/10.2134/age2018.08.0028.

Dodds, W.K., W.W. Bouska, J.L. Eitzmann, T.J. Pilger, K.L. Pitts, A.J. Riley, J.T. Schloesser, and D.J. Thornbrugh. 2009. Eutrophication of U.S. freshwaters: Analysis of potential economic damages. *Environ. Sci. Technol.* 43(1): 12–19.

Drewry, J.J., L.T.H. Newham, and R.S.B. Greene. 2011. Index models to evaluate the risk of phosphorus and nitrogen loss at catchment scales. *J. Environ. Manag.* 92(3): 639–649. http://doi.org/10.1016/j.jenvman.2010.10.001.

Fischer, P., R. Pothig, B. Gücker, and M. Venohr. 2018. Phosphorus saturation and superficial fertilizer application as key parameters to assess the risk of diffuse phosphorus losses from agricultural soils in Brazil. *Sci. Total Environ.* 630: 1515–1527. https://doi.org/10.1016/j.scitotenv.2018.02.070.

Heathwaite, A.L., A. Sharpley, M. Bechmann, and S. Rekolainen. 2005. Assessing the risk and magnitude of agricultural nonpoint source phosphorus pollution. In *Phosphorus: Agriculture and the Environment*, ed. J.T. Sims and A.N. Sharpley, 981–1020. Agron. Monogr. 46. CSSA, ASA, and SSSA, Madison, WI.

Heckrath, G., P.C. Brookes, P.R. Poulton, and K.W.T. Goulding. 1995. Phosphorus leaching from soils containing different phosphorus concentrations in the Broadbalk experiment. *J. Environ. Qual.* 24(5): 904–910. https://doi.org/10.2134/jeq1995.00472425002400050018x.

Hesketh, N., and P.C. Brookes. 2000. Development of an indicator for risk of phosphorus leaching. *J. Environ. Qual.* 29(1): 105–110.

Hooda, P.S., A.R. Rendell, A.C. Edwards, P.J.A. Withers, M.N. Aitken, and V.W. Truesdale. 2000. Relating soil phosphorus indices to potential phosphorus release to water. *J. Environ. Qual.* 29(4): 1166–1171. https://doi.org/10.2134/jeq2000.00472425002900040018x.

Jarvie, H.P., A.N. Sharpley, B. Spears, A.R. Buda, L. May, and P.J. Kleinman. 2013. Water quality remediation faces unprecedented challenges from "legacy phosphorus". *Environ. Sci. Technol.* 47(16): 8997–8998.

Joosse, P.J., and D.B. Baker. 2011. Context for re-evaluating agricultural source phosphorus loadings to the great lakes. *Can. J. Soil Sci.* 91(3): 317–327. https://doi.org/10.4141/cjss10005.

Kamprath, E.J., and M.E. Watson. 1980. Conventional soil and tissue tests for assessing phosphorus status of soils. In *The Role of Phosphorus in Agriculture*, ed. F.E. Khasawneh, E.C. Sample, and E.J. Kamprath, 433–469. CSSA, ASA, and SSSA, Madison, WI.

King, K.W., M.R. Williams, L.T. Johnson, D.R. Smith, G.A. LaBarge, and N.R. Fausey. 2017. Phosphorus availability in western Lake Erie Basin drainage waters: Legacy evidence across spatial scales. *J. Environ. Qual.* 46(2): 466–469. https://doi.org/10.2134/jeq2016.11.0434.

Kingery, W.L., C.W. Wood, D.P. Delaney, J.C. Williams, and G.L. Mullins. 1994. Impact of long-term application of broiler litter on environmentally related soil properties. *J. Environ. Qual.* 23: 139–147.

Kleinman, P.J.A., and A.N. Sharpley. 2002. Estimating soil phosphorus sorption saturation from Mehlich-3 data. *Commun. Soil Sci. Plant Anal.* 33(11–12): 1825–1839. https://doi.org/10.1081/CSS-120004825.

Kleinman, P.J.A., B.A. Needelman, A.N. Sharpley, and R.W. McDowell. 2003. Using soil phosphorus profile data to assess phosphorus leaching potential in manured soils. *Soil Sci. Soc. Am. J.* 67(1): 215–224. https://doi.org/10.2136/sssaj2003.2150.

Koopmans, G.F., R.W. Mcdowell, W.J. Chardon, O. Oenema, and J. Dolfing. 2002. Soil phosphorus quantity-intensity relationships to predict increased soil phosphorus loss to overland and subsurface flow. *Chemosphere* 48(7): 679–687.

Leytem, A., D. Bjorneberg, and D. Tarkalson. 2017. The phosphorus site index: A systematic approach to assess the risk of nonpoint source pollution of Idaho waters by agricultural phosphorus. Idaho State Dep. of Agric. https://agri.idaho.gov/main/wp-content/uploads/2018/01/Phosphorus-Site-Index-reference-2017-revised.pdf (accessed 17 Aug. 2022).

Little, J.L., S.C. Nolan, and J.P. Casson. 2006. Relationships between soil-test phosphorus and runoff phosphorus in small Alberta watersheds. In *Alberta Soil Phosphorus Limits Project*. 2: Field-scale losses and soil limits.150 pp. Alberta Agriculture, Food and Rural Development. Lethbridge, Alberta, Canada.

Maguire, R.O., and J.T. Sims. 2002. Soil testing to predict phosphorus leaching. *J. Environ. Qual.* 31(5): 1601–1609.

Maguire, R.O., W.J. Cardon, and R.R. Simard. 2005. Assessing potential environmental impacts of soil phosphorus by soil testing. In *Phosphorus: Agriculture and the Environment*, ed. J.T. Sims, A.N. Sharpley, 145–180. Agron. Monogr. 46. CSSA, ASA, and SSSA, Madison, WI.

McDowell, R.W., and A.N. Sharpley. 2001. Approximating phosphorus release from soils to surface runoff and subsurface drainage. *J. Environ. Qual.* 30(2): 508–520. https://doi.org/10.2134/jeq2001.302508x.

McIntosh, J.L. 1969. Bray and Morgan soil test extractants modified for testing acid soils from different parent materials. *Agron. J.* 61(2): 259–265.

McKeague, J.A., and J.H. Day. 1966. Dithionate and oxalate extractable Fe and Al as aids in differentiating various classes of soils. *Can. J. Soil Sci.* 46(1): 13–22.

Mehlich, A. 1953. Determination of P, Ca, Mg, K, Na, and NH4. Mimeo. North Carolina Soil Testing Div., Raleigh, NC.

Mehlich, A. 1984. Mehlich 3 soil test extractant: A modification of Mehlich 2 extractant. *Commun. Soil Sci. Plant Anal.* 15(12): 1409–1416.

Morgan, M.F. 1941. Chemical soil diagnosis by the universal soil testing system. Conn. Agric. Stn. Bull. 450. 579-628.

Mukherjee, A., V.D. Nair, M.W. Clark, and K.R. Reddy. 2009. Development of indices to predict phosphorus release from wetland soils. *J. Environ. Qual.* 38(3): 878–886. https://doi.org/10.2134/jeq2008.0230.

Nair, V. D., M.W. Clark, K.R. Reddy, and K.R. Reddy. 2015. Evaluation of legacy phosphorus storage and release from wetland soils. *J. Environ. Qual.* 44(6): 1956–1964. https://doi.org/10.2134/jeq2015.03.0154.

Nair, V. D. 2014. Soil phosphorus saturation ratio for risk assessment in land use systems. *Front. Environ. Sci.* 2. https://doi.org/10.3389/fenvs.2014.00006.

Nair, V. D., and D.A. Graetz. 2002. Phosphorus saturation in spodosols impacted by manure. *J. Environ. Qual.* 31(4): 1279–1285. https://doi.org/10.2134/jeq2002.1279.

Nair, V. D., and W.G. Harris. 2004. A capacity factor as an alternative to soil test phosphorus in phosphorus risk assessment. New zeal. *J. Agric. Res.* 47: 491–497. https://doi.org/10.1080/00288233.2004.9513616.

Nair, V.D., and W.G. Harris. 2014. Soil phosphorus storage capacity for environmental risk assessment. *Adv. Agric.* 2014: 1–9. https://doi.org/10.1155/2014/723064

Nair, V. D., K.M. Portier, D.A. Graetz, and M.L. Walker. 2004 An environmental threshold for degree of phosphorus saturation in sandy soils. *J. Environ. Qual.* 33(1): 107–113. https://doi.org/10.2134/jeq2004.1070.

Nair, V. D., W.G. Harris, D. Chakraborty, and M. Chrysostome. 2010. Understanding soil phosphorus storage capacity. Available online at: http://edis.ifas.ufl.edu/pdffiles/SS/SS54100.pdf. SL336.

Nelson, N.O., and A.L. Shober. 2012. Evaluation of phosphorus indices after twenty years of science and development. *J. Environ. Qual.* 41(6): 1703–1710. http://doi.org/10.2134/jeq2012.0342.

Olsen, S.R., C.V. Cole, F.S. Watanabe, and L.A. Dean. 1954. Estimation of available phosphorus in soils by extraction with sodium bicarbonate. USDA circ. 939. U.S. Gov. Print. Office, Washington, DC.

Pautler, M.C., and J.T. Sims. 2000. Relationships between soil test phosphorus, soluble phosphorus, and phosphorus saturation in Delaware soils. *Soil Sci. Soc. Am. J.* 64(2): 765–773.https://doi.org/10.2136/sssaj2000.642765x.

Pavinato, P.S., M.R. Cherubin, A. Soltangheisi, G.C. Rocha, D.R. Chadwick, and D.L. Jones. 2020. Revealing soil legacy phosphorus to promote sustainable agriculture in Brazil. *Sci. Rep.* 10(1): 15615.

Pease, J., R. Parsons, and K. Kenyon. 1998. Economic and environmental impacts of nutrient loss reduction on dairy and dairy/poultry farms. Virginia Coop. Ext. Publ. 448–231/REAP R033. Virginia Tech, Blacksburg.

Peck, T.R., and P.N. Soltanpour. 1990. The principals of soil testing. In *Soil Testing and Plant Analysis*, ed. R.L. Westerman. and S.S.S.A. Book Ser. 3. 1-9. SSSA, Madison, WI.

Penn, C., J. McGrath, J. Bowen, and S. Wilson. 2014. Phosphorus removal structures: A management option for legacy phosphorus. *J. Soil Water Conserv.* 69(2): 51A–56A.

Pierzynski, G.M. 1991. The chemistry and mineralogy of phosphorus in excessively fertilized soils. *Crit. Rev. Environ. Control* 21(3–4): 265–295.

Pierzynski, G.M., and T.J. Logan. 1993. Crop, soil, and management effects on phosphorus soil test levels: A review. *J. Prod. Agric.* 6(4): 513–520. https://doi.org/10.2134/jpa1993.0513.

Pote, D.H., T.C. Daniel, D.J. Nichols, A.N. Sharpley, P.A. Moore, Jr., D.M. Miller, and D.R. Edwards. 1999. Relationship between phosphorus levels in three ultisols and phosphorus concentrations in runoff. *J. Environ. Qual.* 28(1): 170–175.

Pote, D.H., T.C. Daniels, A.N. Sharpley, P.A. Moore, Jr., D.R. Edwards, and D.J. Nichols. 1996. Relating extractable soil phosphorus to phosphorus losses in runoff. *Soil Sci. Soc. Am. J.* 60(3): 855–859.

Prasad, R., and D. Chakraborty. 2019. Phosphorus basics: Understanding phosphorus forms and their cycling in the soil. ANR-2535 https://www.aces.edu/wp-content/uploads/2019/04/ANR-2535-Phosphorus-Basics_041719L.pdf.

Reid, K., K. Schneider, and B. McConkey. 2018. Components of phosphorus loss from agricultural landscapes, and how to incorporate them into risk assessment tools. *Front. Earth Sci.* 6: 1–15. https://doi.org/10.3389/feart.2018.00135.

Schoumans, O.F. 2009. Determination of the degree of phosphate saturation in noncalcareous soils. In *Methods for Phosphorus Analysis for Soils, Sediments, Residuals, and Water*, 2nd ed., ed. J.L. Kovar and G.M. Pierzynski, 29–32. SERA 17 Bulletin 408. Virginia Tech, Blacksburg. Retrieved from https://sera17dotorg.files.wordpress.com/2015/02/sera-17-methods-for-p-2009.pdf. . .

Schoumans, O.F. 2013. Description of the phosphorus sorption and desorption processes in lowland peaty clay soils. *Soil Sci.* 178(6): 291–300.

Sharpley, A.N., D. Beegle, C. Bolster, L. Good, B. Joern, Q. Ketterings, J. Lory, R. Mikkelsen, D. Osmond, and P. Vadas. 2012. Phosphorus indices: Why we need to take stock of how we are doing. *J. Environ. Qual.* 41(6): 1711–1719.

Sharpley, A., and D. Beegle. 2001. Managing phosphorus for agriculture and the environment. Penn State Ext UC162, 1–16. https://extension.psu.edu/managing-phosphorus-for-agriculture-and-the-environment.

Sharpley, A., H.P. Jarvie, A. Buda, L. May, B. Spears, and P. Kleinman. 2013. Phosphorus legacy: Overcoming the effects of past management practices to mitigate future water quality impairment. *J. Environ. Qual.* 42(5): 1308–1326. https://doi.org/10.2134/jeq2013.03.0098.

Sharpley, A.N. 1995. A dependence of runoff phosphorus on extractable soil phosphorus. *J. Environ. Qual.* 24(5): 920–926.

Sharpley, A.N., K.R. Brye, J.M. Burke, L.G. Berry, M.B. Daniels, and P. Webb. 2020. Can soil phosphorus sorption saturation estimate future potential legacy phosphorus sources? *Agrosyst. Geosci. Environ.* 1–10. https://doi.org/10.1002/agg2.20122.

Sharpley, A.N., R.W. Mcdowell, and P.J.A. Kleinman. 2001. Phosphorus loss from land to water: Integrating agricultural and environmental management. *Plant Soil* 237(2): 287–307. https://doi.org/10.1023/A:1013335814593.

Sharpley, A.N., S.J. Smith, and W.R. Bain. 1993. Nitrogen and phosphorus fate from long-term poultry litter applications to Oklahoma soils. *Soil Sci. Soc. Am. J.* 57(4): 1131–1137.

Sharpley, A.N., T. Daniel, G. Gibson, L. Bundy, M. Cabrera, T. Sims, R. Stevens, J. Lemunyon, P. Kleinman, and R. Parry. 2006. Best management practices to minimize agricultural phosphorus impacts on water quality. U.S. Department of Agriculture, Agricultural Research Service, ARS–163, 50 pp.

Sharpley, A.N., T.C. Daniel, J.T. Sims, and D.H. Pote. 1996. Determining environmentally sound soil phosphorus levels. *J. Soil Water. Conserv.* 51: 160–166.

Sharpley, A.N., T.C. Daniel, J.T. Sims, J.L. Lemunyon, R.G. Stevens, and R. Parry. 2003. *Agricultural Phosphorus and Eutrophication*. 2nd ed. USDA, A.R.S. ARS-149. U.S. Gov. Print. Office, Washington, DC.

Sims, J.T., R.O. Maguire, A.B. Leytem, K.L. Gartley, and M.C. Paulter. 2002. Evaluation of Mehlich3 as an agri-environment soil phosphorus test for the Mid-Atlantic United States of America. *Soil Sci. Soc. Am. J.* 66(6): 2016–2032. https://doi.org/10.2136/sssaj2002.2016.

Singh, G., G. Kaur, K. Williard, J. Schoonover, and K.A. Nelson. 2020. Managing phosphorus loss from agroecosystems of the Midwestern United States: A review. *Agronomy* 10(4): 1–64. https://doi.org/10.3390/agronomy10040561.

Smith, D.R., R.D. Harmel, M. Williams, R. Haney, and K.W. King. 2016. Managing acute phosphorus loss with fertilizer source and placement: Proof of concept. *Agric. Environ. Lett.* 1(1): 150015. https://doi.org/10.2134/ael2015.12.0015.

Soltanpour, P.N., and A.P. Schwab. 1977. A new soil test for simultaneous extraction of macro- and micronutrients in alkaline soils. *Commun. Soil Sci. Plant Anal.* 8(3): 195–207.

Sposito, G. 1986. Distinguishing adsorption from surface precipitation. In *Geochemical Processes at Mineral Surfaces*, ed. J.A. Davies and K.F. Hayes. Am. Chem. Soc. Symp. Ser. 323. Am. Chem. Soc., Washington, DC.

Stevenson, F.J. 1982. *Humus Chemistry.* Wiley Interscience, New York.

Torbert, H.A., T.C. Daniel, J.L. Lemunyon, and R.M. Jones. 2002. Relationship of soil test phosphorus and sampling depth to runoff phosphorus in calcareous and noncalcareous soils. *J. Environ. Qual.* 31(4): 1380–1387.

United States Environmental Protection Agency. (1986). Quality criteria for water. USEPA Report 440/5-86-001. USEPA, Office of Water Regulations and standards, U.S. Gov. Print. Office, Washington, DC, 477pp.

U.S. Environmental Protection Agency. 2000. The quality of our nation's water. A summary of the National water Quality Inventory 1998: Report to Congress. EPA841-S-00-001.

US. Environmental Protection Agency. 2013. Aquatic life ambient water quality criteria for ammonia – Freshwater. EPA-822-R-13-001. USEPA, Office of Water, Office of Sci Technol, Washington, DC.

Vadas, P.A., L.W. Good, P.A. Moore, and N. Widman. 2009. Estimating phosphorus loss in runoff from manure and fertilizer for a phosphorus loss quantification tool. *J. Environ. Qual.* 38(4): 1645–1653. https://doi.org/10.2134/jeq2008.0337.

Vance, C.P., C. Uhde-Stone, and D. Allan. 2003. Phosphorus acquisition and use: Critical adaptation by plants for securing non-renewable resources. *New Phytol.* 15: 423–447.

Watts, D.B., H.A. Torbert, and E.E. Codling. 2019. Poultry production management on the buildup of nutrients in litter. *Int. J. Poult. Sci.* 18(9): 445–453. https://doi.org/10.3923/ijps.2019.445.453.

Zhang, T., Y. Wang, C.S. Tan, and T. Welacky. 2020. An 11-year agronomic, economic, and phosphorus loss potential evaluation of legacy phosphorus utilization in a clay loam soil of the Lake Erie Basin. *Front. Earth Sci.* 8: 115. https://doi.org/10.3389/feart.2020.00115.

2 Molecular Mechanisms Underlying Phosphate Sensing and Signalling in Plants

Viswanathan Satheesh, Abdul Wahab, Muthusamy Ramakrishnan, and Mingguang Lei

2.1 INTRODUCTION

Plants need phosphorus (P) for their normal growth and development, and they absorb P in the form of inorganic phosphate (Pi) ([$H_2PO_4^-$] and [HPO_4^{2-}]), as P is not readily available and less mobile in the soil (Holford 1997; Kochian et al. 2004; Seguel et al. 2013; Marschner 2012; do Nascimento et al. 2015). It is also required for cellular functions such as DNA/RNA and membrane phospho-lipid synthesis, as building blocks of ATP and intracellular signalling. Maintenance of Pi homeostasis is essential and possible only when plants can sense available Pi through concerted activities of acquiring Pi, followed by transport, metabolism, storage, and remobilisation. The Pi starvation response (PSR) is an adaptive mechanism that plants possess to navigate conditions of fluctuating levels of Pi, which helps in the maintenance of Pi homeostasis. Through morphophysiological, biochemical, and molecular changes, the plants are able to mount a response to counter the effects of inadequate Pi nutrition. These responses are then attenuated by the resupply of Pi to the plants (Chiou and Lin 2011; Wu et al. 2013; López-Arredondo et al. 2014). Even though a significant amount of progress has been made in deciphering the intricate details of plant Pi nutrition, recent studies on determining the intracellular Pi signalling molecule(s) and further studies that have built on the discovery, provide deeper insights into how plants cope with low Pi stress at the molecular level (Wu et al. 2013; López-Arredondo et al. 2014; Dong et al. 2019; Zhu et al. 2019). In this chapter, we delve into the latest research on how plants sense Pi and transmit the signal within the system to regulate cellular Pi homeostasis.

2.2 LOCAL PHOSPHATE SENSING

2.2.1 Root System Architectural (RSA) Changes and Adaptation

PSRs are either local or systemic in nature. Under this division, RSA modifications are related to the local response. These modifications are triggered by Pi concentration in the rhizosphere leading to such changes as primary root growth inhibition, increased lateral root growth, and increased root hair density (Bates and Lynch 1996).

2.2.2 Molecular Mechanisms Governing RSA Changes

The key molecular players in the inhibition of primary root growth under conditions that have reduced levels of Pi are the ferroxidases LOW PHOSPHATE ROOT1/2 (LPR1/2) and a P5-type ATPase PHOSPHATE DEFICIENCY RESPONSE2 (PDR2), (Svistonoff et al. 2007; Ticconi et al. 2009). While LPR1/2 are expressed in the meristem and root cap, PDR2 is located in the endoplasmic reticulum. Under Pi-starved conditions, LPR1 is probably trafficked to the plasma membrane from the ER to oxidise Fe^{2+} to Fe^{3+}, and this oxidation leads to the accumulation of the ferric ion in the apoplasm, which forms a complex with malate. Formation of the complex activates reactive oxygen species (ROS) production, promoting callose deposition in the stem cell niche (SCN) of the primary root (Müller et al. 2015). SHORT-ROOT (SHR) is an important transcription factor that governs root development, and the deposition of callose in the SCN restricts SHR movement and the stem cell function is lost, thereby arresting primary root growth (Müller et al. 2015). PDR2, interestingly, regulates SCARECROW (SCR) expression, another key transcription factor in root development, in the root of plants subject to low Pi conditions. This again hampers root patterning and stem cell maintenance (Ticconi et al. 2009). LPR1-mediated regulation of root apical meristem activity involves SENSITIVE TO PROTON RHIZOTOXICITY (STOP1), a transcription factor that regulates ALUMINUM ACTIVATED MALATE TRANSPORTER 1 (ALMT1) expression, which is a plasma membrane-localised malate secretion system (Müller et al. 2015; Mora-Macías et al. 2017; Balzergue et al. 2017; Godon et al. 2019). The CLAVATA3 (CLV3)/ENDOSPERM SURROUNDING REGION14 (CLE14) peptide that acts downstream of the PDR2–LPR1 module is perceived by the CLV2/PEP1 RECEPTOR2 (PEPR2) receptors and contributes to the suppression of the PIN-FORMED (PIN)–auxin and SHR–SCARECROW (SCR) pathways to cause RAM exhaustion that is callose-independent (Gutiérrez-Alanís et al. 2017). Even though these mechanisms could end in the reprogramming of the RSA, blue light alone could be enough to promote primary root growth inhibition under low Pi conditions (Zheng et al. 2019). It can do so by activating the malate-mediated photo-Fenton reaction producing hydroxyl radicals, which then cause primary root growth inhibition (Zheng et al. 2019). It was also observed that cryptochromes and their downstream signalling components SPA1, COP1, and HY5 are involved in RSA modification. The translocation of shoot-derived HY5 to the root requires blue light, which autoactivates root HY5, which in turn is required for LPR1 activation to inhibit primary root growth under Pi-depleted conditions (Gao et al. 2021).

2.2.3 Hormonal Control

Hormones are significant players in RSA modifications, and in this section, we discuss their influence on root growth under Pi starvation. The role of auxin in root development is well studied, and it is well known that PHR1, the central regulator of PSR, directly targets the auxin receptor TIR1 along with other genes from the ARF and Aux/IAA gene families under Pi starvation and therefore is suspected to influence lateral root formation and auxin signalling (López-Bucio et al. 2002; Pérez-Torres et al. 2008; Castrillo et al. 2017; Du and Scheres 2018). Several genes, including *ARF7, ARF19, ARK2, PUB9, SLR/IAA14, OsARF12, OsARF16, OsPht1;8, TAA1, AUX1*, and *RSL2/4* genes (Okushima et al. 2005; Narise et al. 2010; Shen et al. 2013; Wang et al. 2014b; Huang et al. 2018; Deb et al. 2014; Jia et al. 2017), have been implicated in the PSR pathways of both *Arabidopsis* and rice. These genes play significant roles in aspects ranging from lateral root initiation, root hair elongation, and Pi foraging.

Cytokinin (CK) is another important hormone that accentuates the root:shoot ratio under Pi-depleted conditions in favour of root growth (Kuiper and Steingröver 1991; Franco-Zorrilla et al. 2002). There are different types of CK, of which cis-zeatin and cis-zeatin riboside are PHR1-dependent and hence are activated when Pi levels are diminished, while trans-zeatin is attenuated under low Pi conditions (Silva-Navas et al. 2019). Higher levels of cis-zeatin compared to trans-zeatin improve root growth and lateral root formation, and cis-zeatin has also been implicated in root hair elongation and Pi allocation, which is critical under Pi-depleted conditions (Silva-Navas et al. 2019). Biosynthesis of ethylene (ET) under low Pi conditions is enhanced, and the transcription factor ETHYLENE-INSENSITIVE3 (EIN3) activates genes that are important for root hair formation (Song et al. 2016), and it has been previously established that ET does indeed promote root hair growth along with root elongation (Borch et al. 1999; Ma et al. 2003; Zhang et al. 2003). In addition, it is interesting to note that EIN3 binds to the *PHR1* promoter in the form of a complex with FHY3, FAR1, and HY5 to activate it (Liu et al. 2017).

Strigolactones (SLs) typically inhibit shoot branching and control lateral root formation and root hair density under Pi-depleted conditions (Umehara et al. 2008, Kapulnik et al. 2011). Gibberellic acid (GA) is known to affect anthocyanin accumulation and alterations in RSA, and GA signalling and metabolism can be altered by the activity of the MYB62 transcription factor to exert an influence on PSR (Jiang et al. 2007; Devaiah et al. 2009). Jasmonic acid (JA) regulates lateral root growth and root hair development (Wang et al. 2002; Cai et al. 2014). VIH1/2 is known to be required for the biosynthesis of inositol pyrophosphate InsP$_8$, a signal molecule driving PSR in plants (Dong et al. 2019; Zhu et al. 2019; Ried et al. 2021; Zhou et al. 2021; Guan et al. 2022). *OsJAZ11* is involved in the elongation of both primary and seminal roots in rice (Pandey et al. 2021). Hormones and their signalling pathways are interconnected with RSA modifications and, in extension, PSR, and more is yet to be uncovered in terms of their involvement and contributions in Pi sensing.

2.2.4 Epigenetic Control of RSA Changes under Pi Starvation

Studies on PSR *vis-à-vis* epigenetic control have been very few. Epigenetic mechanisms act as an additional layer of control in most, if not all, biological processes, and PSR is one such process that is governed by it (Yong-Villalobos et al. 2015; Secco et al. 2015). H2A.Z, a variant of H2A, regulates chromatin function by modulating nucleosome dynamics and DNA methylation. ARP6 is required for the deposition of H2A.Z at several known PSR genes; the failure of which results in inhibited primary roots and an increased number of root hairs (Smith et al. 2010). Further, a reduced H2A.Z chromatin occupancy led to an increase in Pi starvation inducible genes in the *atipk1-1* mutant (Kuo et al. 2014). Studies on methylation mutants *viz.*, *drm1*, *drm2*, and *cmt3*, individually and in the triple mutant, showed that low Pi resulted in shorter primary roots and lateral root density in the mutants than in the wild type (Yong-Villalobos et al. 2015).

In our latest study, we showed that BRAHMA (BRM), a switch defective/sucrose non-fermentable ATPase, is an important player in the local response mechanism in plants under Pi starvation (Li et al. 2022). Earlier studies showed that histone deacetylases HDA6, HDA9, and HDA19 are involved in the regulation of root cell length (Chen et al. 2015), and HDC1 is involved in influencing primary root growth and iron deposition at the root tips via LPR1/2 mediation under low Pi conditions (Xu et al. 2020). In our study, apart from showing that mutation in BRM leads to elevated LPR1/2 expression and the corresponding accumulation of iron under Pi-starved conditions, we also demonstrated the recruitment of HDA6-HDC1 complex by BRM to facilitate H3 deacetylate histone H3 at LPR loci. These recent studies clearly show the importance of epigenetic regulation of PSR genes, and more studies will be undertaken to enhance our understanding of these mechanisms.

2.3 SYSTEMIC PHOSPHATE SIGNALLING

2.3.1 Intracellular P Sensors: SPX Domains and Their Roles in Pi Signalling and Transport

Pi starvation is a major constraint affecting plant growth and development and hence agricultural productivity. A major breakthrough in this field was the identification of inositol pyrophosphates (PP-InsPs) as the metabolic messengers for Pi signalling. The SPX domain, initially identified in yeast (Syg1, Secco et al. 2012), is usually found in the N-terminus of eukaryotic proteins and is conserved and hydrophilic. The domain is divided into three subdomains separated by low-similarity regions (Liu et al. 2018). In plants, there are four groups of SPX domain proteins. Proteins containing only the SPX domain (AtSPX1, Duan et al. 2008; OsSPX1/2, Wang et al. 2009; and OsSPX4, Lv et al. 2014) are the first group. The second group contains one SPX domain and one EXS domain

(AtPHO1, Wild et al. 2016). The third has an SPX domain and a RING domain (AtNLA, Park et al. 2014). One SPX domain and one MFS domain (OsVPE1 and OsVPE2, Wang et al. 2015) are found in the fourth group.

With the identification of the SPX domain as a Pi sensor that interacts with PHR orthologous proteins to inhibit their function under Pi-replete conditions in the presence of the metabolic messengers inositol pyrophosphates (PP-InsPs) (Shi et al. 2014; Wang et al. 2014b; Puga et al. 2014; Lv et al. 2014; Wild et al. 2016; Dong et al. 2019; Osorio et al. 2019), several avenues have opened to understand in greater detail the Pi sensing and signalling mechanism in plants. Identification of the PP-InsPs as important signalling molecules in Pi sensing has led to structural studies that have shown how the interaction of these molecules with SPX proteins affects the binding of these proteins to the PHR proteins, the master regulators of Pi starvation responses in plants (Ried et al. 2021; Zhou et al. 2021; Guan et al. 2022). In these studies, the focus was on proteins with a single SPX domain.

AtNLA contains a single N-terminal SPX domain and a C-terminal RING domain and plays a key role in the regulation of the AtPHT1 Pi transporter (Lin et al. 2013; Park et al. 2014). Apart from AtPHT1, AtNLA also interacts with AtNRT1.7 and AtORE1 to regulate nitrogen and leaf senescence responses (Lin et al. 2013; Yang et al. 2020; Liu et al. 2017; Park et al. 2018). Given these studies, it would be interesting to study the effect of PP-InsPs on these interactions. The PHO1 protein has an SPX domain and an EXS domain, and interaction with PHO2 results in its degradation (Liu et al. 2012; Wege et al. 2016). Interestingly, when PHO1 lacked the SPX domain or when mutations were introduced in the PBC or KSC residues, PHO1 was unable to perform the root-to-shoot Pi translocation activity in the *pho1* mutant, suggesting the importance of the domain and InsP/PP-InsP interaction in Pi translocation (Wege et al. 2016; Wild et al. 2016).

PHT5-type is another important N-terminal SPX domain-containing protein with an MFS domain in the C-terminal end. It is a major vacuolar Pi import transporter. In yeast, the five-subunit vacuolar VTC complex transports polyP that is synthesised and stored in the vacuole, and four of the subunits contain the SPX domain. As the polymerase activity of the VTC complex is InsP7-dependent, mutating PBC or KSC residues dysregulate the polymerase activity of the complex (Lonetti et al. 2011; Wild et al. 2016). Therefore, the binding of PP-InsPs most probably affects the complex, structurally leading to alterations in its activity.

2.3.2 PP-InsPs: The Metabolic Messengers

The inositol pyrophosphates (PP-InsPs) are high-energy molecules derived from inositol polyphosphates (InsP$_6$) and play key roles as signal molecules in several biological processes in eukaryotes (Azevedo and Saiardi 2017; Shears 2018; Lorenzo-Orts et al. 2020). The PP-InsP, InsP$_7$, is obtained from InsP$_6$ by the catalytic activity of the enzyme inositol phosphate kinase ITPK1 (Riemer et al. 2021) and is further converted to InsP$_8$ by the bifunctional kinase/phosphatase enzymes VIH1 and VIH2 by phosphorylation (Laha et al. 2015; Wild et al. 2016; Dong et al. 2019; Zhu et al. 2019). Along with Hothorn and colleagues (Dong et al. 2019; Zhu et al. 2019; Ried et al. 2021), we showed that SPX domain-containing proteins bind to InsP$_8$, leading to restoration of the SPX1-PHR1 interaction, which is stronger than when 5-InsP$_7$ is involved. Both ITPK1 and VIH2 are interdependent in maintaining Pi homeostasis, as the formation of both InsP$_7$ and InsP$_8$ are dependent on both these enzymes (Riemer et al. 2021). Under conditions of an adequate supply of Pi, Pi regulates the phosphatase activity of VIH1/2, and there is increased production of InsP$_8$. This increased level of the PP-InsP translates into the formation of the InsP$_8$-SPX-PHR complex, thereby inhibiting the activity of PHR homologs (Gu et al. 2017; Zhu et al. 2019). The InsP$_8$-SPX1 complex binds to the CC domain of PHRs as dimers and prevents them from binding to the promoters of their target genes (Ried et al. 2021). In fact, InsPs stabilise the helix α1 structure by binding to the SPX protein enabling the InsPs to decouple the PHR protein dimer allosterically (Zhou et al. 2021). By oligomerisation and DNA binding, the SPX domain proteins inhibit PHR activity. Under low Pi conditions, however, a reduction in kinase activity results in reduced InsP$_8$ levels, and SPX domain proteins can no longer bind to PHR transcription factors (Dong et al. 2019; Zhu et al. 2019). These recent findings have greatly enhanced our understanding of the Pi signalling mechanism in plants and therefore have enabled researchers to identify newer avenues and methods to improve Pi use efficiency in crops.

2.4 PI TRANSPORT UNDER VARYING PI LEVELS

To support shoot growth, Pi is taken up by roots and exported into xylem vessels for long-distance translocation. PHO1 is a membrane-spanning protein localised in the Golgi with both its amino- and carboxy-termini toward the cytosol (Wege et al. 2016). The tripartite SPX domain in the N-terminal of PHO1 can bind with inositol pyrophosphate, indicating its function in Pi sensing (Wild et al. 2016). PHO1 is specifically expressed in the root pericycle and is critical for Pi loading into the xylem (Hamburger et al. 2002; Liu et al. 2012), and *pho1* mutants have significantly lower Pi content in the shoot with severe phosphate starvation response phenotypes (Poirier et al. 1991). PHO1 also plays an important role in transferring Pi from the seed coat to the embryo in developing seeds (Vogiatzaki et al. 2017). *PHO1* is regulated by Pi starvation at the transcriptional level. When Pi is replete, the transcription factor WRKY6 binds to the W-box motif in the *PHO1* promoter and represses its transcription (Chen et al. 2009). When Pi is depleted, PRU1, the E3 ligase, is induced, and it ubiquitinates WRKY6 to direct its degradation by 26S proteasome, which releases the PHO1 repression and enhances Pi translocation from root to shoot (Ye et al. 2018). PHO1 is also regulated at the post-transcriptional level. PHO1 associates with and is targeted by PHO2, a ubiquitin-conjugating E2 enzyme, for degradation through multivesicular body (MVB)-mediated vacuolar proteolysis (Liu et al. 2012). PHO2 not only targets PHO1 in pericycle cells but also mediates the degradation of PHT1 Pi transporters in epidermal and cortex layers to regulate Pi uptake (Huang et al. 2013). The *pho2* mutant absorbs and translocates higher levels of Pi to the shoot when Pi is replete (Dong et al. 1998). PHO2 itself is regulated by a systemic signal under Pi starvation at the post-transcriptional level. When Pi is deficient, PHR1 activates the expression of the microRNA gene, miR399, and the mature miR399 moves through the phloem from shoot to root where it directs the cleavage of *PHO2* mRNA (Aung et al. 2006; Bari et al. 2006; Chiou et al. 2006). Recently, it was demonstrated that PHO2, in concert with NITROGEN LIMITATION ADAPTATION (NLA), also regulates the stability of NAC transcription factor ORE1 in the nucleus to determine leaf senescence during nitrogen deficiency (Park et al. 2018). In our recent study, we report a novel role for SHR in the translocation of Pi from root to shoot (Xiao et al. 2022). We show that under Pi-deprived conditions, SHR protein accumulation is repressed, which ensures that the HD-ZIPIII transcription factor PHB is not inhibited by SHR. PHB activates *PHO2* expression leading to the degradation of PHO1 in the xylem-pole pericycle cells (Xiao et al. 2022).

Pi transporters, found in the plasma membrane, play an essential role in Pi acquisition because there exists a steep concentration gradient between levels of Pi in the plant and soil. PHOSPHATE TRANSPORTER 1 (PHT1) is a root cell transmembrane protein that facilitates the acquisition of Pi and is affected by the availability of Pi in the soil (Shin et al. 2004). The PHT1 proteins are activated under Pi-depleted conditions by the homologous transcription factors AtPHR1/AtPHL1, as they are released from the control of the SPX domain proteins (Rubio et al. 2001; Bustos et al. 2010; Dong et al. 2019; Zhu et al. 2019).

The PHT1 proteins are further regulated by three other important proteins, PHOSPHATE TRANSPORTER TRAFFIC FACILITATOR 1 (PHF1), PHO2 and NLA. While PHF1 is instrumental in controlling the targeting of PHT1 transporter proteins to the plasma membrane (Gonzalez et al. 2005), PHO2, a ubiquitin conjugase, and NLA, a ubiquitin E3 ligase, play key roles in the degradation of these proteins in the ER (Huang et al. 2013; Lin et al. 2013; Park et al. 2014) and plasma membrane (Bayle et al. 2011; Chen et al. 2015), respectively. Once ubiquitinated, the ALG2-interacting protein X (ALIX) in association with the endosomal complex required for transport

(ESCRT)-III transports the marked proteins to the vacuole where they are degraded (Cardona-López et al. 2015). Activation or retention of Pi transporters can be influenced by the phosphorylation, or the lack of it, of a serine residue in both *Arabidopsis* AtPHT1;1 and rice OsPT8 (Bayle et al. 2011; Chen et al. 2015). Phosphorylation ensures that the transporters are retained in the endoplasmic reticulum to attenuate intracellular Pi concentration. On the contrary, OsPT8 dephosphorylation by the type 2C protein phosphatase OsPP95 assists in its exit from the ER (Yang et al. 2020).

2.5 ROLE OF PHR IN PLANT–MICROBE INTERACTIONS

In recent years, the mechanisms regulating PHR1-driven PSR systems have become understood in great detail. How this system impacts plants and their interaction with soil fungi and their associated microbiome to enable the plants to adapt to the environment is another major area of focus that has gained traction currently. In the *phr1* mutant, plant response to other abiotic and biotic stresses is also compromised (Castrillo et al. 2017; Finkel et al. 2019). When subject to Pi starvation, severe morpho-physiological changes in root and shoot are noticed in plants that are not mycorrhizal in nature. These changes relate to depleted Pi stores (Raghothama 1999), lateral root growth, and altered root exudate profiles (Paredes et al. 2018; Ziegler et al. 2016). In *phr1*, *phl1*, and *phf1* mutants, the microbiota are altered in natural as well as synthetic microbial communities (Castrillo et al. 2017). The PHR1 effect on the plant immune system is negative regulatory in nature, resulting in elevated susceptibility to pathogens apart from the changes in the microbiota. To improve Pi uptake, plants establish symbioses with mycorrhizal fungi under Pi starvation. In legumes, symbiosis with arbuscular mycorrhizal (AM) fungi activates Pi transporters such as MtPT4 and LjPT3 (Maeda et al. 2006, Javot et al. 2007). In *Arabidopsis*, where symbiosis with AM fungi is not possible, symbiosis is established with Colletotrichum *tofieldiae*, an endophytic fungus, improving Pi uptake (Hiruma et al. 2016). This symbiosis is developed only under low Pi levels. Therefore, the host plant's Pi status is important in deciding the symbiotic relationship (Hacquard et al. 2016). In *Arabidopsis*, the transporters that are activated are Pht1;2 and Pht1;3 (Hiruma et al. 2016). Beneficial associations between *Arabidopsis* and *Bacillus amyloliquefaciens* have been reported to exist only under normal Pi levels, and under low Pi conditions, the plant response is similar to plants that are affected by pathogens (Morcillo et al. 2020). Activation of the transporters is by PHR1, which also integrates the PSR with responses from the immune system (Castrillo et al. 2017). Therefore, the Pi status of the plant determines symbiotic relationships as well as plant immunity responses.

2.6 CONCLUSIONS

Several significant studies deciphering critical molecular constituents in the Pi sensing and signalling pathway that are essential for the maintenance of Pi homeostasis in plants have been reported in recent years (Figure 2.1). Since the identification of SPX domain proteins as sensors that negatively regulate PHR-mediated activation of gene expression, SPX domain proteins containing only the SPX domain have been shown to regulate PHR transcription factors through interaction with InsP$_8$. The structural details of the exact mechanism by which SPX proteins repress PHR homologs were elucidated recently (Ried et al. 2021; Zhou et al. 2021; Guan et al. 2022). It would be interesting to see how SPX proteins with other associated domains such as SPX-MFS, SPX-RING, and SPX-EXS work to control key processes in the Pi sensing pathway. By disrupting the SPX domain in proteins such as NLA and PHT5-type, the effect on Pi transport can be studied, and to know if these proteins are affected by intracellular Pi concentrations, the PP-InsP binding sites in the SPX domain could be disrupted to study the effects of such mutations. Investigating these aspects of the Pi sensing and signalling pathway should provide us with essential information to improve our understanding of the mechanisms adopted by plants to acquire, allocate, and store Pi, allowing us to develop strategies to enhance P use efficiency in crops.

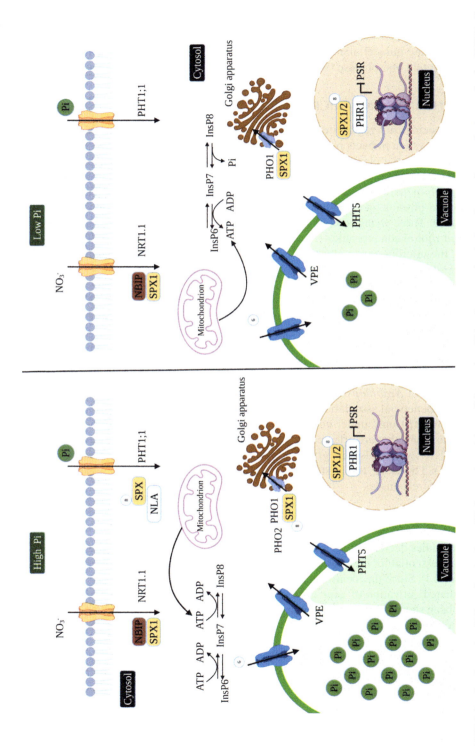

FIGURE 2.1 A schematic of Pi sensing in plants. Under nutrient-rich conditions, when Pi levels are optimal, there is a push towards the synthesis of PP-InsPs, particularly InsP$_8$, which binds to the sensor SPX protein eliciting structural changes that facilitate its binding to PHR1, rendering it inactive. In this bound state, PHR1 cannot drive the expression of genes that are involved in the Pi starvation response. On the contrary, when cellular Pi levels are diminished, InsP$_8$ levels also are reduced, and the SPX domain proteins without the ligand cannot bind to the PHR1 protein, hence allowing the transcription factor to bind to the promoters of the PSR genes to activate their expression.

REFERENCES

Aung K, Lin SI, Wu CC, Huang YT, Su CL, Chiou TJ (2006) pho2, a phosphate overaccumulator, is caused by a nonsense mutation in a microRNA399 target gene. *Plant Physiol* 141(3): 1000–1011.

Azevedo C, Saiardi A (2017) Eukaryotic phosphate homeostasis: The inositol pyrophosphate perspective. *Trends Biochem Sci* 42(3): 219–231.

Balzergue C, Dartevelle T, Godon C, Laugier E, Meisrimler C, Teulon JM, Creff A, Bissler M, Brouchoud C, Hagège A, et al. (2017) Low phosphate activates STOP1-ALMT1 to rapidly inhibit root cell elongation. *Nat Commun.* https://doi.org/10.1038/ncomms15300.

Bari R, Datt Pant B, Stitt M, Scheible WR (2006) PHO2, microRNA399, and PHR1 define a phosphate-signaling pathway in plants. *Plant Physiol* 141(3): 988–999.

Bates TR, Lynch JP (1996) Stimulation of root hair elongation in Arabidopsis thaliana by low phosphorus availability. *Plant Cell Environ* 19(5): 529–538.

Bayle V, Arrighi JF, Creff A, Nespoulous C, Vialaret J, Rossignol M, Gonzalez E, Paz-Ares J, Nussaume L (2011) Arabidopsis thaliana high-affinity phosphate transporters exhibit multiple levels of posttranslational regulation. *Plant Cell* 23(4): 1523–1535.

Borch K, Bouma TJ, Lynch JP, Brown KM (1999) Ethylene: A regulator of root architectural responses to soil phosphorus availability. *Plant Cell Environ* 22(4): 425–431.

Bustos R, Castrillo G, Linhares F, et al. (2010) A central regulatory system largely controls transcriptional activation and repression responses to phosphate starvation in Arabidopsis. *PLOS Genet* 6(9): e1001102.

Cai XT, Xu P, Zhao PX, Liu R, Yu LH, Xiang CB (2014) *Arabidopsis* ERF109 mediates cross-talk between jasmonic acid and auxin biosynthesis during lateral root formation. *Nat Commun* 5: 5833.

Cardona-López X, Cuyas L, Marín E, Rajulu C, Irigoyen ML, Gil E, Puga MI, Bligny R, Nussaume L, Geldner N, et al. (2015) ESCRT-III-associated protein ALIX mediates high-affinity phosphate transporter trafficking to maintain phosphate homeostasis in Arabidopsis. *Plant Cell* 27(9): 2560–2581.

Castrillo G, Lima Teixeira PJP, Paredes SH, et al. (2017) Root microbiota drive direct integration of phosphate stress and immunity. *Nature* 543(7646): 513–518.

Castrillo G, Teixeira PJPL, Paredes SH, Law TF, de Lorenzo L, Feltcher ME, Finkel OM, Breakfield NW, Mieczkowski P, Jones CD, Paz-Ares J, Dangl JL (2017) Root microbiota drive direct integration of phosphate stress and immunity. *Nature* 543(7646): 513–518. https://doi.org/10.1038/NATURE21417.

Chen CY, Wu K, Schmidt W (2015) The histone deacetylase HDA19 controls root cell elongation and modulates a subset of phosphate starvation responses in *Arabidopsis*. *Sci Rep* 5: 15708.

Chen J, Wang Y, Wang F, Yang J, Gao M, Li C, Liu Y, Liu Y, Yamaji N, Ma JF, et al. (2015) The rice CK2 kinase regulates trafficking of phosphate transporters in response to phosphate levels. *Plant Cell* 27(3): 711–723.

Chen Y-F, Li L-Q, Xu Q, Kong Y-H, Wang H, Wu W-H (2009) The WRKY6 transcription factor modulates PHOSPHATE1 expression in response to low Pi stress in Arabidopsis. *Plant Cell* 21(11): 3554–3566.

Chiou T-J, Lin S-I (2011) Signaling network in sensing phosphate availability in plants. *Annu Rev Plant Biol* 62: 185–206. https://doi.org/10.1146/annurev-arplant-042110-103849.

Chiou TJ, Aung K, Lin SI, Wu CC, Chiang SF, Su CL (2006) Regulation of phosphate homeostasis by microRNA in Arabidopsis. *Plant Cell* 18(2): 412–421.

Deb S, Sankaranarayanan S, Wewala G, Widdup E, Samuel MA (2014) The S-Domain Receptor kinase Arabidopsis Receptor Kinase2 and the U Box/Armadillo Repeat-Containing E3 ubiquitin Ligase9 module mediates lateral root development under phosphate starvation in Arabidopsis. *Plant Physiol* 165(4): 1647–1656.

Devaiah BN, Madhuvanthi R, Karthikeyan AS, Raghothama KG (2009) Phosphate starvation responses and gibberellic acid biosynthesis are regulated by the MYB62 transcription factor in Arabidopsis. *Mol Plant* 2(1): 43–58.

do Nascimento C, Pagliari P, Schmitt D, et al. (2015) Phosphorus concentrations in sequentially fractionated soil samples as affected by digestion methods. *Sci Rep* 5: 17967. https://doi.org/10.1038/srep17967.

Dong B, Rengel Z, Delhaize E (1998) Uptake and translocation of phosphate by *pho2* mutant and wild-type seedlings of *Arabidopsis thaliana*. *Planta* 205(2): 251–256.

Dong J, Ma G, Sui L, Wei M, Satheesh V, Zhang R, Ge, S, Li J, Zhang T-E, Wittwer C, et al. (2019) Inositol pyrophosphate InsP8 acts as an intracellular phosphate signal in Arabidopsis. *Mol Plant* 12(11): 1463–1473.

Du Y, Scheres B (2018) Lateral root formation and the multiple roles of auxin. *J Exp Bot* 69(2): 155–167. https://doi.org/10.1093/JXB/ERX223.

Duan K, Yi K, Dang L, Huang H, Wu W, Wu P (2008) Characterization of a sub-family of Arabidopsis genes with the SPX domain reveals their diverse functions in plant tolerance to phosphorus starvation. *Plant J* 54(6): 965–975.

Finkel O, Salas-González I, Castrillo G, Spaepen S, Law TF, Teixeira PJPL., et al. (2019) The effects of soil phosphorus content on plant microbiota are driven by the plant phosphate starvation response. *PLOS Biol* 17(11): e3000534.

Finkel OM, Salas-González I, Castrillo G, et al. (2019) The effects of soil phosphorus content on plant microbiota are driven by the plant phosphate starvation response. *PLOS Biol* 17(11): e3000534.

Franco-Zorrilla JM Martín AC, Solano R, Rubio V, Leyva A, Paz-Ares J (2002). Mutations at CRE1 impair cytokinin-induced repression of phosphate starvation responses in Arabidopsis. *Plant J* 32: 353–360.

Gao Y-Q, Bu L-H, Han M-L, Wang Y-L, Li Z-Y, Liu H-T, Chao D-Y (2021) Long-distance blue light signalling regulates phosphate deficiency-induced primary root growth inhibition. *Mol Plant* 14(9): 1539–1553.

Godon C, Mercier C, Wang X, David P, Richaud P, Nussaume L, Liu D, Desnos T (2019) Under phosphate starvation conditions, Fe and Al trigger accumulation of the transcription factor STOP1 in the nucleus of Arabidopsis root cells. *Plant J* 99(5): 937–949.

Gonzalez E, Solano R, Rubio V, Leyva A, Paz-Ares J (2005) Phosphate transporter traffic facilitator1 is a plant-specific SEC12-related protein that enables the endoplasmic reticulum exit of a high-affinity phosphate transporter in Arabidopsis. *Plant Cell* 17(12): 3500–3512.

Gu C, Nguyen H-N, Hofer A, Jessen HJ, Dai X, Wang H, et al. (2017a) The significance of the bifunctional kinase/phosphatase activities of PPIP5Ks for coupling inositol pyrophosphate cell-signaling to cellular phosphate homeostasis. *J Biol Chem* 292(11): 4544–4555.

Guan Z, Zhang Q, Zhang Z, et al. (2022) Mechanistic insights into the regulation of plant phosphate homeostasis by the rice SPX2 – PHR2 complex. *Nat Commun* 13(1): 1581.

Gutierrez-Alanis D, Yong-Villalobos L, Jimenez-Sandoval P, Alatorre-Cobos F, Oropeza-Aburto A, Mora-Macias J, Sanchez-Rodriguez F, Cruz-Ramirez A, Herrera-Estrella L (2017) Phosphate starvation-dependent iron mobilization induces CLE14 expression to trigger root meristem differentiation through CLV2/PEPR2 signaling. *Dev Cell* 41(5): 555-570.e3.

Hacquard S, Kracher B, Hiruma K, Münch PC, Garrido-Oter R, Thon MR, et al. (2016) Survival trade-offs in plant roots during colonization by closely related beneficial and pathogenic fungi. *Nat Commun* 7: 11362.

Hamburger D, Rezzonico E, MacDonald-Comber Petetot J, Somerville C, Poirier Y (2002) Identification and characterization of the Arabidopsis PHO1 gene involved in phosphate loading to the xylem. *Plant Cell* 14(4): 889–902.

Hiruma K, Gerlach N, Sacristán S, Nakano RT, Hacquard S, Kracher B., et al. (2016) Root endophyte Colletotrichum tofieldiae confers plant fitness benefits that are phosphate status dependent. *Cell* 165(2): 464–474.

Holford ICR (1997) Soil phosphorus: Its measurement, and its uptake by plants. *Aust J Soil Res* 35(2): 227–239.

Huang KL, Ma GJ, Zhang ML, Xiong H, Wu H, Zhao CZ, Liu CS, Jia HX, Chen L, Ren F (2018) The ARF7 and ARF19 transcription factors positively regulate PHOSPHATE STARVATION RESPONSE1 in Arabidopsis roots. *Plant Physiol* 178(1): 413–427.

Huang TK, Han CL, Lin SI, Chen YJ, Tsai YC, Chen YR, Chen JW, Lin WY, Chen PM, Liu TY, et al. (2013) Identification of downstream components of ubiquitin-conjugating enzyme PHOSPHATE2 by quantitative membrane proteomics in Arabidopsis roots. *Plant Cell* 25(10): 4044–4060.

Huang TK, Han CL, Lin SI, Chen YJ, Tsai YC, Chen YR, Chen JW, Lin WY, Chen PM, Liu TY, et al. (2013) Identification of downstream components of ubiquitin-conjugating enzyme PHOSPHATE2 by quantitative membrane proteomics in Arabidopsis roots. *Plant Cell* 25(10): 4044–4060.

Javot H, Penmetsa RV, Terzaghi N, Cook DR, Harrison MJ (2007) A Medicago truncatula phosphate transporter indispensable for the arbuscular mycorrhizal symbiosis. *Proc Natl Acad Sci U S A* 104(5): 1720–1725.

Jia H, Zhang S, Wang L, Yang Y, Zhang H, Cui H, et al. (2017) OsPht1;8, a phosphate transporter, is involved in auxin and phosphate starvation response in rice. *J Exp Bot* 68(18): 5057–5068.

Jiang C, Gao X, Liao L, Harberd NP, Fu X (2007) Phosphate starvation root architecture and anthocyanin accumulation responses are modulated by the gibberellin–DELLA signaling pathway in Arabidopsis. *Plant Physiol* 145(4): 1460–1470.

Kapulnik Y, Delaux PM, Resnick N, Mayzlish-Gati E, Wininger S, Bhattacharya C, et al. (2011) Strigolactones affect lateral root formation and root-hair elongation in Arabidopsis. *Planta* 233(1): 209–216.

Kochian LV, Hoekenga OA, Pinero MA (2004) How do crop plants tolerate acid soils? Mechanisms of aluminum tolerance and phosphorus efficiency. *Annu Rev Plant Biol* 55: 459–493.

Kuiper D, SteingrÖVer E (1991) Responses of growth, shoot to root ratio and cytokinin concentrations in root tissue of two barley varieties, differing if their salt resistance. In *Developments in Agricultural and Managed Forest Ecology* McMichael BL, Persson H eds. Elsevier: NY, USA, 463–471.

Kuo HF, Chang TY, Chiang SF, Wang WD, Charng YY, Chiou TJ (2014) Arabidopsis inositol pentakisphosphate 2-kinase, AtIPK1, is required for growth and modulates phosphate homeostasis at the transcriptional level. *Plant J* 80(3): 503–515.

Laha D, Johnen P, Azevedo C, Dynowski M, Weiss M, Capolicchio S, Mao H, Iven T, Steenbergen M, Freyer M, et al. (2015) VIH2 regulates the synthesis of inositol pyrophosphate InsP8 and jasmonate-dependent defenses in Arabidopsis. *Plant Cell* 27(4): 1082–1097.

Li T, Zhang R, Satheesh V, Wang P, Ma G, Guo J, An GY, Lei M (2022) The chromatin remodeler BRAHMA recruits histone DEACETYLASE6 to regulate root growth inhibition in response to phosphate starvation in Arabidopsis. *J Integr Plant Biol*. Aug 16. https://doi.org/10.1111/jipb.13345 (Epub ahead of print). PMID: 35972795.

Lin WY, Huang TK, Chiou TJ (2013) Nitrogen limitation adaptation, a target of microRNA827, mediates degradation of plasma membrane-localized phosphate transporters to maintain phosphate homeostasis in Arabidopsis. *Plant Cell* 25(10): 4061–4074.

Lin WY, Huang TK, Chiou TJ (2013) Nitrogen limitation adaptation, a target of microRNA827, mediates degradation of plasma membrane-localized phosphate transporters to maintain phosphate homeostasis in Arabidopsis. *Plant Cell* 25(10): 4061–4074.

Liu N, Shang W, Li C, Jia L, Wang X, Xing G, Zheng W (2018) Evolution of the SPX gene family in plants and its role in the response mechanism to phosphorus stress. *Open Biol* 8(1): 170231.

Liu TY, Huang TK, Tseng CY, Lai YS, Lin SI, Lin WY, Chen JW, Chiou TJ (2012) PHO2-dependent degradation of PHO1 modulates phosphate homeostasis in Arabidopsis. *Plant Cell* 24(5): 2168–2183.

Liu W, Sun Q, Wang K, Du Q, Li W-X (2017) Nitrogen limitation adaptation (NLA) is involved in source-to-sink remobilization of nitrate by mediating the degradation of NRT1.7 in Arabidopsis. *New Phytol* 214(2): 734–744.

Liu Y, Xie Y, Wang H, Ma X, Yao W, Wang H (2017) Light and ethylene coordinately regulate the phosphate starvation response through transcriptional regulation of PHOSPHATE STARVATION RESPONSE1. *Plant Cell* 29(9): 2269. https://doi.org/10.1105/TPC.17.00268.

Lonetti A, Szijgyarto Z, Bosch D, Loss O, Azevedo C, Saiardi A (2011) Identification of an evolutionarily conserved family of inorganic polyphosphate endopolyphosphatases. *J Biol Chem* 286 (37): 31966–31974.

López-Arredondo DL, Leyva-González MA, González-Morales SI, López-Bucio J, Herrera-Estrella L (2014) Phosphate nutrition: Improving low-phosphate tolerance in crops. *Annu Rev Plant Biol* 65: 95–123. https://doi.org/10.1146/annurev-arplant-050213-035949.

Lopez-Bucio J, Hernandez-Abreu E, Sanchez-Calderon L, Nieto-Jacobo M, Simpson J, Herrera-Estrella L (2002) Phosphate availability alters architecture and causes changes in hormone sensitivity in the Arabidopsis root system. *Plant Physiol* 129(1): 244–256.

Lorenzo-Orts L, Couto D, Hothorn M (2020) Identity and functions of inorganic and inositol polyphosphates in plants. *New Phytol* 225(2): 637–652.

Lv Q, Zhong Y, Wang Y, Wang Z, Zhang L, Shi J, Wu Z, Liu Y, Mao C, Yi K Wu P (2014) SPX4 negatively regulates phosphate signaling and homeostasis through its interaction with PHR2 in rice. *Plant Cell* 26(4): 1586–1597.

Ma Z, Baskin TI, Brown KM, Lynch JP (2003) Regulation of root elongation under phosphorus stress involves changes in ethylene responsiveness. *Plant Physiol* 131(3): 1381–1390.

Maeda D, Ashida K, Iguchi K, Chechetka SA, Hijikata A, Okusako Y, et al. (2006) Knockdown of an arbuscular mycorrhiza-inducible phosphate transporter gene of Lotus japonicus suppress mutualistic symbiosis. *Plant Cell Physiol* 47(7): 807–817.

Marschner P (2012) *Marschner's Mineral Nutrition of Higher Plants*. 3rd edition. Academic Press: Cambridge, 649 p.

Mora-Macías J, Ojeda-Rivera JO, Gutiérrez-Alanís D, Yong-Villalobos L, Oropeza-Aburto A, Raya-González J, Jiménez-Domínguez G, Chávez-Calvillo G, Rellán-Álvarez R, Herrera-Estrella L (2017) Malate-dependent Fe accumulation is a critical checkpoint in the root developmental response to low phosphate. *Proc Natl Acad Sci* 114(17): E3563–E3572.

Morcillo RJL, Singh SK, He D, An G, Vílchez JI, Tang K, et al. (2020) Rhizobacterium-derived diacetyl modulates plant immunity in phosphate-dependent manner. *EMBO J* 39(2): e102602.

Müller J, Toev T, Heisters M, Teller J, Moore KL, Hause G, Dinesh DC, Bürstenbinder K, Abel S (2015) Iron-dependent callose deposition adjusts root meristem maintenance to phosphate availability. *Dev Cell* 33: 216–230.

Narise T, Kobayashi K, Baba S, Shimojima M, Masuda S, Fukaki H, et al. (2010) Involvement of auxin signaling mediated by IAA14 and ARF7/19 in membrane lipid remodeling during phosphate starvation. *Plant Mol Biol* 72(4–5): 533–544.

Okushima Y, Overvoorde PJ, Arima K, Alonso JM, Chan A, Chang C, Ecker JR, Hughes B, Lui A, Nguyen D, Onodera C, Quach H, Smith A, Yu G, Theologis A (2005) Functional genomic analysis of the AUXIN RESPONSE FACTOR gene family members in Arabidopsis thaliana: Unique and overlapping functions of ARF7 and ARF19. *Plant Cell* 17(2): 444–463. https://doi.org/10.1105/TPC.104.028316.

Osorio MB, Ng S, Berkowitz O, De Clercq I, Mao C, Shou H, Whelan J, Jost R (2019) SPX4 acts on PHR1-dependent and -independent regulation of shoot phosphorus status in Arabidopsis. *Plant Physiol.* Sep; 181(1): 332–352.

Pandey BK, Verma L, Prusty A, Singh AP, Bennett MJ, Tyagi AK, Giri J, Mehra P (2021) OsJAZ11 regulates phosphate starvation responses in rice. *Planta* 254(1): 8.

Paredes SH, Gao T, Law TF, et al. (2018) Design of synthetic bacterial communities for predictable plant phenotypes. *PLOS Biol* 16(2): e2003962.

Park BS, Seo JS, Chua NH (2014) Nitrogen limitation adaptation recruits phosphate2 to target the phosphate transporter PT2 for degradation during the regulation of Arabidopsis phosphate homeostasis. *Plant Cell* 26(1): 454–464.

Park BS, Yao T, Seo JS, Wong ECC, Mitsuda N, Huang CH, Chua NH (2018) Arabidopsis nitrogen limitation adaptation regulates ORE1 homeostasis during senescence induced by nitrogen deficiency. *Nat Plants* 4(11): 898–903.

Pérez-Torres CA, Lopez-Bucio J, Cruz-Ramirez A, Ibarra-Laclette E, Dharmasiri S, Estelle M, Herrera-Estrella L (2008) Phosphate availability alters lateral root development in Arabidopsis by modulating auxin sensitivity via a mechanism involving the TIR1 auxin receptor. *Plant Cell* 20(12): 3258–3272.

Poirier Y, Thoma S, Somerville C, Schiefelbein J (1991) Mutant of Arabidopsis deficient in xylem loading of phosphate. *Plant Physiol* 97(3): 1087–1093.

Puga MI, Mateos I, Charukesi R, Wang Z, Franco-Zorrilla JM, de Lorenzo L, Irigoyen ML, Masiero S, Bustos W, Rodríguez J, Leyva A, Rubio V, Sommer H, Paz-Ares J (2014) SPX1 is a phosphate-dependent inhibitor of Phosphate Starvation Response 1 in Arabidopsis. *Proc Natl Acad Sci USA* 111(41): 14947–14952.

Raghothama KG (1999) Phosphate acquisition. *Annu Rev Plant Physiol Plant Mol Biol* 50: 665–693.

Ried MK, Wild R, Zhu J, Pipercevic J, Sturm K, Broger L, et al. (2021) Inositol pyrophosphates promote the interaction of SPX domains with the coiled-coil motif of PHR transcription factors to regulate plant phosphate homeostasis. *Nat Commun* 12(1): 384.

Riemer E, Qiu D, Laha D, et al. (2021) ITPK1 is an InsP6/ADP pohsphotransferase that controls phosphate signaling in Arabidopsis. *Mol Plant* 14(11): 1864–1880.

Rubio V, Linhares F, Solano R, Martin AC, Iglesias J, Leyva A, et al (2001) A conserved MYB transcription factor involved in phosphate starvation signaling both in vascular plants and in unicellular algae. *Genes Dev* 15(16): 2122–2133.

Secco D, Wang C, Arpat BA, Wang Z, Poirier Y, Tyerman SD, Wu P, Shou H, Whelan J (2012) The emerging importance of the SPX domain-containing proteins in phosphate homeostasis. *New Phytol* 193(4): 842–851.

Secco D, Wang C, Shou H, Schultz MD, Chiarenza S, Nussaume L, Ecker JR, Whelan J, Lister R (2015) Stress induced gene expression drives transient DNA methylation changes at adjacent repetitive elements. *Elife* 4: e09343.

Seguel A, Cumming J, Klugh-Stewart K, Cornejo P, Borie P (2013) The role of arbuscular mycorrhizas in decreasing aluminium phytotoxicity in acidic soils: A review. *Mycorrhiza* 23(3): 167–183.

Shears SB (2018) Intimate connections: Inositol pyrophosphates at the interface of metabolic regulation and cell signaling. *J Cell Physiol* 233(3): 1897–1912.

Shen C, Wang S, Zhang S, Xu Y, Qian Q, Qi Y, et al. (2013) OsARF16, a transcription factor, is required for auxin and phosphate starvation response in rice (Oryza sativaL.). *Plant Cell Environ* 36(3): 607–620.

Shi J, Hu H, Zhang K et al. (2014) The paralogous SPX3 and SPX5 genes redundantly modulate pi homeostasis in rice. *J Exp Bot* 65: 859–870.

Shin H, Shin HS, Dewbre GR, Harrison MJ (2004) Phosphate transport in Arabidopsis: Pht1;1 and Pht1;4 play a major role in phosphate acquisition from both low- and high-phosphate environments. *Plant J* 39(4): 629–642.

Silva-Navas J, Conesa CM, Saez A, Navarro-Neila S, Garcia-Mina JM, Zamarreño AM, Baigorri R, Swarup R, del Pozo JC (2019) Role of cis-zeatin in root responses to phosphate starvation. *New Phytol* 224(1): 242–257. https://doi.org/10.1111/NPH.16020.

Smith AP, Jain A, Deal RB, Nagarajan VK, Poling MD, Raghothama KG, Meagher RB (2010) Histone H2A.Z regulates the expression of several classes of phosphate starvation response genes but not as a transcriptional activator. *Plant Physiol* 152(1): 217–225. https://doi.org/10.1104/pp.109.145532.

Song L, Yu H, Dong J, Che X, Jiao Y, Liu D (2016) The molecular mechanism of ethylene-mediated root hair development induced by phosphate starvation. *PLOS Genet* 12(7): e1006194.

Svistoonoff S, Creff A, Reymond M, Sigoillot-Claude C, Ricaud L, Blanchet A, Nussaume L, Desnos T (2007) Root tip contact with low-phosphate media reprograms plant root architecture. *Nat Genet* 39(6): 792–796.

Ticconi CA, Lucero RD, Sakhonwasee S, Adamson AW, Creff A, Nussaume L, Desnos T, Abel S (2009) ER-resident proteins PDR2 and LPR1 mediate the developmental response of root meristems to phosphate availability. *Proc Natl Acad Sci USA* 106(33): 14174–14179.

Umehara M, Hanada A, Yoshida S, Akiyama K, Arite T, Takeda-Kamiya N, et al. (2008) Inhibition of shoot branching by new terpenoid plant hormones. *Nature* 455(7210): 195–200.

Vogiatzaki E, Baroux C, Jung J-Y, Poirier Y (2017) PHO1 exports phosphate from the chalazal seed coat to the embryo in developing Arabidopsis seeds. *Curr Biol* 27(19): 2893–2900.

Wang C, Yue W, Ying Y, Wang S, Secco D, Liu Y, Whelan J, Tyerman SD, Shou H (2015) Rice SPX-major facility Superfamily3, a vacuolar phosphate efflux transporter, is involved in maintaining phosphate homeostasis in rice. *Plant Physiol* 169(4): 2822–2831.

Wang S, Ichii M, Taketa S, Xu L, Xia K, Zhou X (2002) Lateral root formation in rice (Oryza sativa): Promotion effect of jasmonic acid. *J Plant Physiol* 159(8): 827–832.

Wang S, Zhang S, Sun C, Xu Y, Chen Y, Yu C, et al (2014b) Auxin response factor (OsARF12), a novel regulator for phosphate homeostasis in rice (Oryza sativa). *New Phytol* 201(1): 91–103.

Wang Z, Hu H, Huang H, Duan K, Wu Z, Wu P (2009) Regulation of OsSPX1 and OsSPX3 on expression of OsSPX domain genes and Pi-starvation signaling in rice. *J Integr Plant Biol* 51(7): 663–674.

Wege S, Khan GA, Jung JY, Vogiatzaki E, Pradervand S, Aller I, Meyer AJ, Poirier Y (2016) The EXS domain of PHO1 participates in the response of shoots to phosphate deficiency via a root-to-shoot signal. *Plant Physiol* 170(1): 385–400.

Wild R, Gerasimaite R, Jung JY, Truffault V, Pavlovic I, Schmidt A, Saiardi A, Jessen HJ, Poirier Y, Hothorn M, Mayer A (2016) Control of eukaryotic phosphate homeostasis by inositol polyphosphate sensor domains. *Science* 352(6288): 986–990.

Wu P, Shou H, Xu G, Lian X (2013) Improvement of phosphorus efficiency in rice on the basis of understanding phosphate signaling and homeostasis. *Curr Opin Plant Biol* 16(2): 205–212.

Xiao X, Zhang J, Satheesh V, et al. (2022) Short-root stabilizes phosphate1 to regulate phosphate allocation in Arabidopsis. *Nature Plants* 8(9): 1074–1081.

Xu JM, Wang ZQ, Wang JY, Li PF, Jin JF, Chen WW, Fan W, Kochian LV, Zheng SJ, Yang JL (2020) Low phosphate represses histone deacetylase complex1 to regulate root system architecture remodeling in Arabidopsis. *New Phytol* 225(4): 1732–1745.

Yang Z, Yang J, Wang Y, Wang F, Mao W, He Q, Xu J, Wu Z, Mao C (2020b) Protein Phosphatase95 regulates phosphate homeostasis by affecting phosphate transporter trafficking in rice. *Plant Cell* 32(3): 740–757.

Ye Q, Wang H, Su T, Wu WH, Chen YF (2018) The ubiquitin E3 ligase PRU1 regulates WRKY6 degradation to modulate phosphate homeostasis in response to low-pi stress in Arabidopsis. *Plant Cell* 30(5): 1062–1076.

Yong-Villalobos L, Gonzalez-Morales SI, Wrobel K, Gutierrez-Alanis D, Cervantes-Perez SA, Hayano-Kanashiro C, Oropeza-Aburto A, Cruz-Ramirez A, Martinez O, Herrera-Estrella L (2015) Methylome analysis reveals an important role for epigenetic changes in the regulation of the Arabidopsis response to phosphate starvation. *Proc Natl Acad Sci USA* 112(52): E7293–E7730.

Zhang YJ, Lynch JP, Brown KM (2003) Ethylene and phosphorus availability have interacting yet distinct effects on root hair development. *J Exp Bot* 54(391): 2351–2361.

Zheng Z, Wang Z, Wang X, Liu D (2019) Blue light-triggered Chemical reactions underlie phosphate deficiency-induced inhibition of root elongation of *Arabidopsis* seedlings grown in Petri dishes. *Mol Plant* 12(11): 1515–1523.

Zhou J, Hu Q, Xiao X, Yao D, Ge S, Ye J., Li H., Cai et al. (2021) Mechanism of phosphate sensing and signaling revealed by rice SPX1-PHR2 complex structure. *Nat Commun* 12(1): 7040.

Zhu J, Lau K, Puschmann R, Harmel RK, Zhang Y, Pries V, Gaugler P, Broger L, Dutta AK, Jessen HJ, et al. (2019) Two bifunctional inositol pyrophosphate kinases/phosphatases control plant phosphate homeostasis. *Elife* 8: 1–25.

Ziegler J, Schmidt S, Chutia R, et al (2016) Non-targeted profiling of semi-polar metabolites in Arabidopsis root exudates uncovers a role for coumarin secretion and lignification during the local response to phosphate limitation. *J Exp Bot* 67(5): 1421–1432.

3 Phosphate Homeostasis and Root Development in Crop Plants

Dhriti Singh and Santosh B. Satbhai

3.1 INTRODUCTION

Plants require a plethora of mineral nutrients for growth and development. These nutrients play a vital role in various biochemical, cellular, and physiological processes that help plants to complete their life cycle. Therefore, these nutrients are some of the critical determinants that influence crop yield worldwide. Although soil is rich in various nutrients, it has been shown that at least 17 nutrients are essential for the optimal growth and development of plants (Kumar, Kumar, and Mohapatra 2021; Prathap et al. 2022). Deficiency or excess of any of these nutrients has a severe impact on crop yield. Phosphorus (P) is one of the essential macronutrients limiting plant growth, development, and crop productivity. P is a component of various biomolecules, including nucleic acids, membrane phospholipids, proteins, and sugars (Prathap et al. 2022). This nutrient is also involved in many metabolic processes, such as glycolysis, photosynthesis, redox reactions, etc. as metabolic intermediates. Thus, P is a key element of most of the molecular components involved in the proper functioning of plants.

P is present in the soil in various inorganic and organic forms. However, plants are unable to absorb organic P unless hydrolysed into inorganic phosphates (Pi) via phosphatases (Vincent et al. 2012). Plants can absorb Pi predominantly present in the form of orthophosphates ($H_2PO_4^-$ or HPO_4^{2-}). These orthophosphates are immobile and are heterogeneously distributed in soil. Several environmental factors influence the P availability in the soil, such as the pH of the soil, the presence of cations, and the presence of microorganisms responsible for the rapid conversion of P into organic forms (Alatorre-Cobos, López-Arredondo, and Herrera-Estrella 2009). Due to all these factors, it is estimated that a significant fraction of cultivated soil has low Pi worldwide (Vance 2001). Moreover, low Pi availability can reduce the crop yield by 30–40% (Mo et al. 2022). Therefore, in order to overcome the P limitation in soil and to increase crop yield, a large amount of P fertilisers are used across the globe. However, crops can employ only 10–15% of the P fertilisers, and the remainder is either absorbed and/or fixed due to microbial activity making it unusable (Prathap et al. 2022). Excessive use of P fertilisers leads to environmental pollution such as eutrophication as well as wastage of natural phosphate rock resources (Ju et al. 2007; Veneklaas et al. 2012). Moreover, despite the usage of P fertilisers, P availability in soil often fluctuates during the life cycle of crops because of the high rate of fixation and slow diffusion of P in soil. Therefore, plants have evolved a suite of molecular, biochemical, and physiological strategies to maintain P homeostasis in fluctuating and often limiting P conditions. These adaptations help crops to alleviate the effect of fluctuating P and increase the yield in the stress condition.

Plant roots are indispensable organs that provide anchorage and mechanical support to plants and absorb water and nutrients from the soil. Water and nutrients, including P, are heterogeneously distributed in the soil; consequently, the spatial organisation of roots is imperative to maximise the P absorption and transport, and thereby optimal plant growth and development (Koevoets et al.

2016). Therefore, a comprehensive study of root development and plasticity in response to varying soil P availability is important to better understand plant adaptation to P stress. To better understand root development, we must first understand the root structure. The three-dimensional organisation of the root and its components is described as root system architecture (RSA). Root components include primary roots, lateral roots, and accessory roots. RSA is determined by the length, number, arrangement, and angle of these components (Fromm 2019; Koevoets et al. 2016). These features influence the explored volume of the soil in conjunction with the root surface area and hence regulate the functioning of the aerial parts (Comas et al. 2013). Both monocotyledon (monocot) and dicotyledon (dicot) root systems can be divided into two parts. The embryonically derived primary root and post-embryonically formed lateral roots and adventitious roots are collectively known as secondary roots (Scheres et al. 1994; Fromm 2019). Lateral roots derive from the branching of primary roots and further branching of already formed lateral roots. Adventitious roots are derived from non-root tissues such as adventitious roots may develop from the lower underground stem nodes, andit may develop from the scutellum in the monocots (Bellini, Pacurar, and Perrone 2014; Fromm 2019). There are other differences between the monocot and dicot root systems. For example, dicot root systems consist of one well-developed primary root and several kinds of lateral roots, known as a tap root system, while the monocot root system has embryonic seminal roots and non-embryonic crown and nodal roots forming a fibrous root system (Koevoets et al. 2016; Fromm 2019). The ability of plants to grow roots from non-root tissues provides them with incredible resilience in a wide range of natural conditions and helps in better absorption of water and minerals as well as mechanical support. The typical root system also consists of root hairs, which are epidermal cell extensions. Root hairs develop behind the elongation zone from a group of particular epidermal cells known as trichoblasts. These trichoblast cells are organised in the roots depending upon the species. Root hairs enormously enhance the root surface area and thereby increase the surface area for water and nutrient absorption as well as the interaction between the plant and microbes (Fromm 2019; Salazar-Henao, Vélez-Bermúdez, and Schmidt 2016).

Various genetic, physiological, and environmental factors steer root development in terms of primary root length, root growth direction, and the number and distribution of lateral roots and root hairs. Among other environmental conditions that affect RSA, P is a prominent factor that modulates RSA tremendously.

3.2 PHOSPHATE HOMEOSTASIS AND ROOT SYSTEM ARCHITECTURE OF CROPS

Extensive research has been done to understand the effect of P availability on root system architecture (RSA). P diffuses slowly in the soil solution and is rapidly fixated, usually resulting in the maximum concentration of Pi in the upper layer of soil. In order to enhance the uptake of Pi, plants forage the topsoil (Lynch 2019). Topsoil foraging in crops generally reorganises the RSA with more production of axial roots, shallower axial root angle, greater lateral root density, and longer and denser root hairs, and is hypothesised to be advantageous for P acquisition (D. Liu 2021; Lynch 2019).

A study on the effect of variable P supply on seven crop plants (maize, wheat, canola, white lupin, soybean, faba bean, and chickpea) showed that species with fibrous roots rely more on external P application. However, they showed low reliance on inherent soil residual P. Contrary to this, crops with tap root systems could procure a substantial amount of soil residual P, and they show comparatively low dependency on fertiliser P application (Lyu et al. 2016). For this study, crops were grown on acidic and calcareous soil with or without P fertiliser, and their RSA was examined. Interestingly, different species showed a significant difference in their RSA modification in response to low P conditions. Crops with the fibrous root system (wheat, maize, and canola) showed a significant decrease in total root length in response to P deficiency in both kinds of soil. On the other hand, crops with tap root systems showed either an intermediate reduction in total root length (soybean and chickpea) or no difference in total root length (white lupin and common bean). In addition to this, crops with fibrous roots showed an increase or decrease in their root surface area in response to low P, but the root surface area of crops with tap root systems remained essentially unchanged. Generally, crops with fibrous root systems exhibited high root plasticity in response to variable P levels. Contrary to this, crops with tap root systems increased root exudation including acid phosphatases and carboxylate in response to low P.

To identify adaptive root traits in response to low Pi conditions, Wen et al. (2019) examined the root systems of 16 crops under different P availability. They observed that low Pi induced a more pronounced effect on crops with thin roots in terms of root branching, root surface area, and first-order root length (Table 3.1). On the contrary, crops with thick roots increased the production of P-mobilising exudates and had more significant colonisation by arbuscular mycorrhiza fungi in response to P deficiency (Wen et al. 2019). All in all, in this study, it was found that there is an array of trade-offs between root morphology, mycorrhizal symbioses, and root exudation to optimise P acquisition under P-limited conditions (Wen et al. 2019). All these observations, along with some other studies, establish that the effect of Pi deficiency on RSA is complex, specific to species and cultivar, and plays a pivotal role in plant response to different P conditions in soil. In this chapter, we will summarise how Pi availability in soil shapes root development in major crops and how it affects crop yield.

3.2.1 ROOT DEVELOPMENT OF MAJOR CROPS IN RESPONSE TO VARIABLE PHOSPHATE

Plants generally alter their root-to-shoot growth ratio in response to different nutrient stress, including Pi deficiency. Different species and genotypes show different responses to low P in terms of their root growth. Though topsoil foraging has been hypothesised to be beneficial in response to P deficiency, different crop species and cultivars show different responses to variable Pi concentration. There is no generalisation with regard to the root growth response of crops under P-deficient conditions, so in the following section, we will discuss the root growth of major crops under nutrient stress.

3.2.1.1 Rice (Oryza sativa)

Kirk and Van Du (1997) analysed the effect of low P concentration on rice roots using sand culture supplemented with a nutrient solution with different concentrations of P. They found that under low P conditions, rice plants showed a reduction in shoot dry mass but increased root dry mass and root surface area as compared to P-sufficient conditions. Additionally, low P also increased the proportion of fine roots and, thereby, the total root length in rice plants (Kirk and Van Du 1997). Analysis of the RSA of six genotypes of rice using hydroponics also showed increased root dry weight and lateral roots under Pi deficiency. However, all these genotypes showed P deficiency-mediated RSA modulation to different extents (Li, Xia, and Wu 2001). Similar to this, 62 varieties of rice grown in hydroponic culture showed a broad spectrum of root elongation responses under P deficiency. Out of 62 varieties of rice tested in this study, the majority of the genotypes showed no root elongation, but others exhibited moderated to high elongation in response to P deficiency (Shimizu et al. 2004).

In another study, root architectural, morphological, and anatomical traits of 15 rice genotypes in response to P deficiency were analysed using pots. In this study, authors planted germinated seeds in pots filled with sand and solid-phased buffered P, which led to a slow release of P in the system. This setup was used to mimic the soil P availability, and the RSA of rice plants was studied at the eight-leaf stage. Researchers observed that low P caused a reduction in the length of small as well as large lateral roots. Nevertheless, different genotypes showed varied plasticity and different proportions of allocated root lengths between larger and shorter roots. In addition to this, P deficiency increased root hair length and density across all the tested genotypes of rice. Different growth conditions could be responsible for contradictory results in the studies.

TABLE 3.1
Traits of Crops under Phosphate Deficiency

Crops	Root:shoot ratio LP	Root:shoot ratio HP	Root biomass (g pot-1) LP	Root biomass (g pot-1) HP	RootDiam (mm) LP	RootDiam (mm) HP	RootBr (No. cm-1) LP	RootBr (No. cm-1) HP	FRL (cm) LP	FRL (cm) HP	SRLf (m g-1 RDW) LP	SRLf (m g-1 RDW) HP	SRLw (m g-1 RDW) LP	SRLw (m g-1 RDW) HP
Wheat	0.36±0.01	0.38±0.03	0.75±0.02	2.60±0.29	0.087±0.002	0.085±0.002	3.89±0.14	3.09±0.04	0.62±0.03	0.45±0.02	301±17	269±20	258±10	181±10
Maize	0.29±0.01	0.24±0.01	1.78±0.05	5.51±0.37	0.136±0.007	0.151±0.007	4.66±0.11	3.84±0.09	0.63±0.01	0.56±0.01	146±13	133±12	59±2	49±2
Soybean	0.18±0.01	0.18±0.02	0.39±0.04	1.56±0.17	0.355±0.015	0.359±0.013	3.26±0.06	2.56±0.18	1.34±0.03	1.16±0.05	164±5	151±11	120±4	109±6
Sorghum	0.33±0.02	0.30±0.02	0.51±0.03	3.58±0.26	0.123±0.005	0.145±0.011	5.69±0.24	4.91±0.14	0.54±0.01	0.47±0.02	168±13	135±9	123±10	89±7

Root trait abbreviations: average diameter of absorptive fine roots (first- and second-order roots) (RootDiam), root branching intensity (RootBr), first-order root length (FRL), specific root length of absorptive fine roots (SRLf), and specific root length of whole root system (SRLw).

Several studies have identified many key genes and quantitative trait loci (QTL) involved in the regulation of root growth under Pi-deficient conditions in rice plants. Researchers identified P-deficiency tolerant Indian rice variety "Kasalath" (Wissuwa, Ae, and Jones 2001). A major QTL, *PHOSPHORUS UPTAKE 1 (Pup1)*, linked with low P tolerance was then identified by Kasalath × Nipponbare (intolerant) backcross population (Wissuwa et al. 2002; Matthias Wissuwa and Ae 2001). A decade later, the *Pup1* locus was sequenced and renamed as *PHOSPHORUS STARVATION TOLERANCE1 (PSTOL1)* (Gamuyao et al. 2012). *PSTOL1* encodes a serine/threonine kinase and is not present in the rice reference genome and other P starvation-intolerant modern varieties (Gamuyao et al. 2012; Chin et al. 2011, 2010). Overexpression of *PSTOL1* in two rice cultivars that naturally lack this gene resulted in a considerable increase in grain yield under P deficiency (Gamuyao et al. 2012). Further experiments showed that *PSTOL1* promotes early root growth and increases total root length and root dry weight as well (Figure 3.1). All these adaptive strategies help plants to acquire more P under P deficiency, leading to enhanced yield (Gamuyao et al. 2012). Similarly, among two contrasting near-isogenic lines with (+*Pup1*) and without (−*Pup1*), in the Kasalath *Pup1* locus, the +*Pup1* line exhibited increased root growth in both P-sufficient and deficient conditions

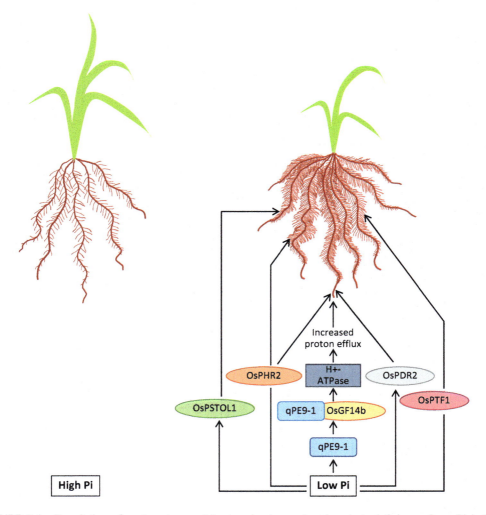

FIGURE 3.1 Regulation of root system architecture in rice under phosphate deficiency. Low Pi induces *qPE9-1* in rice roots, which in turn interacts with OsGF14b. OsGF14b triggers H+ efflux leading to root elongation. Arrows indicate positive regulation, while solid lines indicate no effects observed.

(Gamuyao et al. 2012). At the same time, the knockdown of *PSTOL1* in RNAi lines led to decreased root length and lateral root numbers. The *Pup1*-containing line as well as the rice cultivars overexpressing *PSTOL1* also showed increased nitrogen (N) and potassium (K) accumulation. Overall, these results indicate that *PSTOL1*-mediated modulation of RSA is not specifically in response to low P but a consecutive growth behaviour (Gamuyao et al. 2012).

Shimizu et al. (2004) identified a QTL on chromosome 6 linked to Pi deficiency-mediated root elongation and named it ROOT ELONGATION UNDER PHOSPHORUS DEFICIENCY (REP). Additionally, they found that this QTL or another tightly linked region is partially linked to reduced excess accumulation of Fe in shoots as well.

A basic helix-loop-helix (bHLH) domain-containing novel transcription factor rice *Pi STARVATION-INDUCED TRANSCRIPTION FACTOR 1 (OsPTF1)* is induced in roots by Pi starvation. Transgenic rice plants overexpressing *OsPTF1* showed enhanced tolerance to Pi starvation in terms of increased tiller number, root and shoot biomass, and P accumulation. Overexpression of *OsPTF1* increased overall root length as well as root surface area (Figure 3.1). Subsequently, it enhanced the instantaneous Pi uptake rate compared to wild-type plants in soil pot and field conditions (Yi et al. 2005). Altogether, these observations suggest that *OsPTF1* is involved in the regulation of primary root growth under different Pi conditions along with other traits.

Wu and Wang (2008) identified two homologs of *Arabidopsis PHOSPHATE STARVATION RESPONSE 1 (AtPHR1)* in rice: *OsPHR1* and *OsPHR2*. *AtPHR1* has been shown to play a vital role in P-deficiency response in *Arabidopsis*. The authors found that overexpression of *OsPHR2* induced root elongation, root hair proliferation, and Pi accumulation in Pi sufficient conditions, thus mimicking Pi deficiency response. Later, Guo et al. (2015) showed that rice orthologs (*OSPHR1, OsPHR2,* and *OsPHR3*) of *AtPHR1* are involved in root hair growth under Pi deficiency. Authors found that PHR1, PHR2, and PHR3 redundantly regulate root hair in response to varying P availability (Figure 3.1). Li et al. (2022) introduced *OsPHR2* into the wheat genome and found that the expression was mainly in roots. Furthermore, expression of *OsPHR2* in wheat resulted in enhanced maximum root length, total root length, and root volume under Pi-deficient conditions. Transgenic lines also showed increased crop yield as compared to wild-type in both P-sufficient and P-deficient conditions.

Ding et al. (2018) showed in a study that low P-induced Iron (Fe) formed plaque on the surface of the roots of rice plants. They further showed that although low Pi inhibited the Fe uptake mechanisms I, II, and III, it induced the expression of vacuolar ion transporter (VIT) genes resulting in enhanced storage of Fe in vacuoles and cell walls. Low Pi also increased the expression as well as activity of β-1-3 glucanase, a callose hydrolysis enzyme, and thereby callose accumulation under Pi deficiency is inhibited in rice. However, observations in the study suggested that low Pi-induced primary root elongation in rice is independent of Fe and callose homeostatic changes (Ding et al. 2018). Rice homolog of the *Arabidopsis* gene *PHOSPHATE DEFICIENCY RESPONSE2 (AtPDR2)*, *OsPDR2*, also regulates root growth in rice in response to P-deficient conditions. Contrary to its *Arabidopsis* counterpart, *OsPDR2* positively regulates primary root development under low P (Figure 3.1). RNAi-mediated knockdown of the *OsPDR2* gene led to a wide-spectrum effect on various vegetative and reproductive traits. Some of the most pronounced effects of the down-regulation of the *OsPDR2* gene were lesser root elongation and reduced yield. Additionally, *OsPDR2* is involved in Pi homeostasis and distribution in rice (Cao et al. 2020). In another study, the author showed that rice G protein γ subunit *qPE9-1* regulates root growth under low P conditions (K. Wang et al. 2021). Low P up-regulated the expression of *qPE9-1* in rice roots. Rice varieties carrying the mutant allele *qpe9-1* have shorter, lesser primary root elongation and P accumulation in response to low P as compared to the varieties carrying the *qPE9-1* allele as well as transgenic plants overexpressing the *qPE9-1* allele. Further experiments showed that in rice, *qPE9-1* potentially interacts with OsGF14b, a 14-3-3 protein. Following that, OsGF14b induces the plasma membrane-associated H+-ATPase, thereby increasing H+ efflux, which leads to root elongation and P uptake (Figure 3.1) (K. Wang et al. 2021).

In addition to these reports, several other genes such as *PURPLE ACID PHOSPHATASE 10C (OsPAP10c)* and *NUTRITION RESPONSE AND ROOT GROWTH (NRR)* have been shown to regulate root growth as well as other Pi deficiency responses in rice (S. Deng et al. 2020; J. Liu et al. 2021; Cao et al. 2020; Zhang et al. 2012).

3.2.1.2 Wheat (Triticum aestivum)

In the case of wheat, research has shown low P-induced root elongation similar to rice (Kaur et al. 2021). Sun and Zhang (2000) studied the root growth of wheat in P-deficient conditions using hydroponics. They found that under P deficiency, wheat showed increased primary root and total root length. Further kinetic experiments suggested that reduction in internal P content is the primary response to P deficiency, followed by changes in root growth. They performed a split root experiment to understand whether internal or external P affects primary root growth in wheat. Different proportions of the roots of wheat plants were kept in two compartments with low and high P. They found that the external P level did not affect the primary root length, but the internal level of P did affect the primary root growth, as the primary root length decreased in both compartments.

Teng et al. (2013) analysed the root growth of wheat in soil with six P-fertiliser rates. The growth characteristics of plants were analysed by collecting the root sample at the flowering stage. The authors found that low levels of P reduced the root dry weight and root length density of wheat plants. Shen et al. (2018) investigated the effect of shoot P concentration on root morphology as well as P-mobilising exudation in wheat using soil supplemented with 11 rates of P supply. In this study, authors observed that wheat relies majorly on root biomass and root length maintenance rather than root exudation to survive extreme P-deficient conditions. Low P reduced the root biomass, but wheat plants maintain root growth by allocating more carbon to roots than shoots. In addition, wheat also maintained total root length by producing more fine roots under nutrient-deficient conditions. However, under extreme P-deficiency stress, fine root reduction was more pronounced than thick roots in wheat (Shen et al. 2018). The difference between the observations in these reports can be attributed to the difference in the growth medium, genotype, harvest time, and P levels used in studies.

In order to understand root branching that optimises P uptake in crops under P-deficient conditions, researchers used a combination of mathematical modelling and experimental analysis (Heppell et al. 2015). They simulated P uptake using root parameters from wheat cultivar "Gallant", and afterwards, they also validated the results experimentally using the same cultivar. They found that changing the arrangement of root branching from linear to highly exponential led to a dramatic 147% increase in P uptake when the root mass was constant. In exponential branching, there is a higher number of branches at the top of the soil than in linear branching, where branches are evenly spaced on the root. Exponential branching causes early emergence of roots and maximises topsoil foraging, and could help crops survive under low Pi conditions (Heppell et al. 2015).

With the purpose of developing a method to analyse the RSA of wheat grown in the soil without destroying it, Nguyen and Stangoulis (2019) studied the RSA of two wheat cultivars with different P-use efficiency (PUE). Two wheat cultivars, P-efficient RAC875 and P-inefficient Wyalkatchem, were grown in special rhizoboxes containing sandy soils supplemented with different P rates. Low Pi induced many changes in the RSA of both the cultivars, ranging from decreased convex hull area (CHA) to reduction in total length, surface area, volume, and number of roots. Although Wyalkatchem showed larger CHA than RAC875 under P-sufficient conditions, RAC875 produced considerably larger CHA as compared to Wyalkatchem under low P. Additionally, under low P conditions, the RAC875 cultivar showed greater root hair density than Wyalkatchem (Nguyen and Stangoulis 2019). These observations suggested that increased CHA and root hair density potentially contribute to the plant tolerance to a low P environment.

Deng et al. (2018) studied two wheat cultivars for two years to monitor the response to P deficiency in the field. They found that both cultivars not only exhibited similar growth and yield potential at each P supply, but they also achieved optimal growth at the same P level. However, the two cultivars employed different strategies to achieve optimal growth under stress conditions, wherein one cultivar modified its root morphology and the other one increased its secretion of acid phosphatase and other physiological processes in roots. These observations indicate that adaptive strategies used under low P conditions in wheat are genotype-specific.

A comparison of two wheat cultivars with different tolerance to low Pi revealed root adaptations that contribute to low Pi tolerance in wheat. The Pi-deficient tolerant wheat cultivar Kenong199 (KN199) showed higher soluble Pi and total P in the shoot as compared to the susceptible cultivar Chinese Spring (CS). However, there was no significant difference in P concentration in the roots of the two cultivars (Zheng et al. 2021). Low Pi-induced root growth was significantly higher in KN199 as compared to CS. KN199 cultivar showed a more pronounced increase in total root length as well as total root surface area than the CS cultivar. The increase in root length in KN199 was mostly caused by enhanced fine roots rather than thick root length under Pi stress. Additionally, KN199 showed a significant increase in root biomass and total root volume. Further experiments showed that genes involved in hormone signalling (jasmonate, ethylene, gibberellin) and root development, such as *TaE2f*, *SOMATIC EMBRYOGENESIS RECEPTOR KINASE1 (TaSERK1)*, and *TaEXPB23*, were differentially regulated in two cultivars. These observations suggest that low

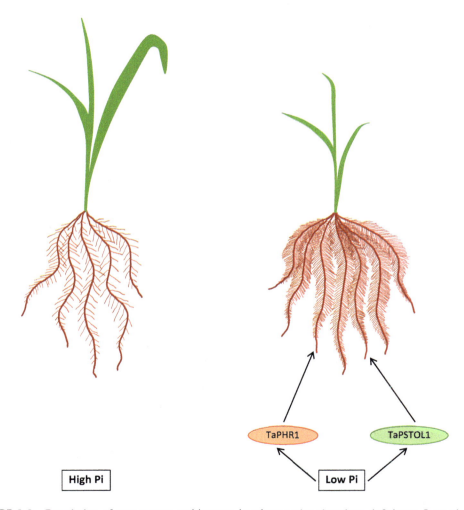

FIGURE 3.2 Regulation of root system architecture in wheat under phosphate deficiency. Low phosphate induces *TaPHR1* and *TaPSTOL1* resulting in better root system architecture under stress conditions in wheat plants.

Pi-mediated changes in RSA in KN199 are regulated by hormone signalling. All these modulations in RSA helped plants acquire more Pi from the soil, thus helping to better tolerate the stress condition (Zheng et al. 2021).

Wang et al. (2013) isolated three orthologs of *AtPHR1* in wheat: *Ta-PHR1-A1*, *B1*, and *D1*. Transgenic wheat plants overexpressing *Ta-PHR1-A1* showed increased expression of a subset of P starvation-responsive genes. Moreover, these transgenic plants showed increased root branching and P uptake under variable P conditions (Figure 3.2). *PSTOL1* has been shown to be involved in the regulation of RSA in crops. Abbas et al. (2022) identified the wheat orthologs of *OsPSTOL1*. They also identified two cultivars of wheat with differential expression patterns of *TaPSTOL1* genes, i.e. Pakistan-13 and Shafaq-6, wherein Pakistan-13 showed enhanced expression of *TaPSTOL1* genes under P deficiency. They also found that Pakistan-13 showed better root growth and increased P uptake under P deficiency as compared to Shafaq-6. This suggests that, similar to its rice ortholog, *TaPSTOL1* regulates RSA under low P conditions in wheat (Figure 3.2).

A recombinant inbred line (RIL) population (H461/CM107) of wheat was used for the identification of QTL associated with root traits under P deficiency at the seedling stage. The authors identified four QTLs in wheat that significantly increased root biomass under P deficiency (X. Yang et al. 2018). In another study, Yang et al. (2021) selected Yangmai 16/Zhongmai 895 derived doubled haploid lines for the identification of QTLs associated with RSA and biomass-related traits under P deficiency. They identified 34 QTLs arranged in seven clusters that regulate root length, root volume, root surface area, root tip number, root dry weight, and shoot dry weight at low P levels.

3.2.1.3 Maize (Zea mays)

A study of maize roots in different P concentrations showed enhanced total root length as well as improved root dry weight in 12-day-old plants under P deficiency (Anghinoni and Barber 1980). Contrary to this, another study showed no difference in root dry weight in response to P deficiency in maize (Khamis, Chaillou, and Lamaze 1990). Mollier and Pellerin (1999) analysed the response of maize root in P-deficient conditions using hydroponics. Maize seedlings of 11 days old were transferred to media, and root growth was marginally enhanced after a few days after P deficiency. However, root growth was substantially reduced after a prolonged period of P deficiency. On the other hand, the elongation rate of primary roots and seminal roots was maintained throughout the span of the experiment. Plants showed a significant reduction in the emergence of new seminal roots and elongation of first-order laterals under low P supply. Nonetheless, the density of lateral roots was not severely affected under the stress condition, suggesting that low P affects the elongation of lateral roots but not the initiation. A similar study was conducted by Li et al. (2012) on maize root response to P starvation. They transferred 10-day-old maize seedlings to hydroponics containing different concentrations of P. Contrary to the earlier report, seedlings showed increased root dry weight and primary root but decreased lateral roots under P deficiency.

In another study on maize roots, different rates of P supplies were used in pot and field experimental setups to understand the effect of soil P level on the ability of roots to take P from the soil. They found that below a certain soil Olsen-P level, root growth first slightly increased and then rapidly decreased, although the level differed in the pot vs field experiments. Moreover, root dry weight remained unchanged to a certain Olsen-P level, and below that, it was sharply reduced (Deng et al. 2014). In a similar study, the effect of shoot P using different P supply rates on maize roots was analysed in a greenhouse (Wen et al. 2017). Experiments showed that when the soil Olsen-P level decreased up to a certain level, several root traits such as total root length, specific root length (ratio of total root length to root biomass), and proportion of fine roots increased. At the same time, root biomass decreased with decreasing P levels. When the soil Olsen-P level was further decreased, leading to extremely low shoot P concentration, these root traits started decreasing with a further decrease in root biomass (Wen et al. 2017). Both of the studies suggest that mild P deficiency induces root growth, but it is inhibited by severe P deficiency.

Gaume et al. (2001) studied the root morphology in four genotypes of maize in both P-sufficient and P-deficient conditions using hydroponics. They selected a low P-tolerant, a low P-susceptible, an acid-tolerant, and a Swiss commercial genotype for this study. All the genotypes showed a similar average length of the three longest roots in the P-sufficient condition. Under the P-deficient condition, the P-tolerant genotype showed a maximum increase in root/shoot ratio among all genotypes. At the same time, the commercial genotype showed the least increase in the root/shoot ratio, with the other genotypes showing an intermediate increase in the root:shoot ratio. In addition to this, all genotypes except for the commercial genotype showed an increase in root development under P-deficient conditions in comparison to P-sufficient conditions (Gaume et al. 2001). Altogether, these observations indicate that changes in root morphology play a significant role in plant tolerance towards P deficiency in maize, and a tolerant genotype modulates the RSA substantially to survive under stress conditions. Liu et al. (2018) analysed the RSA of different maize lines in the field and compared it with their PUE. They found that all parental inbred lines except for one showed reduced total root length in P-deficient conditions. In the case of lateral roots, some parental lines showed increased lateral roots, but the lateral root numbers remain unchanged in other lines under low P. At the same time, P deficiency reduced the number of crown roots, but the seminal roots remained unchanged under the stress condition. These observations indicated that root responses to low P conditions are genotype-specific in maize.

Li et al. (2011) found that the maize ortholog of *OsPTF1*, *ZmPTF1*, is induced in maize root responses to P deficiency. Furthermore, when transgenic plants overexpressing *ZmPTF1* were grown in P-deficient conditions in hydroponics and sand pots, they showed better root growth and increased biomass as compared to wild-type plants (Figure 3.3). In an effort to search the *PSTOL1* homolog of maize, Azevedo et al. (2015) investigated 145 maize RILs using hydroponics. QTL mapping of maize lines showed that *ZmPSTOL* genes co-localised with the QTLs associated with root traits, increased biomass, and P content. Wang et al. (2019) investigated 356 inbred lines of maize in the seedling stage for the identification of single nucleotide polymorphisms (SNPs) and genes associated with root responses under P deficiency. They identified 13 root traits with a significant phenotypic difference in P-sufficient and P-deficient conditions. Out of 356 maize lines, six extremely sensitive as well as seven P-deficiency tolerant lines were selected for further study. The authors identified several significant SNPs and genes associated with the 13 traits using genome-wide association (GWAS) mapping. Out of all, a haplotype Hap5 containing 12 favourable SNPs exhibited a positive correlation with total root length, total root area, and root forks under both P-sufficient and P-deficient conditions in maize seedlings.

In addition to the above-mentioned factors, a recent study by Zhou et al. (2020) revealed that light intensity also has a pronounced effect on the RSA of maize under low P_i conditions. In this study, maize plants were grown in a field in natural light and low light with different Olsen-P supplies. They found that under low Pi stress, maize plants showed higher total root length and total root area, and the percentage of fine roots increased under natural light than in low light. However, total root length and total root area were strongly inhibited under severe P deficiency in both of the light intensities. Under natural light, maize plants allocate more sucrose to the roots, which is a product of photosynthesis. This increased sucrose in roots mimics the low P signal and serves as a carbon and energy source, resulting in increased root growth that helps plants to adapt under low P conditions. However, low light conditions caused low photosynthetic activity, which resulted in less carbon and energy to roots. This led to reduced root growth, thereby decreasing the requirement for P and other nutrients (T. Zhou et al. 2020). In another study on maize seedlings, researchers observed a link between root zone warming and root growth as well as seedling growth and development under variable P conditions (Xia et al. 2021). In this study, the authors used hydroponics with a controlled water temperature to regulate the root zone temperature. Low P, along with moderate temperature in the root zone, significantly induced total root length as well as root surface area in maize (Xia et al. 2021).

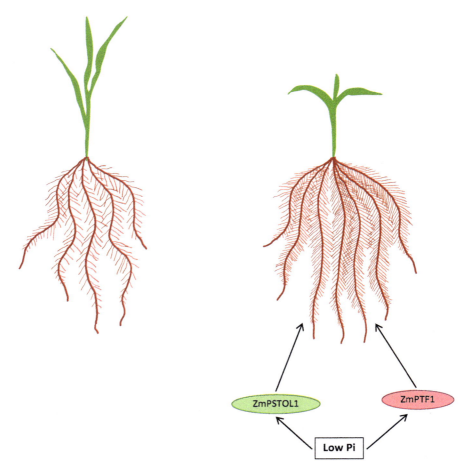

FIGURE 3.3 Regulation of root system architecture in maize under phosphate deficiency. Low phosphate up-regulates the expression of Zm*PSTOL1* and Zm*PTF1* leading to better root system architecture under stress conditions in maize plants.

3.2.1.4 Soybean (Glycine max)

In order to understand the effect of Pi deficiency on the RSA of soybean and the importance of RSA traits in the Pi efficiency of soybean, Zhao et al. (2004) created an "applied core collection" of soybean germplasm. The authors assessed the correlation between RSA and P efficiency of soybean as well as a putative evolutionary pattern of these traits in fields. This study was performed in P-deficient soil in China. The authors used two P levels using two experimental setups; in one setup, P was supplied in the topsoil, and in the other, there was no additional P supply. Soybean with shallow root architecture showed better P efficiency and, thereby, better yield because of the better spatial distribution of root architecture in P-rich topsoil. In addition to this, the authors also found a possible evolutionary relationship among shoot type, RSA, and P efficiency in soybean. In this study, researchers found that cultivated soybeans with bushy shoots had shallow roots and thus better P efficiency. Contrary to this, wild soybean had deep RSA and showed low P efficiency, and semi-wild soybean had an intermediate RSA and moderate P efficiency. Results of this study suggest that during evolution from wild-type soybean to cultivar, RSA became shallower and acquired better P efficiency. This might be because of the addition of fertilisers to the topsoil and domestication of this crop (Zhao et al. 2004).

Guo et al. (2011) investigated the involvement of a Pi starvation-induced β-expansin gene of soybean, *GmEXPB2*. GmEXPB2 was found to be a secretory protein that is primarily expressed in the roots. Transgenic soybean composite plants overexpressing *GmEXPB2* exhibited increased root length as well as increased P content and fresh weight. Similarly, the knockdown of *GmEXPB2* resulted in reduced root length and P content along with lower fresh weight. Overexpression of *GmEXPB2* in *Arabidopsis* led to enhanced root cell division and elongation, as well as heightened plant growth and P uptake in both P-deficient and P-sufficient conditions. Altogether, these observations indicate that *GmEXPB2* influences the P efficiency of soybean plants via modulation of RSA (Guo et al. 2011). Zhou et al. (2014) showed that overexpression of *GmEXPB2* in soybean induced root growth along with leaf expansion. Subsequently, transgenic soybean plants showed increased P efficiency in low P in hydroponics. Similarly, in a recent study, it was shown that transgenic soybean plants overexpressing *GmEXPB2* had increased growth and yield (Yang et al. 2021). Overexpression of *GmEXPB2* promoted root growth and, thereby, the P efficiency of soybean plants similar to the earlier study. Furthermore, in this study, the authors also showed that the expression of *GmEXPB2* under P starvation is mediated via a basic-helix-loop-helix transcription factor, *GmPTF1* (Yang et al. 2021).

Zhou et al. (2016) studied the yield and P-acquisition potentials of 274 genotypes of soybean under variable P availability in the field. Out of these genotypes, five P-efficient and five P-inefficient genotypes were further selected and studied in both P-sufficient and P-deficient conditions using hydroponics. P-efficient genotypes of soybean showed many adaptive traits ranging from organic acid exudation to altered root:shoot ratio. P-efficient genotypes had pronounced differences in the root traits than P-inefficient genotypes. Under low P conditions, P-efficient genotypes showed enhanced total root length, root surface area, specific root length, and root volume (Zhou et al. 2016).

Zhang et al. (2020) compared two soybean genotypes with different P-deficiency tolerance. The authors conducted high-confident linkage and association mapping leading to the identification of *ETHYLENE-OVERPRODUCTION PROTEIN 1 (GmETO1)* underlying a major QTL involved in Pi uptake in soybean. *GmETO1* is a major ethylene-biosynthesis regulator which is up-regulated under P deficiency. Moreover, overexpression of *GmETO1* in transgenic soybean hairy roots increased the proliferation and elongation of hairy roots and thus Pi uptake and PUE. Similarly, silencing of *GmETO1* led to opposite results, suggesting that ,this gene plays a key role in the regulation of root morphology under low P conditions and, thus plant response to nutrient stress (Zhang et al. 2020).

3.2.1.5 Sorghum (Sorghum bicolor)

Nine lines of sorghum were tested in P-sufficient and P-deficient conditions in soil for the identification of root traits beneficial for plant survival and yield under low P (Rocha et al. 2010). In this study, researchers found that root length and root surface area play a more important role in plant adaptation under P deficiency than root diameter or volume. In addition to this, specific root length and tissue density showed a strong correlation with sorghum productivity in low P availability in soil (Rocha et al. 2010). Since *OsPSTOL1* has been shown to regulate RSA in rice under P deficiency, Hufnagel et al. (2014) were interested in understanding the involvement of homologs of *OsPSTOL1* in sorghum in low P conditions. Researchers employed association mapping and analysed two sorghum association panels phenotyped for yield and biomass production in low P conditions in Brazil and Mali, as well as for RSA and P uptake under low P in hydroponics. *SbPSTOL1* alleles underlying smaller root diameters were associated with increased P uptake in P-deficient conditions in hydroponics. At the same time, *SbPSTOL1* alleles associated with enhanced root surface area also enhanced yield under P deficiency in the soil. Multiple *SbPSTOL1* genes contribute to the modulation of the overall RSA of sorghum in response to low P, which assists in enhanced grain yield under the stress condition (Hufnagel et al. 2014).

Parra-Londono et al. (2018) assessed the root traits of 194 Sorghum genotypes in response to different P availability using germination paper and mini-rhizotrons. Sorghum RSA was analysed at the two-leaf stage as well as after the emergence of nodal roots. The authors selected 19 root traits for analysis under P-sufficient and P-deficient conditions. At the same time, only 16 traits of RSA were observed at early development, i.e. the two-leaf stage of sorghum. Based on these observations, the authors allocated genotypes to different RSA clusters using multivariate analysis. Sorghum genotypes showing compact, bushy, and shallow root systems provided a potential adaptive strategy in response to P deficiency in the field (Parra-Londono et al. 2018). Bernardino et al. (2019) employed multi-trait QTL mapping to understand the genetic determinants of P efficiency or PUE of sorghum. Root surface area showed a positive correlation with grain yield in P-deficient conditions. The authors found 14 QTLs associated with root morphology and/or grain yield. Out of these QTLs, three QTLs were related to root morphology and are located in close proximity to the *SbPSTOL1* gene. An allelic variation at the *SbPSTOL1* gene has been shown to regulate root diameter in sorghum under P deficiency similar to that of its maize ortholog (Azevedo et al. 2015; Bernardino et al. 2019).

3.2.1.6 Common Bean (Phaseolus vulgaris)

Bonser, Lynch, and Snapp (1996) investigated the effect of low P on the growth angles of basal roots in common beans using a paper growth pouch supplemented with nutrient solution. The authors studied 16 genotypes of common bean, among which P deficiency reduced the growth angle in six genotypes. Simultaneously, low P enhanced the growth angle in only one genotype of bean, and the root angle of nine genotypes remained unchanged. The effect of global and local P availability on the root angle of common bean was also investigated using a split pouch system. Different parts of the roots were exposed to nutrient solutions with low Pi and high Pi, and both roots showed a growth angle similar to that of the high Pi solution. This therefore suggests that the change in root angle under P deficiency is a response to global P availability. Additionally, a low P-mediated decrease in the root angle of common bean showed a correlation with reduced shoot P (Bonser, Lynch, and Snapp 1996). In a similar study by Liao et al. (2001), the root growth angle of common bean under low P in both two-dimensional paper growth pouch and three-dimensional sand and soil cultures was analysed. The P-efficient genotype of the common bean showed shallower roots under low P in all three media, i.e., pouches, sand, and soil. However, P-inefficient genotypes produced deeper roots in response to P deficiency (Liao et al. 2001). Altogether, these observations suggest that the effect of low P on root angle in the common bean is genotype-specific and regulates plant adaptation to nutrient stress. The same group also investigated QTLs associated with this trait and other root traits, as well as PUE in the common bean using RILs. They identified several QTLs associated with basal root growth angle, shallow basal root length, and relative shallow basal root length, along with QTLs for other root traits such as root hair growth and root length (Hong Liao et al. 2004, 2006; Yan et al. 2004). These QTLs related to shallow root traits also showed an association with P-acquisition efficiency (PAE) in the field.

Yao et al. (2014) investigated the involvement of SPX domain-containing proteins in the low P responses in the common bean, as SPX domains have been reported to be involved in P signalling in yeast, *Arabidopsis*, rice, and rapeseed. The authors cloned three *SPX* members from the common bean. Transgenic bean hairy roots overexpressing *PvSPX1* showed altered root morphology in terms of inhibited root growth as well as enhanced root hair zone. At the same time, overexpression of *PvSPX1* in bean transgenic hairy roots resulted in enhanced root P accumulation. The authors also found that PvSPX1 functions downstream to PvPHR1 in the regulation of P signalling in common beans (Yao et al. 2014).

3.3 CONCLUSIONS

In this chapter, we have summarised the root developmental responses of crops to Pi availability in the soil. As highlighted in this chapter, the root response to Pi is significantly complex and diverse across all plant species. Various factors affect low P-mediated root changes, such as crop species, genotype, and growth medium and conditions. Additionally, as shown in the various reports, the developmental stage of the crop has a pronounced effect on the root responses under Pi deficiency. There have been many comprehensive studies to understand the effect of low Pi on root growth in various crops. Despite this progress, there are still gaps in our knowledge about the root response in crops under low P conditions. However, we can draw a generalised outline of the effect of soil P on the RSA. As shown in many studies, P deficiency triggers root growth in many crops to increase Pi uptake. At the same time, severe P deficiency reduced root growth, which may be the result of reduced metabolic activity because of low P in plants. In addition to this, several other environmental factors, such as light intensity and soil temperature, have also been shown to influence root growth in different P conditions. Altogether, in nature, several factors shape the RSA to optimise the growth and development of plants under low P conditions.

Plants modulate their RSA to maintain Pi homeostasis under P deficiency with the help of various gene regulation networks. In recent years, there have been several studies revealing the key genes involved in the regulation of RSA under P deficiency, such as *PSTOL1* and *PHRs SPX1*. Although our knowledge about the genes involved in the root response of crops under low P is increasing exponentially, still we lack the comprehensive knowledge about the factors acting upstream and downstream of these genes. The underlying molecular machinery of the root growth response in low P conditions is very important. These genes can help in developing transgenic crops with a high PAE with the help of the current genome editing tools. Therefore, in order to achieve sustainable agriculture, further studies need to be undertaken to understand the root response to fluctuating Pi levels in soil as well as the molecular network underlying these responses.

3.4 ACKNOWLEDGEMENT

We thank Pooja Jhakar and Sanskar Mishra for their input in figure preparation. S.B.S. acknowledges intramural funding support from the Indian Institute of Science Education and Research (IISER) Mohali. S.B.S. is a recipient of the Ramalingaswami Fellowship from the Department of Biotechnology (DBT), Govt. of India, and acknowledges funding received from DBT. S.B.S also acknowledges the Science and Engineering Research Board (SERB) for early career research funding (ECR/2018/001580). D.S. acknowledges the postdoctoral fellowship received from IISER Mohali.

REFERENCES

Abbas, Hina, Muhammad Kashif Naeem, Marya Rubab, Emilie Widemann, Muhammad Uzair, Nageen Zahra, Bilal Saleem, et al. 2022. "Role of Wheat Phosphorus Starvation Tolerance 1 Genes in Phosphorus Acquisition and Root Architecture." *Genes* 13(3): 487. https://doi.org/10.3390/GENES13030487/S1.

Alatorre-Cobos, Fulgencio, Damar López-Arredondo, and Luis Herrera-Estrella. 2009. "Genetic Determinants of Phosphate Use Efficiency in Crops." *Genes for Plant Abiotic Stress*, October: 143–65. https://doi.org/10.1002/9780813809380.CH6.

Anghinoni, I., and S. A. Barber. 1980. "Phosphorus Influx and Growth Characteristics of Corn Roots as Influenced by Phosphorus Supply1." *Agronomy Journal* 72(4): 685–88. https://doi.org/10.2134/AGRONJ1980.00021962007200040028X.

Azevedo, Gabriel C., Adriana Cheavegatti-Gianotto, Bárbara F. Negri, Bárbara Hufnagel, Luciano da Costa e Silva, Jurandir V. Magalhaes, Antonio Augusto F. Garcia, Ubiraci G. P. Lana, Sylvia M. de Sousa, and Claudia T. Guimaraes. 2015. "Multiple Interval QTL Mapping and Searching for PSTOL1 Homologs Associated with Root Morphology, Biomass Accumulation and Phosphorus Content in Maize Seedlings Under Low-P." *BMC Plant Biology* 15(1). https://doi.org/10.1186/S12870-015-0561-Y.

Bellini, Catherine, Daniel I. Pacurar, and Irene Perrone. 2014. "Adventitious Roots and Lateral Roots: Similarities and Differences." *Annual Review of Plant Biology* 65(1): 639–66. https://doi.org/10.1146/annurev-arplant-050213-035645.

Bernardino, Karine C., Maria Marta Pastina, Cícero B. Menezes, Sylvia M. De Sousa, Laiane S. Maciel, G. C. Geraldo Carvalho, Claudia T. Guimarães, et al. 2019. "The Genetic Architecture of Phosphorus Efficiency in Sorghum Involves Pleiotropic QTL for Root Morphology and Grain Yield Under Low

Phosphorus Availability in the Soil." *BMC Plant Biology* 19(1): 1–15. https://doi.org/10.1186/S12870-019-1689-Y/FIGURES/5.

Bonser, Amy M., Jonathan Lynch, and Sieglinde Snapp. 1996. "Effect of Phosphorus Deficiency on Growth Angle of Basal Roots in Phaseolus vulgaris." *New Phytologist* 132(2): 281–88. https://doi.org/10.1111/J.1469-8137.1996.TB01847.X.

Cao, Yue, Ajay Jain, Hao Ai, Xiuli Liu, Xiaowen Wang, Zhi Hu, Yafei Sun, et al. 2020. "OsPDR2 Mediates the Regulation on the Development Response and Maintenance of Pi Homeostasis in Rice." *Plant Physiology and Biochemistry : PPB* 149(April): 1–10. https://doi.org/10.1016/J.PLAPHY.2019.12.037.

Chin, Joong Hyoun, Rico Gamuyao, Cheryl Dalid, Masdiar Bustamam, Joko Prasetiyono, Sugiono Moeljopawiro, Matthias Wissuwa, and Sigrid Heuer. 2011. "Developing Rice with High Yield under Phosphorus Deficiency: Pup1 Sequence to Application." *Plant Physiology* 156(3): 1202–16. https://doi.org/10.1104/PP.111.175471.

Chin, Joong Hyoun, Xiaochun Lu, Stephan M. Haefele, Rico Gamuyao, Abdelbagi Ismail, Matthias Wissuwa, and Sigrid Heuer. 2010. "Development and Application of Gene-Based Markers for the Major Rice QTL Phosphorus Uptake 1." *TAG. Theoretical and Applied Genetics. Theoretische und Angewandte Genetik* 120(6): 1073–86. https://doi.org/10.1007/S00122-009-1235-7.

Comas, Louise H., Steven R. Becker, Von Mark V. Cruz, Patrick F. Byrne, and David A. Dierig. 2013. "Root Traits Contributing to Plant Productivity Under Drought." *Frontiers in Plant Science*. Frontiers Research Foundation. https://doi.org/10.3389/fpls.2013.00442.

Deng, Suren, Linghong Lu, Jingyi Li, Zezhen Du, Tongtong Liu, Wenjing Li, Fangsen Xu, Lei Shi, Huixia Shou, and Chuang Wang. 2020. "Purple Acid Phosphatase 10c Encodes a Major Acid Phosphatase That Regulates Plant Growth under Phosphate-Deficient Conditions in Rice." *Journal of Experimental Botany* 71(14): 4321–32. https://doi.org/10.1093/JXB/ERAA179.

Deng, Yan, Keru Chen, Wan Teng, Ai Zhan, Yiping Tong, Gu Feng, Zhenling Cui, Fusuo Zhang, and Xinping Chen. 2014. "Is the Inherent Potential of Maize Roots Efficient for Soil Phosphorus Acquisition?." *PLOS One* 9(3): e90287. https://doi.org/10.1371/JOURNAL.PONE.0090287.

Deng, Yan, Wan Teng, Yi Ping Tong, Xin Ping Chen, and Chun Qin Zou. 2018. "Phosphorus Efficiency Mechanisms of Two Wheat Cultivars as Affected by a Range of Phosphorus Levels in the Field." *Frontiers in Plant Science* 871: 1614. https://doi.org/10.3389/FPLS.2018.01614/BIBTEX.

Ding, Yan, Zegang Wang, Menglian Ren, Ping Zhang, Zhongnan Li, Sheng Chen, Cailin Ge, and Yulong Wang. 2018. "Iron and Callose Homeostatic Regulation in Rice Roots under Low Phosphorus." *BMC Plant Biology* 18(1): 1–14. https://doi.org/10.1186/S12870-018-1486-Z/TABLES/4.

Fromm, Hillel. 2019. "Root Plasticity in the Pursuit of Water." *Plants*. MDPI AG. https://doi.org/10.3390/plants8070236.

Gamuyao, Rico, Joong Hyoun Chin, Juan Pariasca-Tanaka, Paolo Pesaresi, Sheryl Catausan, Cheryl Dalid, Inez Slamet-Loedin, Evelyn Mae Tecson-Mendoza, Matthias Wissuwa, and Sigrid Heuer. 2012. "The Protein Kinase Pstol1 from Traditional Rice Confers Tolerance of Phosphorus Deficiency." *Nature* 488(7412): 535–39. https://doi.org/10.1038/nature11346.

Gaume, Alain, Felix Mächler, Carlos De León, Luis Narro, and Emmanuel Frossard. 2001. "Low-P Tolerance by Maize (Zea Mays L.) Genotypes: Significance of Root Growth, and Organic Acids and Acid Phosphatase Root Exudation." *Plant and Soil* 228(2): 253–64.

Guo, Meina, Wenyuan Ruan, Changying Li, Fangliang Huang, Ming Zeng, Yingyao Liu, Yu Yanan, et al. 2015. "Integrative Comparison of the Role of the Phosphate Response1 Subfamily in Phosphate Signaling and Homeostasis in Rice." *Plant Physiology* 168(4): 1762. https://doi.org/10.1104/PP.15.00736.

Guo, Wenbing, Jing Zhao, Xinxin Li, Lu Qin, Xiaolong Yan, and Hong Liao. 2011. "A Soybean β-Expansin Gene GmEXPB2 Intrinsically Involved in Root System Architecture Responses to Abiotic Stresses." *The Plant Journal* 66(3): 541–52. https://doi.org/10.1111/J.1365-313X.2011.04511.X.

Heppell, J., P. Talboys, S. Payvandi, K. C. Zygalakis, J. Fliege, P. J. A. Withers, D. L. Jones, and T. Roose. 2015. "How Changing Root System Architecture Can Help Tackle a Reduction in Soil Phosphate (P) Levels for Better Plant P Acquisition." *Plant, Cell and Environment* 38(1): 118–28. https://doi.org/10.1111/PCE.12376.

Hufnagel, Barbara, Sylvia M. de Sousa, Lidianne Assis, Claudia T. Guimaraes, Willmar Leiser, Gabriel C. Azevedo, Barbara Negri, et al. 2014. "Duplicate and Conquer: Multiple Homologs of Phosphorus-Starvation Tolerance1 Enhance Phosphorus Acquisition and Sorghum Performance on Low-Phosphorus Soils." *Plant Physiology* 166(2): 659–77. https://doi.org/10.1104/PP.114.243949.

Ju, X. T., C. L. Kou, P. Christie, Z. X. Dou, and F. S. Zhang. 2007. "Changes in the Soil Environment from Excessive Application of Fertilizers and Manures to Two Contrasting Intensive Cropping Systems on the North China Plain." *Environmental Pollution* 145(2): 497–506. https://doi.org/10.1016/J.ENVPOL.2006.04.017.

Kaur, Gazaldeep, Vishnu Shukla, Varsha Meena, Anil Kumar, Deepshikha Tyagi, Jagtar Singh, Pramod Kaitheri Kandoth, Shrikant Mantri, Hatem Rouached, and Ajay Kumar Pandey. 2021. "Physiological and Molecular Responses to Combinatorial Iron and Phosphate Deficiencies in Hexaploid Wheat Seedlings." *Genomics* 113(6): 3935–50. https://doi.org/10.1016/J.YGENO.2021.09.019.

Khamis, S. A. A. D. I., Sylvain Chaillou, and Thierry Lamaze. 1990. "CO2 Assimilation and Partitioning of Carbon in Maize Plants Deprived of Orthophosphate." *Journal of Experimental Botany* 41(12): 1619–25. https://doi.org/10.1093/JXB/41.12.1619.

Kirk, G. J. D., and L. E. Van Du. 1997. "Changes in Rice Root Architecture, Porosity, and Oxygen and Proton Release under Phosphorus Deficiency." *New Phytologist* 135(2): 191–200. https://doi.org/10.1046/J.1469-8137.1997.00640.X.

Koevoets, Iko T., Jan Henk Venema, J. Theo, M. Elzenga, and Christa Testerink. 2016. "Roots Withstanding Their Environment: Exploiting Root System Architecture Responses to Abiotic Stress to Improve Crop Tolerance." *Frontiers in Plant Science*. https://doi.org/10.3389/fpls.2016.01335.

Kumar, Suresh, Santosh Kumar, and Trilochan Mohapatra. 2021. "Interaction Between Macro- and Micronutrients in Plants." *Frontiers in Plant Science* 12(May): 753. https://doi.org/10.3389/FPLS.2021.665583/XML/NLM.

Li, Hai Bo, Ming Xia, and Ping Wu. 2001. "Effect of Phosphorus Deficiency Stress on Rice Lateral Root Growth and Nutrient Absorption." *Acta Botanica Sinica*. 43(11) 1154-1160.

Li, Yan, Yuhui Fang, Chaojun Peng, Xia Hua, Yu Zhang, Xueli Qi, Zhengling Li, Yumin Wang, Lin Hu, and Weigang Xu. 2022. "Transgenic Expression of Rice OsPHR2 Increases Phosphorus Uptake and Yield in Wheat." *Protoplasma* 259(5): 1271–82. https://doi.org/10.1007/S00709-021-01702-5/FIGURES/7.

Li, Zhaoxia, Qiang Gao, Yazheng Liu, Chunmei He, Xinrui Zhang, and Juren Zhang. 2011. "Overexpression of Transcription Factor ZmPTF1 Improves Low Phosphate Tolerance of Maize by Regulating Carbon Metabolism and Root Growth." *Planta* 233(6): 1129–43. https://doi.org/10.1007/S00425-011-1368-1/FIGURES/6.

Li, Zhaoxia, Changzheng Xu, Kunpeng Li, Shi Yan, Xun Qu, and Juren Zhang. 2012. "Phosphate Starvation of Maize Inhibits Lateral Root Formation and Alters Gene Expression in the Lateral Root Primordium Zone." *BMC Plant Biology* 12(1): 1–17. https://doi.org/10.1186/1471-2229-12-89/TABLES/7.

Liao, H., G. Rubio, X. Yan, A. Cao, K. M. Brown, and J. P. Lynch. 2001. "Effect of Phosphorus Availability on Basal Root Shallowness in Common Bean." *Plant and Soil* 232(1–2): 69–79.

Liao, Hong, Xiaolong Yan, Gerardo Rubio, Steve E. Beebe, Matthew W. Blair, and Jonathan P. Lynch. 2004. "Genetic Mapping of Basal Root Gravitropism and Phosphorus Acquisition Efficiency in Common Bean." *Functional Plant Biology: FPB* 31(10): 959–70. https://doi.org/10.1071/FP03255.

Liao, H., X. Yan, G. Rubio, S. E. Beebe, M. W. Blair, and J. P. Lynch. 2006. "Erratum: Genetic Mapping of Basal Root Gravitropism and Phosphorus Acquisition Efficiency in Common Bean (Functional Plant Biology 31:10 (959–970))." *Functional Plant Biology*. https://doi.org/10.1071/FP03255_CO.

Liu, Dong. 2021. "Root Developmental Responses to Phosphorus Nutrition." *Journal of Integrative Plant Biology*. https://doi.org/10.1111/jipb.13090.

Liu, Jianping, Wencheng Liao, Bo Nie, Jianhua Zhang, and Weifeng Xu. 2021. "OsUEV1B, an Ubc Enzyme Variant Protein, Is Required for Phosphate Homeostasis in Rice." *The Plant Journal: for Cell and Molecular Biology* 106(3): 706–19. https://doi.org/10.1111/TPJ.15193.

Liu, Zhigang, Xiangsheng Liu, Eric J. Craft, Lixing Yuan, Lingyun Cheng, Guohua Mi, and Fanjun Chen. 2018. "Physiological and Genetic Analysis for Maize Root Characters and Yield in Response to Low Phosphorus Stress." *Breeding Science* 68(2): 268. https://doi.org/10.1270/JSBBS.17083.

Lynch, Jonathan P. 2019. "Root Phenotypes for Improved Nutrient Capture: An Underexploited Opportunity for Global Agriculture." *New Phytologist* 223(2): 548–64. https://doi.org/10.1111/NPH.15738.

Lyu, Yang, Hongliang Tang, Haigang Li, Fusuo Zhang, Zed Rengel, William R. Whalley, and Jianbo Shen. 2016. "Major Crop Species Show Differential Balance between Root Morphological and Physiological Responses to Variable Phosphorus Supply." *Frontiers in Plant Science* 7 (December 2016): 1939. https://doi.org/10.3389/FPLS.2016.01939/BIBTEX.

Mo, Xiaohui, Guoxuan Liu, Zeyu Zhang, Xing Lu, Cuiyue Liang, and Jiang Tian. 2022. "Mechanisms Underlying Soybean Response to Phosphorus Deficiency through Integration of Omics Analysis." *International Journal of Molecular Sciences* 23(9): 4592. https://doi.org/10.3390/IJMS23094592.

Mollier, A., and S. Pellerin. 1999. "Maize Root System Growth and Development as Influenced by Phosphorus Deficiency." *Journal of Experimental Botany* 50(333): 487–97. https://doi.org/10.1093/JXB/50.333.487.

Nguyen, Van Lam, and James Stangoulis. 2019. "Variation in Root System Architecture and Morphology of Two Wheat Genotypes Is a Predictor of Their Tolerance to Phosphorus Deficiency." *Acta Physiologiae Plantarum* 41(7): 1–13. https://doi.org/10.1007/S11738-019-2891-0/TABLES/6.

Parra-Londono, Sebastian, Mareike Kavka, Birgit Samans, Rod Snowdon, Silke Wieckhorst, and Ralf Uptmoor. 2018. "Sorghum Root-System Classification in Contrasting P Environments Reveals Three Main Rooting Types and Root-Architecture-Related Marker–Trait Associations." *Annals of Botany* 121(2): 267–80. https://doi.org/10.1093/AOB/MCX157.

Prathap, V., Anuj Kumar, Chirag Maheshwari, and Aruna Tyagi. 2022. "Phosphorus Homeostasis: Acquisition, Sensing, and Long-Distance Signaling in Plants." *Molecular Biology Reports* 49(8): 8071–86. https://doi.org/10.1007/S11033-022-07354-9/TABLES/1.

Rocha, M. C. da, G. V. Miranda, M. J. V. Vasconcelos, P. C. Magalhães, G. A. de Carvalho Júnior, L. A. Silva, M. O. Soares, F. R. O. Cantão, F. Rodrigues, and R. E. Schaffert. 2010. "Characterization of Root Morphology in Contrasting Genotypes of Sorghum at Low and High Phosphorus Level." *Revista Brasileira de Milho e Sorgo* 9(1): 65–78.

Salazar-Henao, Jorge E., Isabel Cristina Vélez-Bermúdez, and Wolfgang Schmidt. 2016. "The Regulation and Plasticity of Root Hair Patterning and Morphogenesis." *Development (Cambridge)*. Company of Biologists Ltd. https://doi.org/10.1242/dev.132845.

Scheres, B., H. Wolkenfelt, V. Willemsen, M. Terlouw, E. Lawson, C. Dean, and P. Weisbeek. 1994. "Embryonic Origin of the Arabidopsis Primary Root and Root Meristem Initials." *Development* 120 (9): 2475–2487. https://doi.org/10.1242/dev.120.9.2475

Shen, Qi, Zhihui Wen, Yan Dong, Haigang Li, Yuxin Miao, and Jianbo Shen. 2018. "The Responses of Root Morphology and Phosphorus-Mobilizing Exudations in Wheat to Increasing Shoot Phosphorus Concentration." *AoB Plants* 10(5). https://doi.org/10.1093/AOBPLA/PLY054.

Shimizu, Akifumi, Seiji Yanagihara, Shinji Kawasaki, and Hiroshi Ikehashi. 2004. "Phosphorus Deficiency-Induced Root Elongation and Its QTL in Rice (Oryza Sativa L.)." *Theoretical and Applied Genetics* 109(7): 1361–68. https://doi.org/10.1007/S00122-004-1751-4/FIGURES/6.

Sun, H. G., and F. S. Zhang. 2000. "Growth Response of Wheat Roots to Phosphorus Deficiency." *Acta Botanica Sinica*. 42(9):913-919

Teng, Wan, Yan Deng, Xin Ping Chen, Xiao Feng Xu, Ri Yuan Chen, Yang Lv, Yan Yan Zhao, et al. 2013. "Characterization of Root Response to Phosphorus Supply from Morphology to Gene Analysis in Field-Grown Wheat." *Journal of Experimental Botany* 64(5): 1403. https://doi.org/10.1093/JXB/ERT023.

Vance, C. P. 2001. "Symbiotic Nitrogen Fixation and Phosphorus Acquisition. Plant Nutrition in a World of Declining Renewable Resources." *Plant Physiology* 127(2): 390. https://doi.org/10.1104/pp.010331.

Veneklaas, Erik J., Hans Lambers, Jason Bragg, Patrick M. Finnegan, Catherine E. Lovelock, William C. Plaxton, Charles A. Price, W. R. Scheible, M. W. Shane, P. J. White, and J. A. Raven. 2012. "Opportunities for Improving Phosphorus-Use Efficiency in Crop Plants." *New Phytologist* 195(2): 306–20. https://doi.org/10.1111/J.1469-8137.2012.04190.X.

Vincent, Andrea G., Jürgen Schleucher, Gerhard Gröbner, Johan Vestergren, Per Persson, Mats Jansson, and Reiner Giesler. 2012. "Changes in Organic Phosphorus Composition in Boreal Forest Humus Soils: The Role of Iron and Aluminium." *Biogeochemistry* 108(1–3): 485–99. https://doi.org/10.1007/S10533-011-9612-0/FIGURES/6.

Wang, Jing, Jinghan Sun, Jun Miao, Jinkao Guo, Zhanliang Shi, Mingqi He, Yu Chen, et al. 2013. "A Phosphate Starvation Response Regulator Ta-PHR1 Is Involved in Phosphate Signalling and Increases Grain Yield in Wheat." *Annals of Botany* 111(6): 1139–53. https://doi.org/10.1093/aob/mct080.

Wang, Ke, Feiyun Xu, Wei Yuan, Dongping Zhang, Jianping Liu, Leyun Sun, Liyou Cui, Jianhua Zhang, and Weifeng Xu. 2021. "Rice G Protein γ Subunit QPE9-1 Modulates Root Elongation for Phosphorus Uptake by Involving 14-3-3 Protein OsGF14b and Plasma Membrane H+-ATPase." *The Plant Journal* 107(6): 1603–15. https://doi.org/10.1111/TPJ.15402.

Wang, Qing-Jun, Yibing Yuan, Zhengqiao Liao, Yi Jiang, Qi Wang, Litian Zhang, Shibin Gao, et al. 2019. "Genome-Wide Association Study of 13 Traits in Maize Seedlings under Low Phosphorus Stress." *The Plant Genome* 12(3): 190039. https://doi.org/10.3835/PLANTGENOME2019.06.0039.

Wen, Zhihui, Haigang Li, Jianbo Shen, and Zed Rengel. 2017. "Maize Responds to Low Shoot P Concentration by Altering Root Morphology Rather than Increasing Root Exudation." *Plant and Soil* 416(1–2): 377–89. https://doi.org/10.1007/S11104-017-3214-0/FIGURES/1.

Wen, Zhihui, Hongbo Li, Qi Shen, Xiaomei Tang, Chuanyong Xiong, Haigang Li, Jiayin Pang, Megan H. Ryan, Hans Lambers, and Jianbo Shen. 2019. "Tradeoffs among Root Morphology, Exudation and Mycorrhizal Symbioses for Phosphorus-Acquisition Strategies of 16 Crop Species." *New Phytologist* 223(2): 882–95. https://doi.org/10.1111/NPH.15833.

Wissuwa, M., N. Ae, and S. S. Jones. 2001. "Genotypic Variation for Tolerance to Phosphorus Deficiency in Rice and the Potential for Its Exploitation in Rice Improvement." *Plant Breeding* 120(1): 43–8. https://doi.org/10.1046/j.1439-0523.2001.00561.x.

Wissuwa, M., J. Wegner, N. Ae, and M. Yano. 2002. "Substitution Mapping of Pup1: A Major QTL Increasing Phosphorus Uptake of Rice from a Phosphorus-Deficient Soil." *TAG. Theoretical and Applied Genetics. Theoretische und Angewandte Genetik* 105(6–7): 890–97. https://doi.org/10.1007/S00122-002-1051-9.

Wissuwa, Matthias, and Noriharu Ae. 2001. "Further Characterization of Two QTLs That Increase Phosphorus Uptake of Rice (Oryza sativa L.) under Phosphorus Deficiency." *Plant and Soil* 237(2): 275–86. https://doi.org/10.1023/A:1013385620875.

Wu, Ping, and X. M. Wang. 2008. "Role of OsPHR2 on Phosphorus Homoestasis and Root Hairs Development in Rice (Oryza sativa L.)." 3(9): 674–75. https://doi.org/10.4161/PSB.3.9.5781.

Xia, Zhenqing, Shibo Zhang, Qi Wang, Guixin Zhang, Yafang Fu, and Haidong Lu. 2021. "Effects of Root Zone Warming on Maize Seedling Growth and Photosynthetic Characteristics Under Different Phosphorus Levels." *Frontiers in Plant Science* 12(December): 2903. https://doi.org/10.3389/FPLS.2021.746152/BIBTEX.

Yan, Xiaolong, Hong Liao, Steve E. Beebe, Matthew W. Blair, and Jonathan P. Lynch. 2004. "QTL Mapping of Root Hair and Acid Exudation Traits and Their Relationship to Phosphorus Uptake in Common Bean." *Plant and Soil* 265(1): 17–29. https://doi.org/10.1007/S11104-005-0693-1.

Yang, Mengjiao, Cairong Wang, Muhammad Adeel Hassan, Faji Li, Xianchun Xia, Shubing Shi, Yonggui Xiao, and Zhonghu He. 2021. "QTL Mapping of Root Traits in Wheat under Different Phosphorus Levels Using Hydroponic Culture." *BMC Genomics* 22(1): 1–12. https://doi.org/10.1186/S12864-021-07425-4/TABLES/5.

Yang, Xilan, Yaxi Liu, Fangkun Wu, Xiaojun Jiang, Yu Lin, Zhiqiang Wang, Zhengli Zhang, G. Chen, Y. Wei, and Y. Zheng. 2018. "Quantitative Trait Loci Analysis of Root Traits under Phosphorus Deficiency at the Seedling Stage in Wheat." *Genome* 61(3): 209–15. https://doi.org/10.1139/GEN-2017-0159.

Yang, Zhaojun, Zhi Gao, Huiwen Zhou, Ying He, Yanxing Liu, Yelin Lai, Jiakun Zheng, Xinxin Li, and Hong Liao. 2021. "GmPTF1 Modifies Root Architecture Responses to Phosphate Starvation Primarily through Regulating GmEXPB2 Expression in Soybean." *The Plant Journal* 107(2): 525–43. https://doi.org/10.1111/TPJ.15307.

Yao, Zhu Fang, Cui Yue Liang, Qing Zhang, Zhi Jian Chen, Bi Xian Xiao, Jiang Tian, and Hong Liao. 2014. "SPX1 Is an Important Component in the Phosphorus Signalling Network of Common Bean Regulating Root Growth and Phosphorus Homeostasis." *Journal of Experimental Botany* 65(12): 3299. https://doi.org/10.1093/JXB/ERU183.

Yi, Keke, Zhongchang Wu, Jie Zhou, Liming Du, Longbiao Guo, Yunrong Wu, and Ping Wu. 2005. "OsPTF1, a Novel Transcription Factor Involved in Tolerance to Phosphate Starvation in Rice." *Plant Physiology* 138(4): 2087. https://doi.org/10.1104/PP.105.063115.

Zhang, Hengyou, Yuming Yang, Chongyuan Sun, Xiaoqian Liu, Lingling Lv, Zhenbin Hu, Yu Deyue, and Dan Zhang. 2020. "Up-Regulating GmETO1 Improves Phosphorus Uptake and Use Efficiency by Promoting Root Growth in Soybean." *Plant, Cell and Environment* 43(9): 2080–94. https://doi.org/10.1111/PCE.13816.

Zhang, Yu Man, Yong Sheng Yan, Li Na Wang, Kun Yang, Na Xiao, Yun Feng Liu, Ya. Ping Fu, Zong Xiu Sun, Rong Xiang Fang, and Xiao Ying Chen. 2012. "A Novel Rice Gene, NRR Responds to Macronutrient Deficiency and Regulates Root Growth." *Molecular Plant* 5(1): 63–72. https://doi.org/10.1093/mp/ssr066.

Zhao, Jing, Jiabing Fu, Hong Liao, Yong He, Hai Nian, Yueming Hu, Linjuan Qiu, Yinsan Dong, and Xiaolong Yan. 2004. "Characterization of Root Architecture in an Applied Core Collection for Phosphorus Efficiency of Soybean Germplasm." *Chinese Science Bulletin* 49(15): 1611–20. https://doi.org/10.1007/BF03184131.

Zheng, Lu, Mohammad Rezaul Karim, Yin Gang Hu, Renfang Shen, and Ping Lan. 2021. "Greater Morphological and Primary Metabolic Adaptations in Roots Contribute to Phosphate-Deficiency Tolerance in the Bread Wheat Cultivar Kenong199." *BMC Plant Biology* 21(1): 1–18. https://doi.org/10.1186/S12870-021-03164-6/FIGURES/10.

Zhou, Jia, Jianna Xie, Hong Liao, and Xiurong Wang. 2014. "Overexpression of β-Expansin Gene GmEXPB2 Improves Phosphorus Efficiency in Soybean." *Physiologia Plantarum* 150(2): 194–204. https://doi.org/10.1111/PPL.12077.

Zhou, Tao, Yongli Du, Shoaib Ahmed, Ting Liu, Menglu Ren, Weiguo Liu, and Wenyu Yang. 2016. "Genotypic Differences in Phosphorus Efficiency and the Performance of Physiological Characteristics in Response to Low Phosphorus Stress of Soybean in Southwest of China." *Frontiers in Plant Science* 7 (November 2016): 1776. https://doi.org/10.3389/FPLS.2016.01776/BIBTEX.

Zhou, Tao, Li Wang, Xin Sun, Xiaochun Wang, Yinglong Chen, Zed Rengel, Weiguo Liu, and Wenyu Yang. 2020. "Light Intensity Influence Maize Adaptation to Low P Stress by Altering Root Morphology." *Plant and Soil* 447(1–2): 183–97. https://doi.org/10.1007/S11104-019-04259-8/FIGURES/7.

4 Crosstalk between Phosphate and Other Nutrients

Xianqing Jia, Long Wang, and Keke Yi

4.1 INTRODUCTION

As sessile organisms, plants must cope with constantly changing environments that are often stressful or unfavourable for growth and development, such as fluctuations in the bioavailability of essential mineral nutrients. Plants require many essential nutrients to complete their life cycle and achieve yield, including macronutrients, e.g., phosphorus (P) and nitrogen (N), as well as micronutrients, e.g., iron (Fe) and zinc (Zn) (Fan et al., 2021; Kumar et al., 2021). In the last two decades, remarkable advances have been made in identifying components and working modules for the crosstalk between P and other elements, especially for N, Fe, and Zn. In this chapter, we will review and discuss the physiological effects and molecular mechanisms of P-N and P-Fe crosstalks, as well as some advances made in P-Zn and P-arsenic (As) crosstalks.

4.2 P-N CROSSTALK

Plants preferentially absorb and utilize P as inorganic phosphate (Pi) and plants have evolved elaborate regulatory networks to acquire Pi and maintain cellular Pi homeostasis, which are mediated by the SPX-InsP-PHR complex (reviewed by Jia et al., 2023). Different from phosphate-starvation response (PSR) signalling, N signalling can be divided into two divergent responses: 1) primary nitrate response (PNR), being immediately (within minutes) triggered when nitrate-starved plants are supplied with nitrate; 2) N starvation response (NSR), being triggered when N is removed from the media (reviewed in Wang et al., 2018; Vidal et al., 2020; de Bang et al., 2021). In brief, a dual functional protein, Nitrate Transporter1.1 (NRT1.1), is the core component of the PNR signalling pathway, which works as both a dual affinity nitrate transporter and a nitrate sensor (thus also termed transceptor). NRT1.1 triggers the calcium response likely decoded by calcium-dependent protein kinases (CPK), which phosphorylate the central transcription factors controlling PNR, such as NIN-like proteins (NLPs). NLPs subsequently activate regulatory networks (including Nitrate-Inducible GARP-type Transcriptional Repressor1 (NIGT1) and Regulator of Leaf Inclination1 (RLI1)) and nitrate-responsive genes (NRT2.1 and nitrate reductase (NIR)) to enhance nitrate acquisition (Ruan et al., 2019; Guo et al., 2022). Although core elements involved in NSR signalling are relatively unclear, several TFs, including NIGT1s and RLI1, were disclosed to be also implicated in NSR signalling and P-N-signalling crosstalk.

Intricate interactions between P and N have been suggested in several physiological aspects, including observations that N acts positively on phosphate uptake and phosphate starvation acts negatively on nitrate uptake and assimilation (Smith and Jackson, 1987; Rufty et al., 1990; de Magalhães et al., 1998). It's believed that plants have evolved complex P-N interaction networks to coordinate N and P acquisition and utilisation under fluctuating nutrient conditions for growth optimisation. In recent years, significant advances have been made in molecular mechanisms and regulatory cascades in P-N crosstalk, which will be introduced in the following sections.

Crosstalk between Phosphate and Other Nutrients

4.2.1 Regulating Pi Homeostasis in Plants in an N-Dependent Manner

Numerous studies have shown that N supplements could activate PSR under phosphate starvation; however, N starvation strongly represses PSR, that is, PSR can be regulated by N availability (Smith and Jackson, 1987; Rufty et al., 1990; de Magalhães et al., 1998; Kant et al., 2011). One of the first vital players discovered to be involved in N-dependent PSR is Nitrogen Limitation Adaptation (NLA). NLA, which encodes a RING-type ubiquitin E3 ligase, is a positive regulator for the development of the adaptability of plants to nitrogen limitation, and the *nla* mutant plants display abrupt early senescence (Peng et al., 2007; Peng et al., 2008). The expression of NLA was then found to be regulated by the low-Pi induced microRNA miR827, and the Pi overaccumulation in the *pho2* mutant shows Pi toxicity in a nitrate-dependent manner similar to the *nla* mutant (Kant et al., 2011). Further studies showed that NLA partners with PHOSPHATE2 (PHO2) to direct the ubiquitination of plasma membrane-localised PHT1s, which triggers clathrin-dependent endocytosis followed by endosomal sorting to vacuoles (Lin et al., 2013; Park et al., 2014). Further, Medici et al. found that NRS signalling could downregulate Pi-starvation-induced (PSI) genes via Phosphate Starvation Response1 (PHR1) and PHO2 (Medici et al., 2019). Specifically, upon N starvation, the stability of the core regulator of PSR signalling, PHR1, is reduced, further reducing PSR. PHO2 was further confirmed as an integrator of the N availability into the PSR, which also upregulates NRT 1.1, and, conversely, NRT1.1 downregulates PHO2 (Medici et al., 2019). Together, these findings suggest that Pi homeostasis in plants is regulated in an N-dependent manner via P-signalling components, such as NLA, PHO2, and PHR1.

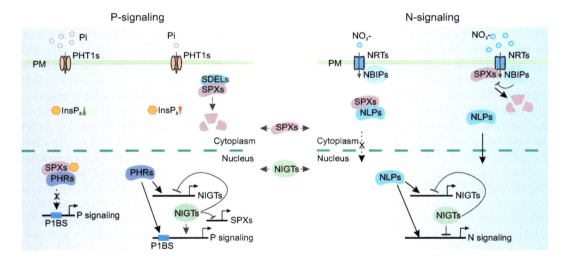

FIGURE 4.1 Crosstalks between Pi signalling and N signalling. SPXs and NIGTs are proposed as two classes of core regulators in P-N-signalling crosstalks. In P signalling, inositol pyrophosphate (InsP8) content is increased under Pi-sufficient conditions. Then InsP8 accumulates and triggers the formation of an SPX–InsP$_8$–PHR (SPX, also known as SYG1/Pho81/XPR1, and PHR, also known as PHOSPHATE STARVATION RESPONSE protein) complex, which blocks the activity of PHRs. Under Pi-deficient conditions, InsP8 content is reduced. Then PHRs are released in the nucleus, activating the expression of P-starvation response (PSR) genes by binding to the P1BS motifs. In addition, the action of PHR converges on the activation of Nitrate-Inducible GARP-type Transcriptional Repressors (NIGTs), which repress SPXs but activate the expression of PSR genes, and NIGTs also repress their own transcriptional expression. In N signalling, high nitrate levels prompt the interaction of NRTs–SPXs (NRTs, also known as Nitrate Transporters) and the degradation of the phosphate sensor involving ubiquitination by NBIPs (NRT1.1B interacting proteins). Under low nitrate conditions, SPXs interact with NLPs (named after NIN-like proteins); thus, nitrate-stimulated degradation of SPXs provokes a coordinated release of NLP TFs to activate the expression of nitrate uptake genes. In addition, NLPs also activate the expression of NIGTs, which repress nitrate uptake genes. PM, plasma membrane.

4.2.2 NIGT1s-Dependent Regulatory Cascades

A class of MYB transcription factors closely related to PHR1/PHL, NIGT1s, are increasingly thought to be core regulators of regulatory cascades in P-N crosstalk (Figure 4.1). NIGT1s are transcriptionally upregulated by NLP and act as repressors of nitrate-responsive genes as well as of their own transcription (Sawaki et al., 2013; Medici et al., 2015). Besides its repressor role in N signalling, NIGT1 also represses the expression of SPXs and, consequently, releases PHR1 to activate the expression of Pi transporters (Kiba et al., 2018; Ueda et al., 2020). On the other side, under nitrate supply, PHR1 induces NIGT1s that, in turn, enhance the Pi absorption via the upregulation of PHT1s (Kiba et al., 2018; Maeda et al., 2018). Another NIGT protein, NIGT1.2, was found to directly upregulate PHT1.1 and PHT1.4 genes by binding the non-canonical NIGT1.2 cis-motifs in their promoters (Wang et al., 2020). Consistently, NIGT1 overexpression alleviated the reduction in phosphate uptake under phosphate-replete conditions, and intricate feedback regulations involving PHR1, NIGT1, and SPX family proteins in the Pi-starvation signalling network were observed (Ueda et al., 2020). In addition, results of mutant protoplast-based assays and in planta analysis using NIGT1 overexpression in the *spx1 spx2* double mutant (Ueda et al., 2020). Thus, the NIGT1-SPX-PHR1 module provides a framework for bidirectional P-N-signalling crosstalk. Specifically, Pi-starvation signalling is activated by nitrate, and NSR signalling is also downregulated by Pi-starvation.

4.2.3 SPXs as Pivotal Nodes in Integrating P-N-Signalling Crosstalk

An SPX domain-containing protein, SPX4, interacts with PHR2 in the cytosol and prevents its translocation into the nucleus in rice (Lv et al., 2014). Besides SPX4, other nuclear-localised SPX proteins, SPX1 and SPX2, can block PHR1 and PHR2 from binding to the promoters of the PSI genes through the formation of the SPX-PHR complex in a phosphate-dependent manner (Wang et al., 2014). Besides its roles in Pi signalling, SPX4 was also found to play critical roles in P-N crosstalk (Hu et al., 2019). In detail, nitrate supply could trigger recruitment of the E2 ubiquitination ligase NBIP1 (named after NRT1.1B interacting protein1) by NRT1.1, which leads to SPX4 degradation to release PHRs to activate Pi-starvation signalling and promote Pi uptake. SPX4 also interacts with the NLP, the master TFs of PNR, endowing SPX4 with a direct role in PNR signalling. Together, this dual role of SPX4 to sequestrate PHR1 and NLPs provides a mechanism for coordinated activation of both PSR and PNR. Inositol pyrophosphates (InsPs) act as "signalling translators" to reflect cellular Pi levels in Pi-signalling networks. Under Pi-sufficient conditions, the InsPs can promote high-affinity interactions between SPXs and PHRs, and then this complex decreases the transcriptional activity of PSR genes mediated by PHRs (Jia et al., 2021). Therefore, whether InsPs play a role in N signalling deserves further study. In addition, another MYB gene involved in PS signalling and modulation of leaf inclination, RLI1, was reported to be a repressor of NS-responsive genes and is highly upregulated by PNR signalling in rice (Ruan et al., 2019). RLI1 expression in roots is proposed to be upregulated by nitrate via PHR1, which is relieved from sequestration by SPX4 in an NRT1.1-NBIP1-dependent manner (Zhang et al., 2021a). RLI1 may act as a weak activator of PHT1, but it mainly serves as a high-affinity interactor of SPX1 and SPX2, relieving the SPX inhibition of PHR1 (Zhang et al., 2021a).

4.2.4 P-N Crosstalk Regulating Root Growth

Nutritional crosstalks modulate plant root growth and development and have been widely reported. It has been well documented that plants use different strategies to regulate their root growth under N-deficiency compared with Pi-deficiency conditions. Compared with single nutrient starvation stress, the combined N and Pi deficiency potentiates the root system architecture in *Arabidopsis*

(Kellermeier et al., 2014). Serval molecular players have been identified to be implicated with P-N crosstalk regulating root growth. The *Arabidopsis* NO^{3-} transporter NPF7.3/NRT1.5 modulates the root architecture in response to Pi-starvation conditions, which is substantially induced by Pi-starvation (Cui et al., 2019). The *atnrt1.5* mutants displayed conspicuously longer primary roots and a significantly reduced lateral root density under Pi-deficient conditions than the wild-type plants. The ethylene synthesis antagonist eliminated these morphological differences in the roots to a certain extent (Cui et al., 2019). NIGT1 (also termed HRS1 or HHO1) is another well-documented crucial molecular regulator connecting N and P signalling to control the primary root response. NIGT1 represses the primary root growth in response to P deficiency conditions but only when NO3- is present (Medici et al., 2015).

A key regulatory protein, STOP1 (also known as Sensitive To Proton Rhizotoxicity1), was also reported to be a central regulator which links external ammonium with efficient Pi acquisition from insoluble Pi sources to shape root growth (Tian et al., 2021). In detail, Pi deficiency promotes ammonium uptake mediated by ammonium transporters (AMTs) and causes rapid acidification of the root surface. Rhizosphere acidification-triggered STOP1 accumulation activates the excretion of organic acids, which help to solubilise Pi from insoluble iron or calcium phosphates. Ammonium uptake by AMTs is downregulated by a CBL-Interacting Protein Kinase 23 (CIPK23) whose expression is directly modulated by STOP1 when ammonium reaches toxic levels. Another independent study also showed that, in *Arabidopsis* roots, the STOP1 in the nucleus was enriched by low pH in a nitrate-independent manner, with the spatial expression pattern of NRT1.1 established by low pH requiring the action of STOP1 (Ye et al., 2021). The *nrt1.1* and *stop1* mutants and the *nrt1.1 stop1* double mutant had a similar hypersensitive phenotype to low pH, indicating that STOP1 and NRT1.1 function in the same pathway for H$^+$ tolerance. Molecular assays revealed that STOP1 is directly bound to the promoter of NRT1.1 to activate its transcription in response to low pH, indicating that STOP1-NRT1.1 is a crucial module for plants to optimise N utilisation and ensure better root growth in acidic conditions. In addition, as a comprehensive core regulator in nutrient crosstalks, STOP1 is also involved in the interactions between Pi and other elements, which will be described below.

4.3 P-FE CROSSTALK

Crosstalk between Pi signalling and Fe signalling has also been widely documented. The chemistry of Pi and Fe is deeply interconnected, as Pi naturally associates with ferric oxides to create insoluble complexes that plant roots can no longer directly take up in the soil. Physiologically, Pi deficiency resulted in significantly increased Fe availability within the plants. Pi availability has been demonstrated to alter the transcriptional regulation of genes involved in the uptake and transport of Fe in the plant (Hirsch et al., 2006; Ward et al., 2008; Zheng et al., 2009). Fe deficiency-induced chlorosis depends on Pi availability (Paz-Ares et al., 2022). In turn, Fe deficiency can promote Pi accumulation in plants (Zheng et al., 2009). Additionally, the low-Pi-triggered root tip growth inhibition could be compromised under Fe deficiency. Over the past few decades, great advances have been made in molecular mechanisms underlying this antagonistic interaction between iron and P, which will be summarised in the following subsections.

4.3.1 P-FE-SIGNALLING INTERPLAY MEDIATED BY A PHR-HRZ MODULE

A class of type E3 ligases, HRZs (also known as Hemerythrin motif-containing Really Interesting New Gene- and Zinc-finger proteins), which participate in the negative control of the stability of some Fe-signalling TFs to mediate Fe starvation response (FSR) (Spielmann and Vert, 2021), are also important players for the signalling crosstalk between P and Fe (Figure 4.2). It has been reported that HRZs could interact with and trigger the degradation of PHR2 (Guo et al., 2022). The transcript and protein abundances of HRZs and PHRs are maintained at moderate levels to

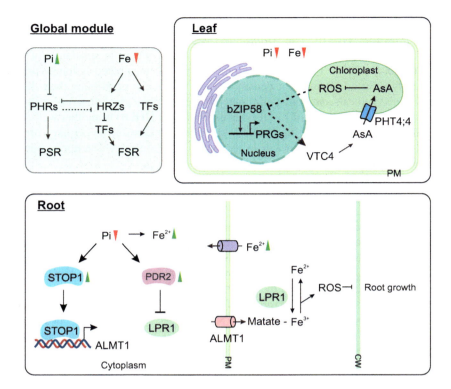

FIGURE 4.2 Crosstalks between Pi signalling and Fe signalling. A global P-Fe interplay was reported recently, which is mediated by a reciprocally inhibitory module formed between rice PHRs and HRZs (also known as Hemerythrin motif-containing Really Interesting New Gene- and Zinc-finger proteins). HRZs are thought to act as Fe sensors and repressors of FSR genes. PHR downregulates HRZ, and HRZ mediates PHR degradation. Thus, P-starvation signalling and Pi accumulation are reduced by Fe supply, and P starvation enhances Fe accumulation. In leaves, under Fe and P deficiency, bZIP58 modulates the expression of the gene encoding VITAMIN C 4 (VTC4), which produces ascorbic acid (AsA) to inhibit reactive oxygen species (ROS) that repress bZIP58 expression; bZIP58 prevents leaf chlorosis through the regulation of photosynthesis-related genes (PRGs) and is suppressed under singular Fe deficiency. In *Arabidopsis* roots, P starvation inhibits root growth by promoting Fe accumulation in the root tip via the PHOSPHATE DEFICIENCY RESPONSE (PDR2)-LOW PHOSPHATE ROOT 1 (LPR1) and Sensitive To Proton Rhizotoxicity1 (STOP1)-Aluminum-activated Malate Transporter 1 (ALMT1) signalling modules. LPR1 localises to the apoplast and expresses ferroxidase activity, which is thought to generate ROS via Fe redox cycling, leading to the induction of callose deposition via unknown ROS signalling and impaired symplastic connectivity. ALMT1-dependent release of malate into the apoplast of interior root tip cell layers is thought to mobilise Fe^{3+} from Fe–Pi complexes for reduction by ascorbate, subsequent re-oxidation by LPR1, and Fe redox cycling. CW, cell wall; PM, plasma membrane.

allow the basal expression of FSR and PSR genes. When the availability of Pi becomes low, the Pi deficiency prompts repression of HRZs, which in turn increases PHRs protein abundance to further inhibit the expression of HRZs. The increased accumulation of PHRs and reduced accumulation of negative regulators of iron signalling, HRZs, will activate the expression of FSR- and PSR-related genes, ultimately causing iron accumulation and promoting low Pi stress adaptation. Under Fe deficiency conditions, the transcript and protein level of HRZs increases, thus facilitating the degradation of PHRs. When both Fe and Pi levels are low, the protein abundances of HRZs and PHRs are maintained at relatively high levels, promoting better adaptation of plants to these two stresses. Together, HRZs and PHRs form a global reciprocal inhibitory module that coordinates the crosstalk between Pi signalling and Fe signalling in plants.

4.3.2 P-Fe Influence on Plant Root Growth

Root tips sense Pi together with Fe to locally inform root development. The local response of roots to Pi limitation depends on external Fe availability (Svistoonoff et al., 2007; Ward et al., 2008), which suggests indirect monitoring of Pi via its chemical interactions with the transition metal. Recent works uncovered a central role of LOW PHOSPHATE ROOT 1 (LPR1) and PHOSPHATE DEFICIENCY RESPONSE 2 (PDR2) in root Pi sensing (Ticconi et al., 2009; Müller et al., 2015; Ziegler et al., 2016). PDR2 codes for the single P5-type ATPase of undefined transport specificity, AtP5A, which maintains essential endoplasmic reticulum functions (Müller et al., 2015). LPR1 encodes a cell wall-resident multicopper oxidase with ferroxidase activity, which was identified as a major QTL in accessions that show opposite root responses to low Pi (Svistoonoff et al., 2007; Müller et al., 2015). The Pi-insensitive *lpr1 lpr2* mutations, which cause available primary root extension on low Pi, suppress the hypersensitive *pdr2* short root phenotype in the triple mutant background (Müller et al., 2015). The redox activity of LPR1-derived Fe^{3+} likely depends on apoplast chemistry and the availability of Fe ligands (Grillet et al., 2014). Together, the antagonistic Fe–Pi interaction is mediated by the PDR2–LPR1 function framework, which likely determines reactive oxygen species (ROS) production and activates callose synthesis to adjust cell-to-cell communication and root growth.

Further studies found that a STOP1-ALMT1 (also known as Aluminum-activated Malate Transporter1) module parallel to the LPR1–PDR2 module, is involved in the low-Pi-triggered inhibition of primary root growth and shapes root system architecture (Figure 4.2). Under low Pi and high Fe conditions, STOP1 protein but not STOP1 transcription accumulated in the nucleus and upregulated the expression of the ALMT1 (Balzergue et al., 2017). The ALMT1 transported the malate across the plasm member, together with LPR1 ferroxidase, promoting the Fe^{3+} aggregation in the apoplast of root cells, which triggered ROS formation (Svistoonoff et al., 2007; Müller et al., 2015; Balzergue et al., 2017). The ROS induced the callose deposition in the meristem and the elongation zone and then displayed enhanced cell wall stiffening mediated by the action of peroxidases that cause inhibition of root elongation (Svistoonoff et al., 2007; Müller et al., 2015; Balzergue et al., 2017). Mutations in STOP1 and ALMT1 restore cell elongation under Pi-starvation conditions, but cell divisions are still inhibited (Balzergue et al., 2017), also indicating the synergistic regulation roles of the two modules in shaping root architecture.

4.3.3 P-Fe Influence on Photosynthesis

Besides regulating root growth, P-Fe crosstalk also influences plant photosynthesis. As Pi can interact with iron to form insoluble complexes, Pi starvation can promote iron accumulation and utilisation in plants, while high Pi content provokes iron deficiency, causing chlorosis, even when the total leaf iron level is sufficient (Zheng et al., 2009; Kumar et al., 2021). Similarly, iron deficiency can promote Pi accumulation in plants (Zheng et al., 2009). Moreover, the plants that grow under a combined Fe and Pi deficiency do not exhibit a chlorotic phenotype and present a stay-green phenotype. These observations indicate that changes in Pi availability will strongly influence iron homeostasis. Recently, a study showed that the mechanism includes ascorbic acid (AsA) and the gene coding for nuclear proteins (bZIP58) that prevent the repression of a core set of photosynthesis genes and associated chlorosis under Fe and P deficiency (Nam et al., 2021). In detail, the nuclear transcription factor bZIP58 appeared to regulate the expression of an AsA biosynthesis gene VITAMIN C 4 (VTC4) and appeared to be a key regulator of PRGs. Its absence could alter chlorophyll accumulation regardless of Fe and P availability. AsA, which VTC4 synthesised, was transported into the chloroplast by PHT4;4. During Fe deprivation conditions, the increase of plastidic ROS levels is perceived at the nucleus as a signal to downregulate the expression of bZIP58 and downstream photosynthesis genes, which results in chlorosis. Under Pi-Fe deprivation, the nuclear transcription factor bZIP58 appeared to regulate the expression of VTC4 to synthesise AsA. The AsA was transported into the chloroplast by PHT4;4 and prevented ROS accumulation, thus maintaining the expression of bZIP58 and its downstream photosynthesis genes and leading to the "stay-green" phenotype. Together, chloroplastic ascorbate transport prevents the downregulation of photosynthesis genes under iron-phosphorus combined deficiency through modulation of ROS homeostasis.

4.4 INTERACTIONS BETWEEN P AND OTHER ELEMENTS

Interactions between P and Zinc (Zn) have also been widely observed and reported. Physiologically, Zn deficiency treatments could increase the P accumulation in the *Arabidopsis* shoot, and Pi-starvation treatments reciprocally cause Zn accumulation (Khan et al., 2014). However, the increase in shoot Pi content in the wild type was observed in the *pho2* mutant but not in the *phr1* or *pho1* mutants, indicating that PHR1 and PHO1 participate in the coregulation of Zn and Pi homeostasis. Further results showed that mutations in PHO1;H3 cause enhanced Zn-deficiency-triggered Pi accumulation, which was not observed in a *pho1 pho1;h3* double mutant. However, unlike PHO1, PHO1;H3 could not export Pi, suggesting that PHO1;H3 restricts root-to-shoot Pi transfer requiring PHO1 function for Pi homeostasis in response to Zn deficiency (Khan et al., 2014). LPCAT1, encoding a lysophosphatidylcholine acyltransferase, is involved in the conversion of lysophosphatidylcholine (LPC) to phosphatidylcholine (PC) (Kisko et al., 2018), which is regulated by bZIP23 leading to a constitutive transcriptional Zn deficiency response (Kisko et al., 2018; Lilay et al., 2021). The *lpcat1* mutants exhibit increased LPC/PC ratio and upregulated expression of the Pi transporter PHT1;1, consequentially leading to Pi accumulation in shoots (Kisko et al., 2018). Most recently, it was found that relatively higher Pi and Zn accumulations were detected in the low phytic acid (LPA) mutant line under standard and also deficient conditions of Pi and Zn, but the activation of shoot Pi accumulation that occurs in WT in response to Zn depletion was not observed in the *lpa* mutant (Belgaroui et al., 2022).

Arsenic (As) is a powerfully toxic compound to nearly all plants. The typical chemical forms of As in the biosphere are As(V) and its reduced form arsenite (As(III)). Due to the structural similarity of the anions with Pi, As(V) can be absorbed into cells via the Pi transporter PHT1, and its detoxification involves the reduction to As(III) by High Arsenic Content 1 (HAC1), followed by fast sequestration in the vacuoles complexed with phytochelatins or by extrusion mediated by Nod-26-like aquaporin (Paz-Ares et al., 2022). An arsenate-responsive WRKY transcription factor, WRKY6, could downregulate PHT1 and PHO1, whose expression is highly induced by As(V) to restrict arsenate-induced transposon activation (Castrillo et al., 2013). On the other hand, As-repression of the phosphate transporter PHT1;1 is associated with the degradation of the PHR1. Once arsenic is sequestered into the vacuole, PHR1 stability is restored, and PHT1;1 expression is recovered. An As-responsive SKP1-like protein and a PHR1 Interactor F-box1 (PHIF1) form an SCF complex responsible for PHR1 degradation (Navarro et al., 2021). Given As's toxic effects and threats in agricultural practices, more work is needed to uncover the molecular pathways underlying As-sensing systems to foster plant tolerance to As toxicity and develop bioremediation strategies for plant-mediated removal of As pollution from the environment for bio-safe food production.

4.5 CONCLUSION AND PROSPECTS

As sessile organisms, land plants are faced with variable environmental stresses, including soil nutrient deficiency and high concentrations of metals or heavy metals, which significantly restrict plant survival and growth. Nutrients in the soil are always coupled together, such as P-Fe and P-Al. In addition, plant cell nutrients are frequently implicated in several metabolism pathways, such as P and N involved in DNA metabolism and P and Fe involved in photosynthesis. Therefore, land plants have evolved intricate mechanisms to coregulate nutrient assimilation and maintain nutrient homeostasis. For decades, the complicated crosstalks between P and other micro/mineral elements in plants have been recognised at morphological and physiological levels, and the molecular bases, regulatory networks, and biological significance of these interactions in plants are starting to be uncovered.

With growing evidence that multiple-nutrient uptake and utilisation are intertwined, and that intercellular nutrient coordination is vital for nutrient utilisation (reviewed in Fan et al., 2021; Jia et al., 2021; Paz-Ares et al., 2022), in the future, more integrative studies are needed to reveal the underlying mechanisms by which plants coordinate these multiple-nutrient stresses at whole-plant systemic levels. Examples include gene co-expression analysis, single-cell transcriptome, multi-omics approaches, efficient nutrient tracing assay, and gene editing. In addition, different from land plants, green algae usually live in water (freshwater or saltwater) with low nutrient content, thus, green algae have evolved a different nutrient utilisation system. Taking P utilisation as an example, green algae always perform sustained luxury P uptake (ability to take up more P than necessary for immediate growth) and store this excess P as polyphosphate (polyP) in vacuoles (land plants store P as Pi in vacuoles) (Solovchenko et al., 2019). Thus, more attention to the nutrient-use strategies of lower green algae would provide us with a more comprehensive understanding of nutrient crosstalks. Recent studies have shown that land plants largely depend upon their associated microbes for growth promotion and nutrient availability under various stresses (reviewed in Arif et al., 2020; Klimasmith and Kent, 2022; Singh et al., 2020; J. Zhang et al., 2021). Thus, further investigations are proposed to explore the roles of plant-associated microbiota (rhizospheric or endophytic) in plant crosstalks between P and other nutrients. These approaches will help us further comprehensively understand the biological significance of these crosstalks, promote nutrient-efficient breeding, and develop new strategies for field nutrient management.

4.6 ACKNOWLEDGEMENTS

This work was supported by the National Natural Science Foundation of China (No: 32130096, 32202593 and 32102478).

REFERENCES

Arif, I., Batool, M., and Schenk, P. M. (2020). Plant microbiome engineering: Expected benefits for improved crop growth and resilience. *Trends Biotechnol.* 38(12):1385–1396.

Balzergue, C., Dartevelle, T., Godon, C., Laugier, E., Meisrimler, C., Teulon, J.-M., Creff, A., Bissler, M., Brouchoud, C., Hagège, A., et al. (2017). Low phosphate activates STOP1-ALMT1 to rapidly inhibit root cell elongation. *Nat. Commun.* 8:15300.

Belgaroui, N., Elifa, W., and Hanin, M. (2022). Phytic acid contributes to the phosphate-zinc signaling crosstalk in Arabidopsis. *Plant Physiol. Biochem.* 183:1–8.

Castrillo, G., Sánchez-Bermejo, E., de Lorenzo, L., Crevillén, P., Fraile-Escanciano, A., TC, M., Mouriz, A., Catarecha, P., Sobrino-Plata, J., Olsson, S., et al. (2013). WRKY6 transcription factor restricts arsenate uptake and transposon activation in Arabidopsis. *Plant Cell* 25(8):2944–2957.

Cui, Y.-N., Li, X.-T., Yuan, J.-Z., Wang, F.-Z., Wang, S.-M., and Ma, Q. (2019). Nitrate transporter NPF7.3/ NRT1.5 plays an essential role in regulating phosphate deficiency responses in Arabidopsis. *Biochem. Biophys. Res. Commun.* 508(1):314–319.

de Bang, T. C., Husted, S., Laursen, K. H., Persson, D. P., and Schjoerring, J. K. (2021). The molecular–physiological functions of mineral macronutrients and their consequences for deficiency symptoms in plants. *New Phytol.* 229(5):2446–2469.

de Magalhães, J. V., Alves, V. M. C., de Novais, R. F., Mosquim, P. R., Magalhães, J. R., Filho, A. F. C. B., and Hubert, D. M. (1998). Nitrate uptake by corn under increasing periods of phosphorus starvation. *J. Plant Nutr.* 21(8):1753–1763.

Fan, X., Zhou, X., Chen, H., Tang, M., and Xie, X. (2021). Cross-talks between macro- and micronutrient uptake and signaling in plants. *Front. Plant Sci.* 12:2076.

Grillet, L., Ouerdane, L., Flis, P., Hoang, M. T. T., Isaure, M.-P., Lobinski, R., Curie, C., and Mari, S. (2014). Ascorbate efflux as a new strategy for iron reduction and transport in plants*. *J. Biol. Chem.* 289(5):2515–2525.

Guo, M., Ruan, W., Zhang, Y., Zhang, Y., Wang, X., Guo, Z., Wang, L., Zhou, T., Paz-Ares, J., and Yi, K. (2022). A reciprocal inhibitory module for Pi and iron signaling. *Mol. Plant* 15(1):138–150.

Hirsch, J., Marin, E., Floriani, M., Chiarenza, S., Richaud, P., Nussaume, L., and Thibaud, M. C. (2006). Phosphate deficiency promotes modification of iron distribution in Arabidopsis plants. *Biochimie* 88(11):1767–1771.

Hu, B., Jiang, Z., Wang, W., Qiu, Y., Zhang, Z., Liu, Y., Li, A., Gao, X., Liu, L., Qian, Y., et al. (2019). Nitrate–NRT1.1B–SPX4 cascade integrates nitrogen and phosphorus signalling networks in plants. *Nat. Plants* 5(4):401–413.

Jia, X., Wang, L., Nussaume, L., and Yi, K. (2023). Cracking the code of plant central phosphate signaling. *Trends Plant Sci.* 28:267–270.

Jia, X., Wang, L., Zeng, H., and Yi, K. (2021). Insights of intracellular/intercellular phosphate transport and signaling in unicellular green algae and multicellular land plants. *New Phytol.* 232(4):1566–1571.

Kant, S., Peng, M., and Rothstein, S. J. (2011). Genetic regulation by NLA and MicroRNA827 for maintaining nitrate-dependent phosphate homeostasis in Arabidopsis. *PLOS Genet.* 7(3):e1002021.

Kellermeier, F., Armengaud, P., Seditas, T. J., Danku, J., Salt, D. E., and Amtmann, A. (2014). Analysis of the root system architecture of Arabidopsis provides a quantitative readout of crosstalk between nutritional signals. *Plant Cell* 26(4):1480–1496.

Khan, G. A., Bouraine, S., Wege, S., Li, Y., de Carbonnel, M., Berthomieu, P., Poirier, Y., and Rouached, H. (2014). Coordination between zinc and phosphate homeostasis involves the transcription factor PHR1, the phosphate exporter PHO1, and its homologue PHO1;H3 in Arabidopsis. *J. Exp. Bot.* 65(3):871–884.

Kiba, T., Inaba, J., Kudo, T., Ueda, N., Konishi, M., Mitsuda, N., Takiguchi, Y., Kondou, Y., Yoshizumi, T., Ohme-Takagi, M., et al. (2018). Repression of nitrogen starvation responses by members of the Arabidopsis GARP-type transcription factor NIGT1/HRS1 subfamily. *Plant Cell* 30(4):925–945.

Kisko, M., Bouain, N., Safi, A., Medici, A., Akkers, R. C., Secco, D., Fouret, G., Krouk, G., Aarts, M. G., Busch, W., et al. (2018). LPCAT1 controls phosphate homeostasis in a zinc-dependent manner. *eLife* 7:e32077.

Klimasmith, I. M., and Kent, A. D. (2022). Micromanaging the nitrogen cycle in agroecosystems. *Trends Microbiol.* 0(11).

Kumar, S., Kumar, S., and Mohapatra, T. (2021). Interaction between macro- and micro-nutrients in plants. *Front. Plant Sci.* 12.

Lilay, G. H., Persson, D. P., Castro, P. H., Liao, F., Alexander, R. D., Aarts, M. G. M., and Assunção, A. G. L. (2021). Arabidopsis bZIP19 and bZIP23 act as zinc sensors to control plant zinc status. *Nat. Plants* 7(2):137–143.

Lin, W.-Y., Huang, T.-K., and Chiou, T.-J. (2013). Nitrogen limitation adaptation, a target of microRNA827, mediates degradation of plasma membrane–localized phosphate transporters to maintain phosphate homeostasis in Arabidopsis. *Plant Cell* 25(10):4061–4074.

Lv, Q., Zhong, Y., Wang, Y., Wang, Z., Zhang, L., Shi, J., Wu, Z., Liu, Y., Mao, C., Yi, K., et al. (2014). SPX4 negatively regulates phosphate signaling and homeostasis through its interaction with PHR2 in rice. *Plant Cell* 26(4):1586–1597.

Maeda, Y., Konishi, M., Kiba, T., Sakuraba, Y., Sawaki, N., Kurai, T., Ueda, Y., Sakakibara, H., and Yanagisawa, S. (2018). A NIGT1-centred transcriptional cascade regulates nitrate signalling and incorporates phosphorus starvation signals in Arabidopsis. *Nat. Commun.* 9(1):1376.

Medici, A., Marshall-Colon, A., Ronzier, E., Szponarski, W., Wang, R., Gojon, A., Crawford, N. M., Ruffel, S., Coruzzi, G. M., and Krouk, G. (2015). AtNIGT1/HRS1 integrates nitrate and phosphate signals at the Arabidopsis root tip. *Nat. Commun.* 6:1–11.

Medici, A., Szponarski, W., Dangeville, P., Safi, A., Dissanayake, I. M., Saenchai, C., Emanuel, A., Rubio, V., Lacombe, B., Ruffel, S., et al. (2019). Identification of Molecular integrators shows that nitrogen actively controls the phosphate starvation response in plants. *Plant Cell* 31(5):1171–1184.

Müller, J., Toev, T., Heisters, M., Teller, J., Moore, K. L., Hause, G., Dinesh, D. C., Bürstenbinder, K., and Abel, S. (2015). Iron-dependent callose deposition adjusts root meristem maintenance to phosphate availability. *Dev. Cell* 33:216–230.

Nam, H.-I., Shahzad, Z., Dorone, Y., Clowez, S., Zhao, K., Bouain, N., Lay-Pruitt, K. S., Cho, H., Rhee, S. Y., and Rouached, H. (2021). Interdependent iron and phosphorus availability controls photosynthesis through retrograde signaling. *Nat. Commun.* 12(1):7211.

Navarro, C., Mateo-Elizalde, C., Mohan, T. C., Sánchez-Bermejo, E., Urrutia, O., Fernández-Muñiz, M. N., García-Mina, J. M., Muñoz, R., Paz-Ares, J., Castrillo, G., et al. (2021). Arsenite provides a selective signal that coordinates arsenate uptake and detoxification through the regulation of PHR1 stability in Arabidopsis. *Mol. Plant* 14(9):1489–1507.

Park, B. S., Seo, J. S., and Chua, N.-H. (2014). Nitrogen limitation adaptation recruits phosphate2 to target the phosphate transporter PT2 for degradation during the regulation of Arabidopsis phosphate homeostasis. *Plant Cell* 26(1):454–464.

Paz-Ares, J., Puga, M. I., Rojas-Triana, M., Martinez-Hevia, I., Diaz, S., Poza-Carrión, C., Miñambres, M., and Leyva, A. (2022). Plant adaptation to low phosphorus availability: Core signaling, crosstalks, and applied implications. *Mol. Plant* 15(1):104–124.

Peng, M., Hannam, C., Gu, H., Bi, Y.-M., and Rothstein, S. J. (2007). A mutation in NLA, which encodes a RING-type ubiquitin ligase, disrupts the adaptability of Arabidopsis to nitrogen limitation. *Plant J.* 50(2):320–337.

Peng, M., Hudson, D., Schofield, A., Tsao, R., Yang, R., Gu, H., Bi, Y.-M., and Rothstein, Steven. J. (2008). Adaptation of Arabidopsis to nitrogen limitation involves induction of anthocyanin synthesis which is controlled by the NLA gene. *J. Exp. Bot.* 59(11):2933–2944.

Ruan, W., Guo, M., Wang, X., Guo, Z., Xu, Z., Xu, L., Zhao, H., Sun, H., Yan, C., and Yi, K. (2019). Two RING-finger ubiquitin E3 ligases regulate the degradation of SPX4, an internal phosphate sensor, for phosphate homeostasis and signaling in rice. *Mol. Plant* 12:1060–1074.

Rufty, T. W., Jr., MacKown, C. T., and Israel, D. W. (1990). Phosphorus stress effects on assimilation of nitrate 1. *Plant Physiol.* 94(1):328–333.

Sawaki, N., Tsujimoto, R., Shigyo, M., Konishi, M., Toki, S., Fujiwara, T., and Yanagisawa, S. (2013). A nitrate-inducible GARP family gene encodes an auto-repressible transcriptional repressor in rice. *Plant Cell Physiol.* 54(4):506–517.

Singh, B. K., Trivedi, P., Egidi, E., Macdonald, C. A., and Delgado-Baquerizo, M. (2020). Crop microbiome and sustainable agriculture. *Nat. Rev. Microbiol.* 18(11):601–602.

Smith, F. W., and Jackson, W. A. (1987). Nitrogen enhancement of phosphate transport in roots of Zea mays L: II. Kinetic and inhibitor studies. *Plant Physiol.* 84(4):1319–1324.

Solovchenko, A. E., Ismagulova, T. T., Lukyanov, A. A., Vasilieva, S. G., Konyukhov, I. V., Pogosyan, S. I., Lobakova, E. S., and Gorelova, O. A. (2019). Luxury phosphorus uptake in microalgae. *J. Appl. Phycol.* 31(5):2755–2770.

Spielmann, J., and Vert, G. (2021). The many facets of protein ubiquitination and degradation in plant root iron-deficiency responses. *J. Exp. Bot.* 72(6):2071–2082.

Svistoonoff, S., Creff, A., Reymond, M., Sigoillot-Claude, C., Ricaud, L., Blanchet, A., Nussaume, L., and Desnos, T. (2007). Root tip contact with low-phosphate media reprograms plant root architecture. *Nat. Genet.* 39(6):792–796.

Tian, W. H., Ye, J. Y., Cui, M. Q., Chang, J. B., Liu, Y., Li, G. X., Wu, Y. R., Xu, J. M., Harberd, N. P., Mao, C. Z., et al. (2021). A transcription factor STOP1-centered pathway coordinates ammonium and phosphate acquisition in Arabidopsis. *Mol. Plant* 14(9):1554–1568.

Ticconi, C. A., Lucero, R. D., Sakhonwasee, S., Adamson, A. W., Creff, A., Nussaume, L., Desnos, T., and Abel, S. (2009). ER-resident proteins PDR2 and LPR1 mediate the developmental response of root meristems to phosphate availability. *Proc. Natl. Acad. Sci.* 106(33):14174–14179.

Ueda, Y., Kiba, T., and Yanagisawa, S. (2020). Nitrate-inducible NIGT1 proteins modulate phosphate uptake and starvation signalling via transcriptional regulation of SPX genes. *Plant J.* 102(3):448–466.

Vidal, E. A., Alvarez, J. M., Araus, V., Riveras, E., Brooks, M. D., Krouk, G., Ruffel, S., Lejay, L., Crawford, N. M., Coruzzi, G. M., et al. (2020). Nitrate in 2020: Thirty years from transport to signaling networks. *Plant Cell* 32(7):2094–2119.

Wang, X., Wang, H.-F., Chen, Y., Sun, M.-M., Wang, Y., and Chen, Y.-F. (2020). The transcription factor NIGT1.2 modulates both phosphate uptake and nitrate influx during phosphate starvation in Arabidopsis and maize. *Plant Cell* 32(11):3519–3534.

Wang, Y.-Y., Cheng, Y.-H., Chen, K.-E., and Tsay, Y.-F. (2018). Nitrate transport, signaling, and use efficiency. *Annu. Rev. Plant Biol.* 69:85–122.

Wang, Z., Ruan, W., Shi, J., Zhang, L., Xiang, D., Yang, C., Li, C., Wu, Z., Liu, Y., Yu, Y., et al. (2014). Rice SPX1 and SPX2 inhibit phosphate starvation responses through interacting with PHR2 in a phosphate-dependent manner. *Proc. Natl. Acad. Sci.* 111(41):14953–14958.

Ward, J. T., Lahner, B., Yakubova, E., Salt, D. E., and Raghothama, K. G. (2008a). The effect of iron on the primary root elongation of Arabidopsis during phosphate deficiency. *Plant Physiol.* 147(3):1181–1191.

Ye, J. Y., Tian, W. H., Zhou, M., Zhu, Q. Y., Du, W. X., Zhu, Y. X., Liu, X. X., Lin, X. Y., Zheng, S. J., and Jin, C. W. (2021). STOP1 activates NRT1.1-mediated nitrate uptake to create a favorable rhizospheric pH for plant adaptation to acidity. *Plant Cell* 33(12):3658–3674.

Zhang, J., Cook, J., Nearing, J. T., Zhang, J., Raudonis, R., Glick, B. R., Langille, M. G. I., and Cheng, Z. (2021b). Harnessing the plant microbiome to promote the growth of agricultural crops. *Microbiol. Res.* 245:126690.

Zhang, Z., Li, Z., Wang, W., Jiang, Z., Guo, L., Wang, X., Qian, Y., Huang, X., Liu, Y., Liu, X., et al. (2021a). Modulation of nitrate-induced phosphate response by the MYB transcription factor RLI1/HINGE1 in the nucleus. *Mol. Plant* 14(3):517–529.

Zheng, L., Huang, F., Narsai, R., Wu, J., Giraud, E., He, F., Cheng, L., Wang, F., Wu, P., Whelan, J., et al. (2009). Physiological and transcriptome analysis of iron and phosphorus interaction in rice seedlings. *Plant Physiol.* 151(1):262–274.

Ziegler, J., Schmidt, S., Chutia, R., Müller, J., Böttcher, C., Strehmel, N., Scheel, D., and Abel, S. (2016). Non-targeted profiling of semi-polar metabolites in Arabidopsis root exudates uncovers a role for coumarin secretion and lignification during the local response to phosphate limitation. *J. Exp. Bot.* 67(5):1421–1432.

5 Hormonal Control of Phosphate Uptake and Assimilation

Shreya Gupta and Amar Pal Singh

5.1 INTRODUCTION

Among the minerals, phosphate (Pi) is one of the major determinants whose availability modulates plant growth-related processes (Jiang et al. 2007; López-Bucio, Cruz-Ramírez, and Herrera-Estrella 2003; Pé et al. 2011; Pandey, Devi, and Singh 2020; Devi, Pandey, and Singh 2020; Kumari et al. 2022). In a natural ecosystem, generally, the level of soluble Pi is low to support plant's growth, development, and metabolic activity (Negi et al. 2016; Cui et al. 2020). Plants' intricate signalling networks have evolved throughout their life span to help them deal with Pi deficiency, which leads to changes in plant physiology, genetic circuitry, and metabolism, and to maintain intracellular Pi homeostasis such as Pi uptake from the rhizosphere and its assimilation and recycling at cellular and tissue levels (Jia, Giehl, and von Wirén 2022; Rouached, Arpat, and Poirier 2010; Bouain et al. 2019; Pandey, Devi, and Singh 2020). To improve the growth and yield of plants in Pi-limiting soils, only phosphorus (Pi) based fertilisers are commonly used, which reduce Pi-use efficiency, deplete Pi reserves, and cause environmental pollution (Bindraban, Dimkpa, and Pandey 2020; Ajmera, Hodgman, and Lu 2019; Vinod and Heuer 2012). Environmental factors such as soil pH and differential availability of mineral elements, e.g. Ca, Fe, Zn, and Al, in the soils, are generally associated with altered Pi uptake and use efficiency, thus leading to compromised growth (Bouain et al. 2019; Cho et al. 2021; Mora-Macías et al. 2017; Müller et al. 2015; Naumann et al. 2022; Blackwell, Darch, and Haslam 2019). Minerals such as Ca, Fe, and Al form complexes with Pi, thus limiting soluble Pi availability to the roots (Meyer et al. 2020; Müller et al. 2015; Singh et al. 2018; Heuer et al. 2017; Xiaoyue Wang et al. 2019; Dong et al. 2017; Satbhai et al. 2017). Soil pH is a crucial factor that regulates the availability of soluble Pi. Acidic (< 5.5) and basic (7.5–8.5) soils are generally associated with Pi fixation with the elements Ca, Fe, and Al, thus making it unavailable for the plants (Meyer et al. 2020; Blackwell, Darch, and Haslam 2019; Heuer et al. 2017).

A change in Pi concentrations in the rhizosphere is sensed by roots (Svistoonoff et al. 2007). Upon sensing low Pi, plants activate a series of signalling events that ultimately result in changes in root traits, organic acid extrusion, and secretion of acid phosphatases to increase Pi uptake efficiency (Balzergue et al. 2017; Paz-Ares et al. 2022; Péret et al. 2011; Heuer et al. 2017; Pantigoso et al. 2020). Simultaneously, plants remobilise Pi within the plant tissues and organelles through phosphate transporters (PHTs) to utilise the internal P pool, such as P remobilisation from senescent to younger leaves to overcome the detrimental changes induced by Pi deficiency (Ye et al. 2015; Nagarajan et al. 2011; Chang et al. 2019; Smith et al. 2003; Irfan et al. 2020). Therefore, for reducing fertiliser use in agriculture without detrimental effects to yield, it is highly desirable to develop plants with efficient Pi uptake capacity from the rhizosphere and with internal P remobilisation and re-utilisation (Heuer et al. 2017).

Many genes encoding functional high and low-affinity *PHOSPHATE TRANSPORTERS* (*PHTs*) have been identified across the plant species that are involved in modulating the Pi uptake and homeostasis (Poirier and Bucher 2002). For example, root-localised *PTHs* (*PHT1* family of transporters) are mainly involved in Pi uptake from soils, whereas root-to-shoot Pi translocation involves PHO family transporters (PHO1 and PHO1;H1) (Heuer et al. 2017). In addition, many intracellular organelle-specific Pi transporters such as *PHT5;1*, *PHT5;2*, *PHT5;3*, *AtPHT3;1*, *AtPHT3;2*, *AtPHT3.3*, and *AtPHT4;6* are also involved in Pi translocation under its limited availability (Młodzińska and Zboińska 2016).

Pi-deficiency-triggered changes in root system architecture (RSA), such as in primary root (PR), lateral root (LR), and root hair (RH) growth, which are the key adaptation mechanisms in plants. Phytohormones (auxin, brassinosteroids (BRs), gibberellic acid (GA), abscisic acid (ABA), strigolactones (SLs), cytokinins (CKs), and ethylene) have been implicated as modulating Pi-deficiency-induced responses in plants. Systemic responses are associated with long-distance Pi transport, tissue, and cellular level translocation, while the local response of the plant to Pi deficiency leads to a change in root traits for its absorption. Pi-availability-dependent regulation of hormone biosynthesis and signalling has been correlated with enhanced Pi-foraging capacity of the plants. Emerging evidence suggests that Pi-induced local responses are plausibly associated with auxin, BRs, and GA, which are involved in modulating root foraging response. However, ABA, SLs, and CKs appear to be associated with low-Pi-dependent systemic responses also (Jiang et al. 2007; Singh et al. 2014; Zhang et al. 2022; Silva-Navas et al. 2019; Czarnecki et al. 2013; Mayzlish-Gati et al. 2012). For example, Pi deficiency increases the activity of the auxin receptor, TRANSPORT INHIBITOR RESPONSE 1 (TIR1), which in turn increases the auxin signalling output during LR growth under limited Pi availability (Pérez-Torres, López-Bucio, and Herrera-Estrella 2009). Auxin biosynthesis and its transport from the root tip to the elongation/differentiation zone (EDZ) have been shown to regulate RH elongation under Pi deficiency (Bhosale et al. 2018). In addition high Pi represses transcription factors *ROOT HAIR DEFECTIVE 6-LIKE 2* and *4* (*RSL2* and *RSL4*), which are involved in polar RH elongation and linked with auxin signalling and responses (Mangano et al. 2018). In rice, the auxin-responsive transcription factor *OsARF16* has been demonstrated to regulate the Pi homeostasis by modulating CK response. Increased Pi content in the *osarf16* mutant is associated with increased levels of genes involved in Pi uptake and solubilisation, such as *PHT1* and *Acid phosphatases* (Shen et al. 2014). Additionally, *OsPHT1;8* has been shown to modulate both auxin and Pi deficiency responses in rice (Jia et al. 2017). Thus, auxin has been shown to modulate both root developmental changes and Pi uptake-related genes for improving plant performance during Pi deficiency. Recently, the function of the stress hormone ABA in Pi uptake and homeostasis has been demonstrated in *Arabidopsis*. Low Pi enhances ABA biosynthesis and signalling, thereby leading to Pi acquisition by enhancing *PHT1;1* levels in an ABSCISIC ACID INSENSITIVE5 (ABI5)-dependent manner (Zhang et al. 2022). The function of ABA in Pi re-utilisation and root-to-shoot translocation has also been demonstrated (Fang Zhu et al. 2018). CKs, which are known to regulate several aspects of plant growth and development by controlling cell division and differentiation processes, are also regulated by altered Pi levels (Dello Ioio et al. 2008). Pi deficiency represses CK levels and the expression of its receptor, CYTOKININ RESPONSE 1 (CRE1) (José M. Franco-Zorrilla et al. 2002; José Manuel Franco-Zorrilla et al. 2005). High levels of CK repress the Pi-deficiency-induced genes likely by modulating the intracellular Pi pool (Xuming Wang et al. 2006). Taken together, the recent advancements in Pi uptake, remobilisation, and intracellular homeostasis suggested a complex crosstalk of hormonal signalling pathways that are essential for plant adaptation under low P nutrition.

In this chapter, we review the roles of phytohormones on plants' ability to adapt under low Pi conditions and the underlying molecular mechanisms associated with low Pi sensing and acquisition, which are considered necessary for improving Pi uptake efficiency. Further, the effects of hormones on Pi homeostasis and remobilisation in plants are also highlighted in relation to Pi distribution among the plant tissues to sustain growth-related processes during its deficiency or low availability.

Hormonal Control of Phosphate Uptake and Assimilation　　　　　　　　　　　　　　　　　　　　61

5.2 PHOSPHATE DEFICIENCY IMPACTS OVERALL PLANT GROWTH AND DEVELOPMENT

Pi deficiency impacts overall plant growth and development and leads to changes in both physiology and metabolism. Pi-deficiency-induced changes in plant growth-related traits vary among plant species and genotypes. The interplay of hormonal signalling pathways has been shown to modulate both underground (root traits) and above-ground (shoot or aerial traits) organs as a result of the adaptive strategy. Inhibition of BR and GA under Pi deficiency could be a beneficial metabolic adjustment that in turn inhibits PR growth to improve topsoil Pi foraging to sustain growth. Additionally, reduced GA levels enhance anthocyanin content in the leaves/shoot in a DELLA-dependent manner, which is a hallmark feature of leaf adaptation to Pi deficiency. ABA, which is involved in regulating several developmental aspects of plants from seed germination to flowering, has been implicated in root growth and Pi utilisation and uptake during its deficiency (Karimi et al. 2021; Zhang et al. 2022; Fang Zhu et al. 2018). The ABA signalling factor ABI5 improves plant growth and Pi content by directly regulating the phosphate transporter gene *PHT1.1* in *Arabidopsis* (Zhang et al. 2022). Auxins are involved in modulating almost every aspect of plant growth, and their role in reprogramming RSA during Pi deficiency from the model plant Arabidopsis to crop plants such as rice has been well characterised (Bhosale et al. 2018; H. Jia et al. 2017; Meng et al. 2013; Giri et al. 2018; Nacry et al. 2005). Increased RH density under Pi deficiency is considered an adaptive response that is associated with root Pi-foraging capacity. Low-Pi-dependent increased RH density in *Arabidopsis* and rice leads to enhanced auxin signalling and transport activity (Bhosale et al. 2018; Giri et al. 2018). In addition, LR formation during Pi deficiency depends on TIR1-mediated signalling events in *Arabidopsis* (Pérez-Torres et al. 2008; Malamy and Ryan 2001). Therefore, auxin regulates many adaptive developmental features of the plant during Pi deficiency. SLs are involved in regulating several growth-related processes in plants (Umehara et al. 2008; Kohlen et al. 2011). Low Pi increases SLs levels in the roots and has been demonstrated to modulate the Pi-deficiency-mediated root and shoot growth as an adaptive response (Umehara et al. 2008; Czarnecki et al. 2013; Umehara et al. 2010; Kohlen et al. 2011).

The effect of Pi deficiency on the biosynthesis and signalling of these phytohormones and hormonal signalling-dependent regulation of plant growth has been summarised in Figure 5.1.

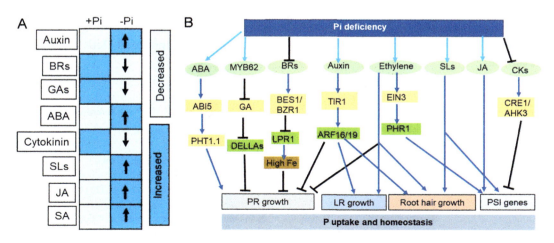

FIGURE 5.1 Figure summarising the effect of Pi deficiency on the biosynthesis and signalling response of growth hormones (inference from the findings in *Arabidopsis*). (A) Represents the levels of hormones regulated by Pi deficiency. (B) Diagram showing the components of hormonal signalling pathways involved in regulating the root system architecture (RSA).

Additionally, the involvement of Pi uptake and utilisation has been discussed in detail in the following sub-sections.

5.2.1 LOW PI SENSING AND HORMONAL CONTROL OF PI-DEFICIENCY RESPONSE

Pi deficiency in the rhizosphere is sensed by the plant roots. Upon sensing, plants undergo several morpho-physiological changes including compromised root and shoot growth, which involves a series of signalling arrays for adaptation in a Pi-limiting environment (Liu 2021; Devi, Pandey, and Singh 2020; Xiaoyue Wang et al. 2019; Svistoonoff et al. 2007). Low Pi sensing through roots involves *LOW PHOSPHATE ROOT1* and *2* (*LPR1/2*), which are expressed at the root tips upon physical contact of roots with the Pi-deficient medium. LPR1 encodes a multicopper oxidase (MCO), and mutants lacking functional *LPR1* and its homolog *LPR2* (*lpr1/lpr2*) showed improved root growth during Pi deficiency (Svistoonoff et al. 2007; Müller et al. 2015; Ticconi et al. 2009). Pi-deficiency-induced changes, such as inhibited root meristem zone length, loss of quiescent centre (QC) activity, reduced cell length, and increased RH density, were blocked in *lpr1/lpr2* mutants (Müller et al. 2015; Naumann et al. 2022; Svistoonoff et al. 2007; Sánchez-Calderón et al. 2005). A P5-type ATPase, *PHOSPHATE DEFICIENCY RESPONSE 2* gene (*PDR2*), regulates root meristematic activity by maintaining the root patterning gene, *SCARECROW* (*SCR*), which functions at the root tip. It was proposed that both PDR2 and LPR1 may act together to regulate root developmental changes during Pi deficiency (Ticconi et al. 2009; Rouached, Arpat, and Poirier 2010). Pi deficiency leads to the activation of both local and systemic signals. In *Arabidopsis*, local signalling responses lead to modification of root traits such as increased LR and RH density, and inhibition of primary root length (PRL), thus leading to a shallow root system to improve topsoil Pi foraging. Systemic signalling responses lead to changes in the cellular Pi pool via long-distance transport and signalling (Pé et al. 2011; Thibaud et al. 2010). In the past few years, systemic Pi response signalling has been much characterised, and components involved in it have been described (Table 5.1). Both root and shoot-derived systemic signals such as sugars, hormones and peptides, RNAs, and Pi itself have been shown to regulate the plant metabolic and growth response at the whole-plant level (Lin et al. 2014). Plant hormones, SLs, and CKs are proposed to act as systemic signals during Pi deficiency. Shoot branching inhibition during Pi deficiency is thought to reduce Pi utilisation efficiency as an adaptive response. Low Pi sensing increases SL biosynthesis in roots and moves acropetally,

TABLE 5.1
Summary of Demonstrated Systemic Signals Involved in Modulation of Pi Deficiency Responses

Signal	Movement of Signal (Root to Shoot or Shoot to Root)	Response	Reference
Sugar	Shoot to root	Inhibits P inducible genes in P deficient roots	(Lin et al. 2014; Liu et al. 2005)
miR399	Shoot to root	Cleaves PHO2 mRNA, increases P uptake and translocation	Pant et al. (2008)
Phosphate	Root to shoot	Inhibition of P-responsive genes	Balzergue et al. (2011); Ticconi et al. (2001)
SLs	Root to shoot	Repression of shoot branching	Beveridge et al. (2000); Kohlen et al. (2011)
CKs (tZ)	Root to shoot	Inhibition of P-responsive genes in shoots	Kuroha et al. (2002); Matsumoto-Kitano et al. (2008)
Ca2+/H+ transporter (CAX mediated)	Shoot to root	Pi uptake	

Hormonal Control of Phosphate Uptake and Assimilation 63

thereby inhibiting shoot branching (Umehara et al. 2008; Kohlen et al. 2011). Nonetheless, SLs are also synthesised in shoots and may act locally to inhibit branching phenotype (Sorefan et al. 2003; Lin et al. 2014). Pi deficiency inhibits CK biosynthesis and response genes, while exogenously applied CKs inhibited Pi-starvation-induced (PSI) genes, suggesting the role of CKs during Pi deficiency (José Manuel Franco-Zorrilla et al. 2005; Xuming Wang et al. 2006; Kuroha et al. 2002). Grafting experiments supported the evidence that root-derived CKs act as a systemic signal (Matsumoto-Kitano et al. 2008). Effects of Pi deficiency on growth hormones and their effect on root traits are summarised in Figures 5.1 A and B. Apart from hormonal signals, sugars and RNAs have been demonstrated to act as systemic signals. For example, *miR399* moves from shoot to root where it cleaves *PHOSPHATE2* (*PHO2*) mRNA and increases Pi uptake and transport (Pant et al. 2008; Lin et al. 2008; Bari et al. 2006). The *PHO2* gene encodes a ubiquitin E2 conjugase, and plants lacking functional *PHO2* gene (*pho2*) over-accumulate Pi in shoots (Lin et al. 2008; Aung et al. 2006; Fujii et al. 2005). Systemic signals with the demonstrated function have been summarised in Table 5.1. Altogether, these findings suggest that coordinated signalling activity of local and systemic responses is crucial for plant fitness and adaptation during Pi deficiency. Major genes involved in hormone-dependent phosphate starvation response (PSR) are summarised in Table 5.2.

5.2.2 Effect of Ethylene Signalling and Response on Plant Growth during Pi Deficiency

The gaseous plant hormone ethylene is synthesised from methionine. In the initial step, *S*-adenosylmethionine (AdoMet) synthetase converts methionine to AdoMet, which is further

TABLE 5.2
Summary of Hormone Signalling and Response-Related Genes Involved in Modulating Plant Growth during Pi Deficiency

Gene Name	Plant Species	Regulation by Growth Hormones	Regulation of Pi-Dependent Growth Processes	Reference
TIR1/ARF7/ARF19	*Arabidopsis*	Auxin signalling	Lateral root development	Pérez-Torres et al. (2008)
TAA1/AUX1	*Arabidopsis*	Auxin biosynthesis/transport	Root hair growth	Bhosale et al. (2018)
PLT1/PLT2	*Arabidopsis*	Auxin signalling and homeostasis	Primary root growth inhibition in *plt1/plt2* mutants	Aida et al. (2004)
SUMO E3 ligase (SIZ1)	*Arabidopsis*	Auxin signalling and homeostasis	Primary root growth inhibition	Miura et al. (2005)
BZR1/BES1	*Arabidopsis*	Brassinosteroid signalling	Primary root growth via LPR1	Singh et al. (2018, 2014)
GA1,3	*Arabidopsis*	Gibberellic acid signalling	Shorter root hair	Jiang et al. (2007)
ETHYLENE INSENSITIVE3 (EIN3)	*Arabidopsis*	Ethylene signalling	Root hair elongation	Xiao et al. (2021)
ABI5	*Arabidopsis*	Abscisic acid signalling	Primary root growth	Zhang et al. (2022)
OsJAZ11	Rice	Jasmonic acid signalling	Primary and seminal root elongation	Pandey et al. (2021)
OsCKX2	Rice	Cytokinin catabolism	Improves shoot biomass	Yan et al. (2022)
OsPIN2	Rice	Auxin signalling/transport	Modulates seminal root growth, LR and RH density	Sun et al. (2019)
OsAUX1	Rice	Auxin signalling/transport	Root hair elongation	Giri et al. (2018)

converted to 1-aminocyclopropane-1-carboxylic acid (ACC) via ACC synthase enzyme (ACS) activity. In the cells, by the action of the ACC oxidase (ACO) enzyme, ACC is converted to ethylene (Bleecker and Kende 2003; Xu and Zhang 2015). These two biosynthetic genes, *ACC* and *ACO*, are critical for ethylene production, and their levels are regulated at both transcriptional and post-transcriptional levels (Bleecker and Kende 2003). Once ethylene is synthesised in the cells, it is perceived by its receptors, ETHYLENE RESPONSE 1 and 2, ETHYLENE RESPONSE SENSOR1 and ETHYLENE INSENSITIVE4 (ETR1/2, ERS1/2, and EIN4), which leads to the inactivation of these receptors. This disrupts Raf-like Ser/Thr protein kinase, CONSTITUTIVE TRIPLE RESPONSE (CTR1) from the receptors and inactivates CTR1, and releases ETHYLENE-INSENSITIVE 2 (EIN2). The C-terminus (EIN2-C) moves to the nucleus and increases the abundance of ethylene transcription factors *EIN3* and *EIN3-LIKE1* (*EIL1*). EIN3/EIL1 mediates ethylene response by regulating several downstream target effectors such as *ETHYLENE RESPONSE FACTORS* (*ERFs*) (Xiao et al. 2021; Wang, Li, and Ecker 2002; Binder and Jez 2020).

The effect of Pi deficiency on ethylene biosynthesis and signalling has been evaluated in many plants including *Arabidopsis thaliana*, *Medicago falcata*, *Lupinus albus*, *Phaseolus vulgaris*, and *Oryza sativa* (Gilbert et al. 2000; Song and Liu 2015; Li et al. 2009; Lei et al. 2011; Borch et al. 1999). Many studies have shown the Pi-availability-dependent regulation of ethylene biosynthesis genes *ACS2/4/6*) and *ACO* in *Arabidopsis* (Lei et al. 2011; Thibaud et al. 2010). Increased levels of *ACS2* and *ACS6* under low Pi were repressed when seedlings were transferred to an adequate Pi medium, suggesting a correlative effect of Pi availability on ethylene (Thibaud et al. 2010; Song and Liu 2015). Genetic screening in *Arabidopsis* led to identifying a hypersensitive mutant to Pi starvation (*hps2*), which is a new allele of the classical ethylene signalling pathway gene, *CTR1*. *hps2* mutant showed increased levels of PSI genes (*AT4*, *IPS1*, *ACP5*, *miR399D*, and *RNS1*), suggesting that compromised ethylene signalling alters PSR (Lei et al. 2011).

The role of ethylene in governing RH and LR development under Pi deficiency has been extensively studied (Borch et al. 1999; Lei et al. 2011; Song and Liu 2015). Chemical inhibition of ethylene biosynthesis by aminoethoxyvinylglycine (AVG) or treatment of plants with ethylene promoted and repressed LR density, respectively (Borch et al. 1999). Mutants with compromised ethylene response also showed a change in LR development/density under Pi deficiency. The ethylene-insensitive mutants *etr1*, *ein2*, and *ein3* showed an increased number and density of lateral roots under the Pi-limiting condition (Pérez-Torres et al. 2008; López-Bucio et al. 2002). In white lupin, cluster root (CR) densely packed with RHs under Pi deficiency increases the soil surface area for Pi exploitation. This hallmark feature under Pi deficiency is associated with ethylene biosynthesis and response. Inhibiting ethylene biosynthesis by Co^{2+} inhibited CR formation and thus led to a compromised Pi-deficiency-induced response (Wang et al. 2015). In addition, the ethylene-insensitive "Never-ripe" (Nr) tomato line failed to promote the formation of adventitious roots under Pi deficiency, suggesting a wider role of ethylene biosynthesis and signalling in governing root traits across the plant species (Kim, Lynch, and Brown 2008).

Increased RH elongation/density under Pi deficiency correlated with the Pi uptake efficiency in many plants (Figure 5.2). It was demonstrated that ethylene increases RH formation/density in *EIN3* dependent manner. The ethylene transcription factor *EIN3* increases RH formation by regulating the expression of cell wall-related genes (Song et al. 2016). A mutation in the ethylene receptor ERS1 (*hps5* mutant) leads to the accumulation of EIN3 protein levels. Increased levels of EIN3 lead to the increased transcription of the genes (direct binding with the promoters) involved in cell wall remodelling and root hair growth, such as *ROOT HAIR SPECIFIC 15* (*RHS15*), *RHS19*, *PROLINE-RICH PROTEIN 1* (*PRP1*), and *EXTENSIN* (*EXT14*). Plants lacking functional EIN3 (*ein3* and *ein3eil1* mutants) repressed the low-Pi-dependent induction of these genes, suggesting that ethylene is a key player in RH formation (Song et al. 2016). Additionally, the authors suggested that the cooperative action of EIN3 and the transcription factor involved in RH formation, RSL4, potentially regulates RH elongation in response to low Pi (Song et al. 2016; Qiu et al. 2021). An R2R3-MYB family transcription factor, *MYB30*, represses RH formation in *Arabidopsis*. Overexpression of *MYB30* inhibits

Hormonal Control of Phosphate Uptake and Assimilation

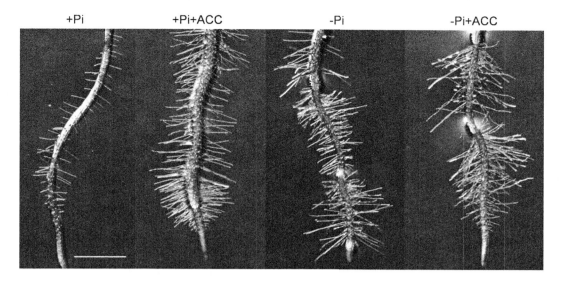

FIGURE 5.2 Effect of ethylene precursor, 1-Aminocyclopropane-1-carboxylic acid (ACC) on root hair elongation/formation under sufficient (+Pi) and low Pi (−Pi) conditions in wild-type (Col0) seedlings.

RH elongation while mutant showed longer RH in response to low Pi. MYB30 physically interacts with EIN3. Both these transcriptional effectors antagonistically regulate root hair formation/elongation where the former represses and later promotes it (Xiao et al. 2021). Altogether, studies suggested a key role of ethylene in modulating root traits as an adaptive response under Pi deficiency.

5.2.3 Role of Auxins and Low-Pi-Derived Signals in Modulating Plant Growth

The role of auxins under low Pi is well explored. PSR and auxin responses are coupled in mediating plant developmental processes. Under low Pi, *Arabidopsis* roots display architectural plasticity and lead to a change in PR growth, LRs, and RH emergence and elongation for maximum Pi utilisation from the top layers of the soil. These responses are well correlated with auxin biosynthesis, signalling, and response (Jia, Giehl, and von Wirén 2022; Gruber et al. 2013; Malamy and Ryan 2001; D. Liu 2021; Lin et al. 2008). Fine-tuning of root architecture by auxin signalling and response pathways in several plants has been demonstrated under altered nutrient availability. Pi deficiency increases auxin sensitivity and signalling, which in turn represses PR elongation and simultaneously promotes LR organogenesis/formation (López-Bucio et al. 2002; Nacry et al. 2005; Devi, Pandey, and Singh 2020). In line, low-Pi-grown roots show increased sensitivity to exogenous auxin than sufficient Pi-grown roots (Pérez-Torres et al. 2008). Low Pi promotes the activity of auxin receptor TIR1, which degrades auxin repressor AUX/IAA (IAA17) and activates transcription factor *AUXIN RESPONSE FACTOR19* (*ARF19*), which in turn regulates genes involved in LR formation and emergence. In line with the same evidence, the *tir1-1* mutant showed a reduction in LR formation under low Pi, suggesting that the TIR1 function is essential for LR formation (Pérez-Torres et al. 2008; Okushima et al. 2005). In addition, the auxin biosynthesis genes *TRYPTOPHAN AMINOTRANSFERASE 1* (*TAA1*), *YUC1*, *YUC2*, and *YUC4* were differentially regulated in a heterogenous Pi environment (Liu et al. 2013). Auxin-regulated genes associated with PR growth and LR formation under low Pi conditions have been extensively studied in many plants. For example, the genes *SUMO E3 ligase* (*SIZ1*), *Inositol Phosphate Kinase* (*IPK1*), *TIR1*, *ARFs*, and *LPR1/2* have been shown to regulate root architecture in response to low Pi (Svistoonoff et al. 2007; Miura et al. 2005; Pérez-Torres et al. 2008; López-Bucio et al. 2005). LR formation is strictly controlled by intrinsic and extrinsic signals, and auxin levels and transport activity appear essential factors in governing LR formation under low Pi conditions. The auxin transport inhibitor 2,3,5-triiodobenzoic acid (TIBA) inhibited LR growth/formation under differential Pi availability, suggesting an important role of auxin distribution under low Pi conditions (Nacry et al. 2005). Recently, the role of *OsAUX1* in RH elongation under low Pi in rice has been demonstrated (Giri et al. 2018). Using X-ray micro CT imaging as a tool, the authors suggested that shallow root angle in *osaux1* is not associated with enhanced Pi uptake due to poor RH elongation under Pi deficiency. Further, the authors demonstrated that AUX1 transports auxin to the differentiation zone (DZ) from the root tip, thus promoting RH elongation in this zone (Giri et al. 2018). In *Arabidopsis*, mutants defective in either auxin biosynthesis or transport regulate low-Pi-dependent RH elongation. The expression of auxin transporter gene *AUX1* in the cells of the LR cap restores low-Pi-dependent RH formation in the *aux1* mutant. These findings further suggested an essential role of auxin transporters in modulating RH elongation during Pi deficiency (Bhosale et al. 2018). In line with the role of auxin transporters in modulating Pi-dependent RH formation/growth, it was demonstrated that phosphatidic acid suppresses auxin efflux carrier PIN-FORMED2 (PIN2) degradation under the low Pi condition to promote RH development (Lin et al. 2020). Overexpression of *OsPIN2* regulates auxin balance and RH formation and growth in response to low Pi, along with increased levels of the genes involved in RH growth (Sun et al. 2019). Altogether, auxin biosynthesis, signalling, transport, and transcriptional responses regulate root developmental processes as an adaptation mechanism in the plants to overcome Pi starvation responses.

5.2.4 Strigolactones (SLs) in Regulating Plant Architecture during Pi Deficiency

The terpenoid lactones, SLs, act as a signalling molecule that controls many developmental features of plants including shoot branching. Emerging pieces of evidence suggested that the action of SLs may be local or systemic depending on their synthesis and environmental conditions (Umehara et al. 2008; Kohlen et al. 2011). It has been demonstrated that Pi deficiency increases SL biosynthesis in the roots of many plants (Umehara et al. 2010; Kapulnik and Koltai 2016; Czarnecki et al. 2013). Increased SL in root exudates has been correlated with mycorrhizal colonisation and nodulation during Pi deficiency (Yoneyama et al. 2007). In *Solanum lycopersicum* and *Trifolium pratense*, a negative correlation between SL exudation and Pi supply has been shown. SL exudation in the rhizosphere was reduced within 24h in Pi-starved plants when transferred to optimal Pi conditions (López-Ráez and Bouwmeester 2008; López-Ráez et al. 2008). Additionally, *Arabidopsis* and rice plants when grown under Pi-deficient conditions showed increased SL levels in the roots and led to inhibition of auxiliary bud outgrowth and tiller numbers respectively. In contrast, mutants impaired with SL biosynthesis and response were unresponsive to such kind of adaptation (Umehara et al. 2010; Kohlen et al. 2011) (Figure 5.3).

In *Arabidopsis*, Pi deficiency increases RH formation/density to maximise Pi exploitation from the rhizosphere (Pandey, Devi, and Singh 2020; Z. Jia, Giehl, and von Wirén 2022). SLs have been shown to regulate this local response associated with Pi-deficiency-dependent root architecture modification (Madmon et al. 2016; Mayzlish-Gati et al. 2012). Thus, RH elongation during early Pi deficiency depends on SL biosynthesis and signalling in *Arabidopsis* (Mayzlish-Gati et al. 2012; Kapulnik and Koltai 2016). Mutants defective in either SL signalling, *more axillary growth 2* (*max2-1*) or biosynthesis (*max4-1*), showed reduced response to RH development/formation during the early phase of Pi deficiency. *max2-1* and *max4-1* mutants showed reduced RH density as compared to the control plants, suggesting that the part of Pi-deficiency-induced local response depends on SL biosynthesis and signalling (Mayzlish-Gati et al. 2012). In addition to RH density, these mutants showed reduced levels of phosphate starvation-induced genes such as *PHOSPHATE TRANSPORTERS* (*PHT1;2, PHT1;4, PHT1;5*) and *ACID PHOSPHATASE* (*ACP5*), suggesting a compromised Pi deficiency response in these mutants (Mayzlish-Gati et al. 2012). Additionally, phosphate (P) content in wild type and *max2-1* remain unchanged during early Pi deficiency (48 hrs). Further, the authors suggested crosstalk of SL and auxin signalling pathways through MAX2 in modulating RH growth

Hormonal Control of Phosphate Uptake and Assimilation

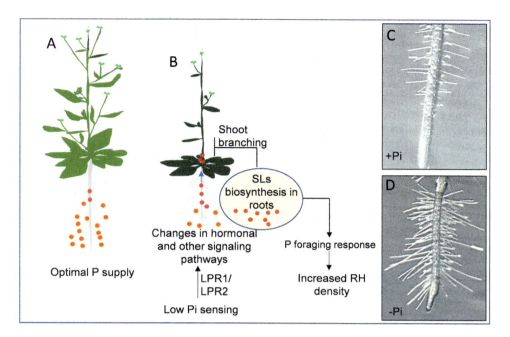

FIGURE 5.3 Illustration of strigolactone (SL) functions in modulating plant architecture during phosphate (Pi) deficiency. (A) Plant growth during optimal Pi supply. (B) Low-Pi sensing mediated by LPR1/LPR2 upon physical contact of the root tip to the Pi-deficient medium. Low-Pi sensing leads to changes in hormonal and peptide signalling pathways. SL biosynthesis increases during Pi deficiency, which promotes root hair (RH) density to improve the root Pi-foraging capacity of the plants. Simultaneously increased SL levels inhibit shoot branching as an adaptive response to potentially reduce Pi utilisation efficiency. (C & D) Wild-type roots grown on +Pi and −Pi conditions showing root hair growth.

during early Pi deficiency response. Endogenous Pi levels activate MAX2, which in turn regulates TIR1-dependent signalling to control RH formation/growth. Additionally, the authors suggested a TIR1 parallel pathway that might be involved in controlling the low Pi effect on RH formation and growth (Mayzlish-Gati et al. 2012). Taken together, the recent findings in the SL field demonstrate the involvement of SLs in modulating both local and systemic Pi-deficiency-induced changes for improving plant growth and Pi foraging and utilisation efficiency (Figure 5.2).

5.2.5 Abscisic Acid (ABA)-Mediated Regulation of Plant Growth and Pi Homeostasis

ABA is a well-characterised stress hormone that governs many physiological and growth-related processes of the plant (Wang, Yu, and Xie 2020; Karimi et al. 2021; Fang Zhu et al. 2018). ABA signalling starts upon the perception of the ligand by Pyrabactin resistance1/Pyr1-like receptor components (PYR/PYL) that relieve phosphatase 2C dependent repression of Snf1-related kinases 2 (SnRK2s). This in turn activates downstream transcription factors such as *ABI5* by phosphorylation-dependent mechanisms (Ng et al. 2014; Zhao et al. 2020). ABI5 is regulated at both transcriptional and post-transcriptional levels to modulate plant growth in response to external and internal cues. ABI5 activity has been shown to regulate many genes involved in seed germination, flowering and seedling establishment, and nutrient stress (Ng et al. 2014; Zhao et al. 2020; Karimi et al. 2021; Fang Zhu et al. 2018). In rice (*Oryza sativa*), early Pi deficiency (1h) leads to a rapid decline in ABA levels, suggesting the interaction of this hormone with Pi signalling components (Fang Zhu et al. 2018). Exogenous ABA-treated plants showed further reduction in root and shoot Pi content. Moreover, ABA treatment represses pectin methylesterase enzyme activity and pectin content, suggesting a putative role of ABA in cell wall Pi utilisation and/or remobilisation during the early phase of Pi deficiency (Fang

Zhu et al. 2018). Recently, the function of ABA in Pi uptake and plant growth during Pi deficiency has been characterised in *Arabidopsis* (Zhang et al. 2022). By using an ABA-responsive, *6×ABRE-R:erGFP* reporter line, the authors showed that low Pi activates ABA signalling. Further, an increase in ABA levels in response to low Pi depends on the ABA-glucosyl ester (ABA-GE) deconjugation gene, *BG1*, and low Pi failed to promote ABA levels in the *bg1* mutant. The authors demonstrated that ABA increases Pi levels where ABI5 directly binds with the phosphate transporter gene (*PHT1;1*) promoter and increases its levels, which in turn increases Pi uptake and root growth under low Pi conditions (Zhang et al. 2022). Recent studies have highlighted the role of ABA in Pi uptake and homeostasis in plants, and it appears essential in modulating the Pi balance of plants. However, further studies are required to understand the function of ABA in growth coordination and its impact on systemic Pi signalling responses across plant species and genotypes.

5.2.6 Impact of Cytokinin (CK) Biosynthesis and Signalling on Plant Growth during Pi Deficiency

CKs are necessary for governing many plant processes that are essential to sustain growth under normal and stressful conditions. The role of CKs has been implicated in modulating source/sink balance, apical dominance, cell division, anthocyanin synthesis, and senescence (Matsumoto-Kitano et al. 2008; Hutchison and Kieber 2002; Argyros et al. 2008). Trans-zeatin (tZ) and cis-zeatin (cZ) CK forms are abundant in *Arabidopsis*, and the synthesis of the tZ form is mediated by ADP/ATP isopentenyl transferases (IPTs) (Miyawaki et al. 2006). Many studies have shown the involvement of CKs during nutrient deficiency in plants (Silva-Navas et al. 2019; Z. Jia, Giehl, and von Wirén 2022). For example, nitrate promotes CK biosynthesis whereas low Pi and potassium repress its levels (Silva-Navas et al. 2019; Samuelson and Larsson 1993; Takei et al. 2001).

CKs have been shown to negatively regulate PSI genes. Studies from *Arabidopsis* suggested that exogenous CKs (6-benzyl amino purine or kinetin) treatment repressed the expression of Pi-deficiency-induced genes such as *PHT1;1*, *IPS1*, and *ACP5* (Xuming Wang et al. 2006; Martín et al. 2000). In addition to repression of the PSI response, CKs also represses LR formation/development under low Pi conditions, suggesting a negative role of CKs in modulating Pi deficiency responses (José M. Franco-Zorrilla et al. 2002). Low Pi represses CK levels and its receptor *CYTOKININ RESPONSE 1* (*CRE1*), thus leading to compromised biosynthesis and signalling activity of hormone under a limiting Pi environment (José M. Franco-Zorrilla et al. 2002). The natural CK form, tZ, shows an accumulation in the xylem and phloem tissues, suggesting that it may be transported to distant organs and tissues (Miyawaki et al. 2006; Matsumoto-Kitano et al. 2008). The levels of CK forms (tZ and iP) are repressed in the CK biosynthesis mutants which show retarded shoot growth. Grafting experiments (mutant onto WT rootstock) further suggested that root-derived CKs promote shoot growth and act as a systemic signal (Matsumoto-Kitano et al. 2008; Kiba et al. 2013; Chien et al. 2018). Recently, it was shown that low Pi regulates the ratio of CK forms tZ/cZ by repressing the synthesis of tZ (Silva-Navas et al. 2019). Similar to *Arabidopsis*, rice seedlings, when treated with 6-benzyl amino purine for three days, repressed the expression of PSI genes and also increased the endogenous Pi pool, suggesting a role of CKs in proving Pi nutrition during its deficiency (Xuming Wang et al. 2006). Knockout lines of *OsCKX2* (*OsCKX2-KO*) showed increased tolerance to Pi deficiency in rice, which led to increased Pi content with improved biomass. Increased tolerance of *OsCKX2*-KO plants to low Pi was found to be associated with increased levels of PHTs (Yan et al. 2022). Thus, CK biosynthesis and signalling-dependent mechanisms are associated with modulating Pi balance and PSI gene expression in plants.

5.2.7 Jasmonic Acid (JA) and Pi-Deficiency Responses in Plants

JA is a well-known hormone involved in plant defence and growth (Vijayan et al. 1998; Y. Yan et al. 2012). The biosynthesis of JA and its forms is a multistep process and starts from the precursor

Hormonal Control of Phosphate Uptake and Assimilation

α-linolenic acid (α-LeA) and leads to the synthesis of JA precursor, 12-oxophytodienoic acid (12-OPDA) (Ahmad et al. 2016; Stintzi and Browse 2000). 12-OPDA further undergoes β-oxidation and different metabolic activities that lead to the formation of active JAs. The 12-Oxophytodienoate reductase 3 (OPR3) function is essential for converting JA precursor to active JAs (Stintzi and Browse 2000). OPR3-independent pathways also operate to modulate JA biosynthesis in plants and may be linked with organ type and environmental conditions, which needs further investigation (Stintzi and Browse 2000; Chini et al. 2018; Ahmad et al. 2016). JA is perceived by its receptor, *CORONATINE-INSENSITIVE1* (*COI1*), which encodes a leucine-rich repeat F-box protein, and the receptor mutants showed compromised response to exogenous JA (Xie et al. 1998; Katsir et al. 2008). JA perception promotes the association of COI1 with the JASMONATE ZIM DOMAIN (JAZ) proteins (a negative regulator of the pathway) and leads to the degradation of JAZs through the 26S proteasome pathway thus, activating the JA response by transcription factors like *MYC2/3/4* (Pauwels and Goossens 2011; Kazan and Manners 2013).

The role of JA biosynthesis and signalling during Pi deficiency has been elucidated in plants. Low Pi increases levels of JA and its form JA-Ile in both root and shoot tissues. *PHOSPHATE 1* (*PHO1*) gene mutant (*pho1*), which is defective in Pi loading to xylem showed a strong shoot Pi deficiency response and also leads to increased levels of JA and JA-Ile in the shoot tissue, suggesting both endogenous and exogenous Pi levels are sufficient to modulate JA biosynthesis and signalling (Khan et al. 2016; Lin et al. 2008; Aung et al. 2006). A transcriptomic analysis of wild-type and *low phosphorus insensitive* mutant (*lpi4*) under Pi-limiting conditions in *Arabidopsis* suggested transcriptional repression of genes involved in JA biosynthesis and signalling in *lpi4* mutant under Pi deficiency. ZIM domain-containing genes (*JAZ1/2/6*) were also repressed in this mutant in low Pi conditions. Among the other biosynthesis genes, *OPR3* is also repressed, suggesting a compromised JA pathway that may be associated with the unresponsiveness of the mutant to Pi deficiency (Stintzi and Browse 2000; Chacón-López et al. 2011). In line with this evidence, the *OPR3* function has been implicated in root growth during Pi deficiency in *Arabidopsis* (Zheng et al. 2016). A knockout mutant of the *OPR3* gene (*opr3*) promotes root growth under Pi deficiency, suggesting a direct role of *OPR3*-dependent JA biosynthesis in regulating root adaptation under a limited Pi environment (Zheng et al. 2016).

In rice, Pi deficiency promotes JA biosynthesis and signalling to protect plants from *Xanthomonas oryzae pv. oryzae* infection. PHOSPHATE STARVATION RESPONSE protein, PHR1/2, regulates a wide range of phosphate-responsive genes in plants (Bari et al. 2006; Kong et al. 2021). It was demonstrated that *OsPHR2* directly binds with the *OsMYC2* promoter and activates it and enhances JA response (Kong et al. 2021). Recently, the role of the ZIM domain protein *OsJAZ11*, which is induced by low Pi in rice, has been implicated in regulating plant architecture during low Pi conditions (Pandey et al. 2021). Overexpression of *OsJAZ11* in a low-Pi-sensitive rice genotype promoted root growth and Pi content. Further, the authors showed that it is under the control of *OsPHR2* (Pandey et al. 2021). Thus, components of PSR are shown to regulate JA signalling and response in crop plants, which could be an important target for developing plants with improved performance during Pi deficiency.

5.2.8 Role of Gibberellic Acid (GA) Biosynthesis and Signalling during Pi Deficiency

Insights from studies of *Arabidopsis* have shed light on the role of GA biosynthesis and signalling in modulating growth during Pi deficiency (Jiang et al. 2007; Y. Xie et al. 2016; Devaiah et al. 2009). The negative regulators of GA signalling, DELLA proteins, accumulate in response to low Pi. The *Arabidopsis* genome contains five DELLA/TVHYNP motif proteins (GAI, repressor of rga1-3 (RGA), RGA-like 1, 2, and 3) which are degraded in response to GA application (Lee et al. 2002; Locascio, Blázquez, and Alabadí 2013). Mutants lacking DELLA proteins (*ga1-3, gai-t6, rga-t2, rgl1-1* and *rgl2-1*) showed reduced PR growth inhibition under the low Pi conditions. The GA-deficient *ga1-3* mutant showed smaller RH, suggesting the role of GA in RH elongation under low Pi conditions. In line with this, low-Pi-grown seedlings showed reduced GA levels, and exogenous GA, promotes growth, and represses anthocyanin content but does not impact total Pi content. This evidence suggested that GA biosynthesis and signalling both are compromised under low Pi conditions (Jiang et al. 2007). The same line of evidence was reported by Devaiah et al. (2009), where a *MYB62* transcription factor regulated a subset of PSR genes and modulated GA biosynthesis and signalling by repressing genes involved in GA synthesis in the shoot/inflorescence tissues under Pi deficiency. Enhanced levels of *MYB62* lead to a compromised RSA and reduce shoot Pi content, suggesting a key role of *MYB62* in GA metabolism and PSR during Pi deficiency (Devaiah et al. 2009).

5.2.9 ROLE OF BRASSINOSTEROIDS UNDER LOW PI CONDITIONS

Brassinosteroids play a key role in various stress responses and regulate plant growth (Pandey, Devi, and Singh 2020; Planas-Riverola et al. 2019; Devi et al. 2022; Singh and Savaldi-Goldstein 2015; Chaiwanon et al. 2016). BR signalling becomes activated upon binding of the ligand with the membrane-bound receptor complex BRASSINOSTEROID INSENSITIVE1 (BRI1), its homologs, BRI1-LIKE1 and 3 (BRL1/3), along with BRI1-ASSOCIATED RECEPTOR KINASE1 (BAK1) (Caño-Delgado et al. 2004; Wang et al. 2001; Russinova et al. 2004). Upon activation of the pathway through a series of phosphorylation and de-phosphorylation events, the BR signalling transcription factors *BRASSINOSTEROID INSENSITIVE1-ETHYLMETHANESULFONATE-SUPPRESSOR 1 (BES1)* and *BRASSINAZOLE RESISTANT1 (BZR1)* become activated to regulate the diverse developmental and stress-related responses in plants (Chen et al. 2019; He et al. 2005; Li et al. 2012; Wang et al. 2019; Wang et al. 2002; H.-S. Lee et al. 2015; Yin et al. 2002; He et al. 2002; Jaillais et al. 2011; Gupta, Devi, and Singh 2022). Low Pi inhibits BR biosynthesis and signalling. The active form of BRs, 28-nor-castasterone and BR biosynthesis genes, were repressed under low Pi conditions (Devi et al. 2020; Singh et al. 2014). A decrease in the nuclear (n)/cytoplasmic (c) ratio of BES1/BZR1 proteins is associated with reduced BR signalling output (Yin et al. 2002; Wang et al. 2002; Singh et al. 2014; 2018). Low Pi reduces the nuclear abundance of BES1, which in turn leads to reduced nuclear BR signalling activity (Singh et al. 2014). Additionally, the constitutively active dominant mutant of BR signalling, *bzr1-1D*, was largely insensitive to low-Pi-dependent root growth inhibition by regulating meristem size and cell length (Singh et al. 2014). LPR1 is known to regulate the Pi-deficiency response by modulating Fe accumulation in the transition/elongation zone of roots (Müller et al. 2015; Naumann et al. 2022). *bzr1-1D* constitutively represses *LPR1* and, thus, low Pi response on root growth (Singh et al. 2018). In line with the findings, it was later shown that BR-regulated U-box40 (PUB40) leads to proteasomal degradation of BZR1 in the roots. The root-specific activity of PUB40 is controlled by increased stability of PUB40, caused by BIN2-mediated phosphorylation. The authors showed that *pub39:pub40:pub41* triple mutants showed increased BZR1 protein accumulation in root tissues, and both *bzr1-1D* and *pub39:pub40:pub41* triple mutants showed reduced response to low-Pi-mediated root growth inhibition. Therefore, root-specific BZR1 stability appears essential for regulating low-Pi-dependent root growth (Kim et al. 2019) (Figures 5.4 A and B).

Other than root development, BRs are also known to regulate shoot developmental responses such as rice leaf inclination. Zou et al. 2020 showed that BRs and low Pi modulate leaf inclination responses in rice. The control of leaf inclination under low Pi conditions is known to be mediated by SPX1, a key regulatory component of the phosphorus signalling system. Regulator of Leaf Inclination 1 (RLI1) and SPX domain-containing protein of PHOSPHATE STARVATION RESPONSE1 (PHR1) subfamily SPX1 (SYG1/Pho81/XPR1) can interact in low phosphorus environments, thus inhibiting control of RLI1 on the transcriptional expression of BRASSINISTEROID UPREGULATED1 (BU1) and BU1-LIKE1 (BUL1) complex1 (BC1) in lamina joint cells and leading to altered leaf inclination (Ruan et al. 2018). By disrupting the Pi-deficit-induced transcription factor BU1 expression, it has been shown that BR deficiency can reduce leaf inclination. Additionally,

Hormonal Control of Phosphate Uptake and Assimilation

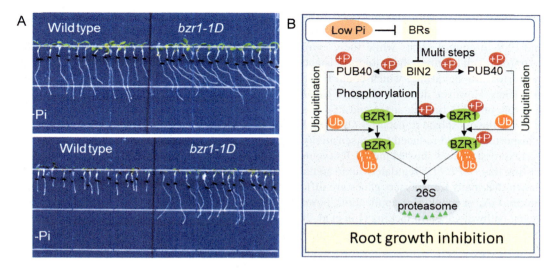

FIGURE 5.4 Effect of Pi deficiency on wild-type and *bzr1-1D* seedlings. (B) Model showing the BZR1 degradation *via* PUB40 ubiquitination mechanism. Low Pi represses BRs, which in turn may promote BIN2-dependent PUB40 activation. PUB40 protein ubiquitinates BZR1 in the root tissues and leads to its degradation, thus inhibiting the BZR1 effect on root growth. Integrated model based on the findings described by Singh et al. (2014, 2018); and Kim et al. (2019).

it was shown that constitutively active or suppressed BR signalling rendered Pi-deficiency-induced leaf inclination insensitive (Zou et al. 2020). Therefore, BR signalling and response intensity in a tissue-specific manner could be an important factor that potentially regulates this Pi-deficiency-induced response.

5.3 EMERGING EVIDENCE OF HORMONAL CROSSTALK AT THE LEVEL OF GENES INVOLVED IN PI UPTAKE AND BALANCE

Recently, the role of ABA in modulating the phosphate transporter gene *PHT1;1* and Pi balance has been identified in *Arabidopsis*. Low Pi increases ABA biosynthesis, which in turn regulates ABI5. ABI5 transcriptionally activates the *PHT1.1* gene, which regulates Pi balance and growth under Pi deficiency (Zhang et al. 2022). Another transporter, *PHT1;5*, which is essential for Pi translocation from older leaves to growing tissue, shows interaction with ethylene. Overexpression lines of this transporter showed increased RH formation and elongation independent of Pi availability. Seedlings treated with either ethylene biosynthesis inhibitor AVG or perception inhibitor (silver nitrate, [AgNO3]) inhibit RH elongation in PHT1;5 overexpression lines, suggesting a possible interaction of ethylene at *PHT1;5* levels, which may be an important target for modulating Pi translocation (Nagarajan et al. 2011).

In rice, *auxin response factor 16* (*OsARF16*), which is involved in auxin signalling and response, regulates the iron and Pi balance of the plants. The *osarf16* mutant showed reduced sensitivity to exogenous auxin and iron deficiency (Shen et al. 2015; 2014). It was demonstrated that *OsARF16* is induced by both exogenous auxins and low Pi. Plants lacking functional *OsARF16* (*osarf16*) showed compromised response to the PR, LR, and RH growth under Pi deficiency (Shen et al. 2013). The *osarf16* mutant showed increased levels of Pi and auxin content with reduced sensitivity to exogenous auxins, suggesting that higher levels of auxin in this mutant may contribute to improved Pi uptake, thereby enhancing Pi content. In the same line of evidence, expression of PSR genes (*OsOPS1*, *OsIPS2*, *OsSQD2*, *OsPT2*, *OsPT8* etc.) was also reduced in the *osarf16* mutant under low Pi conditions, suggesting an overall compromised low Pi response (Shen et al. 2013). Later, it was shown that cytokinin modulates *OsARF16* to balance phosphate transport activity in rice (Shen et al. 2014). Cytokinin (6-BA) treatment increases *OsARF16* levels and also represses Pi uptake and PSR in control plants while these cytokinin-dependent responses were compromised in the *osarf16*. Taken together, these findings suggested a key role of *OsARF16* in plausibly linking auxin and cytokinin signalling, which could be an important target for improving plant response to Pi deficiency. However, direct evidence of *OSARF16* in regulating PHTs and PSI genes is still lacking. The role of PHTs linking auxin and Pi deficiency responses has also been established. In rice and tobacco, the phosphate transporter *OsPHT1;8* mediates such responses. Overexpression lines of this transporter showed reduced auxin sensitivity and increased Pi content (Jia et al. 2017). *OsPHT1;8* overexpression lines showed more LRs (young), presumably due to higher auxin levels, further suggesting that *OsPHT1;8* regulated auxin signaling in regulating lateral root formation. In rice, it was shown that many transporters genes are differentially regulated by exogenous hormones treatment (auxin, GA, or cytokinin) application; however, mechanisms underlying their regulation by hormonal pathways are still lacking (Liu et al. 2011). *PHT1* gene family member *AsPT5*, in *Astragalus sinicus*, is induced by auxin under moderate Pi conditions. It was shown that under moderately high Pi (300 µM) conditions, *AsPT5* alters the distribution of auxin in lateral roots. Additionally, overexpression lines showed enhanced RH formation and growth of lateral roots. However, the effect of AsPT5 on Pi and auxin content and its function under low Pi conditions is still unknown (Fan et al. 2020).

Emerging pieces of evidence suggest that hormonal cues are involved in regulating PHTs; however, further research is required to understand the precise mechanisms of transporter genes regulation (both transcriptional and post-transcriptional) by hormonal pathways at the cellular and tissue levels to modulate Pi uptake, distribution, and its long-distance transport.

5.4 CONCLUSIONS

The chapter highlighted the role of different phytohormones – auxin, BRs, SLs, GAs, CKs, ABA, and ethylene – under low phosphorus availability and how these hormones regulate various development aspects in the plants to ensure survival. The molecular mechanisms describing the roles of different plant hormones under low Pi conditions could be important in understanding plant developmental traits necessary for crop improvement. The role of different plant growth regulators (PGRs) in various abiotic stress conditions is well explored to reprogram RSA as well as regulation of overall plant development. Despite a lot of progress, there is still a gap in understanding the root foraging responses during the scarcity of nutrients including nitrogen, phosphorus, and others in crop plants. It would be beneficial to comprehend the involvement of these hormonal signalling cues in crop improvement by enhancing nutrient foraging response from recent results concerning the impact of various phytohormones in regulating the root Pi-foraging responses. However, information surrounding the understanding of the tissue and organ-specific activity of hormone signalling components, as well as the biosynthesis mechanisms in regulating nutrient, homeostasis, and transport in plants, is still vague. Understanding the processes of cell/tissue-specific control of phytohormones linking nutrient uptake is challenging due to limitations in phenotyping and the lack of mutants in crop plants. Therefore, developing crops with effective adaptive features for increased nutrient uptake may need the use of genome editing technologies like CRISPR-Cas9 (Clustered Regularly Interspaced Short Palindromic Repeats-CRISPR-associated protein 9). The study of the various nutrient stresses is also limited to laboratory conditions, which might not be useful in understanding the realistic adaptive strategies of plants in nature, where there are chances of multiple stress conditions. Therefore, the current scenario of plant development research only focuses on a single nutrient stress condition, which may not explore the holistic framework and multi-level complexity of various stresses to develop crops that can sustain harsh environmental stresses. Further, many stresses in nature are coupled; for example, nutrient scarcity and salinity stress (Kawa et al. 2016). Future work in this area of study could be the construction of a multi-dimensional interactome between different hormone networks and their corresponding regulatory genes under combined stress conditions. Furthermore, the concentration-dependent control of hormones and their mechanisms of modulating adaptive responses in plants are yet to be explored. Another question that emerges is how the interacting hormones and their gene complexes quantitatively control root and shoot growth. Unravelling the knowledge of genetics and the complex hormonal symphony at the cellular level can be manipulated to improve stress tolerance. Despite a lot of progress made in the understanding of adaptive root and shoot traits during harsh environmental conditions and the role of various PGRs, a lot of challenges still lie ahead.

5.5 ACKNOWLEDGEMENTS

We acknowledge DBT e-library Consortium (DeLCON) for providing access to e-resources. APS acknowledges funding support from DST-SERB, India (ECR/2018/000526). SG acknowledges the University Grants Commission of India (UGC, India) for the Junior/Senior Research Fellowship (JRF/SRF).

REFERENCES

Ahmad, Parvaiz, Saiema Rasool, Alvina Gul, Subzar A. Sheikh, Nudrat A. Akram, Muhammad Ashraf, A. M. Kazi, and Salih Gucel. 2016. "Jasmonates: Multifunctional Roles in Stress Tolerance." *Frontiers in Plant Science* 7 (June 2016): 813. https://doi.org/10.3389/FPLS.2016.00813/BIBTEX.

Aida, Mitsuhiro, Dimitris Beis, Renze Heidstra, Viola Willemsen, Ikram Blilou, Carla Galinha, Laurent Nussaume, Yoo Sun Noh, Richard Amasino, and Ben Scheres. 2004. "The PLETHORA Genes Mediate Patterning of the Arabidopsis Root Stem Cell Niche." *Cell* 119 (1). doi:10.1016/j.cell.2004.09.018.

Ajmera, Ishan, T. Charlie Hodgman, and Chungui Lu. 2019. "An Integrative Systems Perspective on Plant Phosphate Research." *Genes* 10 (2): 139. https://doi.org/10.3390/genes10020139.

Argyros, Rebecca D., Dennis E. Mathews, Yi Hsuan Chiang, Christine M. Palmer, Derek M. Thibault, Naomi Etheridge, D. Aaron Argyros, Michael G. Mason, Joseph J. Kieber, and G. Eric Schallera. 2008. "Type B Response Regulators of Arabidopsis Play Key Roles in Cytokinin Signaling and Plant Development." *Plant Cell* 20 (8): 2102–16. https://doi.org/10.1105/TPC.108.059584.

Aung, Kyaw, Shu I. Lin, Chia Chune Wu, Yu Ting Huang, Chun Lin Su, and Tzyy Jen Chiou. 2006. "Pho2, a Phosphate Overaccumulator, Is Caused by a Nonsense Mutation in a MicroRNA399 Target Gene." *Plant Physiology* 141 (3): 1000. https://doi.org/10.1104/PP.106.078063.

Balzergue, Coline, Virginie Puech-Pags, Guillaume Bécard, and Soizic F. Rochange. 2011. "The Regulation of Arbuscular Mycorrhizal Symbiosis by Phosphate in Pea Involves Early and Systemic Signalling Events." *Journal of Experimental Botany* 62 (3). doi:10.1093/jxb/erq335.

Balzergue, Coline, Thibault Dartevelle, Christian Godon, Edith Laugier, Claudia Meisrimler, Jean Marie Teulon, Audrey Creff, et al. 2017. "Low Phosphate Activates STOP1-ALMT1 to Rapidly Inhibit Root Cell Elongation." *Nature Communications* 8 (1): 1–16. https://doi.org/10.1038/ncomms15300.

Bari, Rajendra, Bikram Datt Pant, Mark Stitt, and Wolf Rüdiger Scheible. 2006. "PHO2, MicroRNA399, and PHR1 Define a Phosphate-Signaling Pathway in Plants." *Plant Physiology* 141 (3): 988–99. https://doi.org/10.1104/PP.106.079707.

Beveridge, Christine A. 2000. "Long-Distance Signalling and a Mutational Analysis of Branching in Pea." In *Plant Growth Regulation*. Vol. 32. doi:10.1023/A:1010718020095.

Bhosale, Rahul, Jitender Giri, Bipin K. Pandey, Ricardo F. H. Giehl, Anja Hartmann, Richard Traini, Jekaterina Truskina, et al. 2018. "A Mechanistic Framework for Auxin Dependent Arabidopsis Root Hair Elongation to Low External Phosphate." *Nature Communications* 9 (1). https://doi.org/10.1038/s41467-018-03851-3.

Binder, Brad M., and Joseph M. Jez. 2020. "Ethylene Signaling in Plants." *Journal of Biological Chemistry* 295 (22): 7710–25. https://doi.org/10.1074/JBC.REV120.010854.

Bindraban, Prem S., Christian O. Dimkpa, and Renu Pandey. 2020. "Exploring Phosphorus Fertilizers and Fertilization Strategies for Improved Human and Environmental Health." *Biology & Fertility of Soils* 56 (3): 299–317. https://doi.org/10.1007/S00374-019-01430-2/FIGURES/4.

Blackwell, Martin, Tegan Darch, and Richard Haslam. 2019. "Phosphorus Use Efficiency and Fertilizers: Future Opportunities for Improvements." *Frontiers of Agricultural Science & Engineering* 6 (4): 332–40. https://doi.org/10.15302/J-FASE-2019274.

Bleecker, A. B., and H. Kende. 2003. "Ethylene: A Gaseous Signal Molecule in Plants." 16 (November): 1–18. https://doi.org/10.1146/ANNUREV.CELLBIO.16.1.1.

Borch, K., T. J. Bouma, J. P. Lynch, and K. M. Brown. 1999. "Ethylene: A Regulator of Root Architectural Responses to Soil Phosphorus Availability." *Plant, Cell & Environment* 22 (4): 425–31. https://doi.org/10.1046/J.1365-3040.1999.00405.X.

Bouain, Nadia, Arthur Korte Id, Santosh B. Satbhai Id, Hye-In Nam, Seung Y. Rhee Id, Wolfgang Busch Id, and Hatem Rouached Id. 2019. "Systems Genomics Approaches Provide New Insights into Arabidopsis thaliana Root Growth Regulation under Combinatorial Mineral Nutrient Limitation." https://doi.org/10.1371/journal.pgen.1008392.

Caño-Delgado, Ana, Yanhai Yin, Yu Cong, Dionne Vefeados, Santiago Mora-García, Jin Chen Cheng, Kyoung Hee Nam, Jianming Li, and Joanne Chory. 2004. "BRL1 and BRL3 Are Novel Brassinosteroid Receptors That Function in Vascular Defferentiation in Arabidopsis." *Development* 131 (21): 5341–51. https://doi.org/10.1242/dev.01403.

Chacón-López, Alejandra, Enrique Ibarra-Laclette, Lenin Sánchez-Calderón, Dolores Gutiérrez-Alanís, and Luis Herrera-Estrella. 2011. "Global Expression Pattern Comparison between Low Phosphorus Insensitive 4 and WT Arabidopsis Reveals an Important Role of Reactive Oxygen Species and Jasmonic Acid in the Root Tip Response to Phosphate Starvation." *Plant Signaling & Behavior* 6 (3): 382–92. https://doi.org/10.4161/PSB.6.3.14160.

Chaiwanon, Juthamas, Wenfei Wang, Jia Ying Zhu, Eunkyoo Oh, and Zhi Yong Wang. 2016. "Information Integration and Communication in Plant Growth Regulation." *Cell*. Cell Press. https://doi.org/10.1016/j.cell.2016.01.044.

Chang, Ming Xing, Mian Gu, Yu Wei Xia, Xiao Li Dai, Chang Rong Dai, Jun Zhang, Shi Chao Wang, et al. 2019. "OsPHT1;3 Mediates Uptake, Translocation, and Remobilization of Phosphate under Extremely Low Phosphate Regimes." *Plant Physiology* 179 (2): 656–70. https://doi.org/10.1104/PP.18.01097.

Chen, Weiyue, Minghui Lv, Yanze Wang, Ping An Wang, Yanwei Cui, Meizhen Li, Ruoshi Wang, Xiaoping Gou, and Jia Li. 2019. "BES1 Is Activated by EMS1-TPD1-SERK1/2-Mediated Signaling to Control Tapetum Development in Arabidopsis thaliana." *Nature Communications* 10 (1). https://doi.org/10.1038/s41467-019-12118-4.

Chien, Pei Shan, Chih Pin Chiang, Shang Jye Leong, and Tzyy Jen Chiou. 2018. "Sensing and Signaling of Phosphate Starvation: From Local to Long Distance." *Plant & Cell Physiology* 59 (9): 1714–22. https://doi.org/10.1093/PCP/PCY148.

Chini, Andrea, Isabel Monte, Angel M. Zamarreño, Mats Hamberg, Steve Lassueur, Philippe Reymond, Sally Weiss, et al. 2018. "An OPR3-Independent Pathway Uses 4,5-Didehydrojasmonate for Jasmonate Synthesis." *Nature Chemical Biology* 14 (2): 171–78. https://doi.org/10.1038/NCHEMBIO.2540.

Cho, Huikyong, Nadia Bouain, Luqing Zheng, and Hatem Rouached. 2021. "Plant Resilience to Phosphate Limitation: Current Knowledge and Future Challenges." *Critical Reviews in Biotechnology* 41 (1): 63–71. https://doi.org/10.1080/07388551.2020.1825321.

Cui, Naxin, Min Cai, Xu Zhang, Ahmed A. Abdelhafez, Li Zhou, Huifeng Sun, Guifa Chen, Guoyan Zou, and Sheng Zhou. 2020. "Runoff Loss of Nitrogen and Phosphorus from a Rice Paddy Field in the East of China: Effects of Long-Term Chemical N Fertilizer and Organic Manure Applications." *Global Ecology & Conservation* 22 (June): e01011. https://doi.org/10.1016/J.GECCO.2020.E01011.

Czarnecki, Olaf, Jun Yang, David J. Weston, Gerald A. Tuskan, and Jin Gui Chen. 2013. "A Dual Role of Strigolactones in Phosphate Acquisition and Utilization in Plants." *International Journal of Molecular Sciences* 14 (4): 7681. https://doi.org/10.3390/IJMS14047681.

Devaiah, Ballachanda N., Ramaiah Madhuvanthi, Athikkattuvalasu S. Karthikeyan, and Kashchandra G. Raghothama. 2009. "Phosphate Starvation Responses and Gibberellic Acid Biosynthesis Are Regulated by the MYB62 Transcription Factor in Arabidopsis." *Molecular Plant* 2 (1): 43–58. https://doi.org/10.1093/MP/SSN081.

Devi, Loitongbam Lorinda, Anshika Pandey, Shreya Gupta, and Amar Pal Singh. 2022. "The Interplay of Auxin and Brassinosteroid Signaling Tunes Root Growth under Low and Different Nitrogen Forms." *Plant Physiology*, April. https://doi.org/10.1093/PLPHYS/KIAC157.

Devi, Loitongbam Lorinda, Anshika Pandey, and Amar Pal Singh. 2020. "Root Plasticity under Low Phosphate Availability : A Physiological and Molecular Approach to Plant Adaptation under Limited Phosphate Availability." *Improving Abiotic Stress Tolerance in Plants*, May, 13–30. https://doi.org/10.1201/9780429027505-2.

Dong, Jinsong, Miguel A. Piñeros, Xiaoxuan Li, Haibing Yang, Yu Liu, Angus S. Murphy, Leon V. Kochian, and Dong Liu. 2017. "An Arabidopsis ABC Transporter Mediates Phosphate Deficiency-Induced

Remodeling of Root Architecture by Modulating Iron Homeostasis in Roots." *Molecular Plant* 10 (2): 244–59. https://doi.org/10.1016/j.molp.2016.11.001.

Fan, Xiaoning, Xianrong Che, Wenzhen Lai, Sijia Wang, Wentao Hu, Hui Chen, Bin Zhao, Ming Tang, and Xianan Xie. 2020. "The Auxin-Inducible Phosphate Transporter AsPT5 Mediates Phosphate Transport and Is Indispensable for Arbuscule Formation in Chinese Milk Vetch at Moderately High Phosphate Supply." *Environmental Microbiology* 22 (6): 2053–79. https://doi.org/10.1111/1462-2920.14952.

Zhu, Fang, Xu Sheng Zhao Xiao, Qi Wu, and Ren Fang Shen. 2018. "Abscisic Acid Is Involved in Root Cell Wall Phosphorus Remobilization Independent of Nitric Oxide and Ethylene in Rice (Oryza sativa)." *Annals of Botany* 121 (7): 1361–68. https://doi.org/10.1093/AOB/MCY034.

Franco-Zorrilla, José M., Ana C. Martin, Roberto Solano, Vicente Rubio, Antonio Leyva, and Javier Paz-Ares. 2002. "Mutations at CRE1 Impair Cytokinin-Induced Repression of Phosphate Starvation Responses in Arabidopsis." *Plant Journal* 32 (3): 353–60. https://doi.org/10.1046/J.1365-313X.2002.01431.X.

Franco-Zorrilla, José Manuel, Ana Carmen Martín, Antonio Leyva, and Javier Paz-Ares. 2005. "Interaction Between Phosphate-Starvation, Sugar, and Cytokinin Signaling in Arabidopsis and the Roles of Cytokinin Receptors CRE1/AHK4 and AHK3." *Plant Physiology* 138 (2): 847–57. https://doi.org/10.1104/PP.105.060517.

Fujii, Hiroaki, Tzyy Jen Chiou, Shu I. Lin, Kyaw Aung, and Jian Kang Zhu. 2005. "A MiRNA Involved in Phosphate-Starvation Response in Arabidopsis." *Current Biology : CB* 15 (22): 2038–43. https://doi.org/10.1016/J.CUB.2005.10.016.

Gilbert, Glena A., J. Diane Knight, Carroll P. Vance, and Deborah L. Allan. 2000. "Proteoid Root Development of Phosphorus Deficient Lupin Is Mimicked by Auxin and Phosphonate." *Annals of Botany* 85 (6): 921–28. https://doi.org/10.1006/ANBO.2000.1133.

Giri, Jitender, Rahul Bhosale, Guoqiang Huang, Bipin K. Pandey, Helen Parker, Susan Zappala, Jing Yang, et al. 2018. "Rice Auxin Influx Carrier OsAUX1 Facilitates Root Hair Elongation in Response to Low External Phosphate." *Nature Communications* 9 (1): 1–7. https://doi.org/10.1038/s41467-018-03850-4.

Gruber, Benjamin D., Ricardo F. H. Giehl, Swetlana Friedel, and Nicolaus von Wirén. 2013. "Plasticity of the Arabidopsis Root System under Nutrient Deficiencies." *Plant Physiology* 163 (1): 161–79. https://doi.org/10.1104/pp.113.218453.

Gupta, Shreya, Loitongbam Lorinda Devi, and Amar Pal Singh. 2022. "Nitric Oxide." *Nitric Oxide in Plants*, May, 230–47. https://doi.org/10.1002/9781119800156.CH12.

He, Jun Xian, Joshua M. Gendron, Yu Sun, Srinivas S. L. Gampala, Nathan Gendron, Catherine Qing Sun, and Zhi Yong Wang. 2005. "BZR1 Is a Transcriptional Repressor with Dual Roles in Brassinosteroid Homeostasis and Growth Responses." *Science* 307 (5715): 1634–38. https://doi.org/10.1126/science.1107580.

He, Jun Xian, Joshua M. Gendron, Yanli Yang, Jianming Li, and Zhi Yong Wang. 2002. "The GSK3-Like Kinase BIN2 Phosphorylates and Destabilizes BZR1, a Positive Regulator of the Brassinosteroid Signaling Pathway in Arabidopsis." *Proceedings of the National Academy of Sciences of the United States of America* 99 (15): 10185–90. https://doi.org/10.1073/pnas.152342599.

Heuer, Sigrid, Roberto Gaxiola, Rhiannon Schilling, Luis Herrera-Estrella, Damar López-Arredondo, Matthias Wissuwa, Emmanuel Delhaize, and Hatem Rouached. 2017. "Improving Phosphorus Use Efficiency: A Complex Trait with Emerging Opportunities." *The Plant Journal* 90 (5): 868–85. https://doi.org/10.1111/TPJ.13423.

Hutchison, Claire E., and Joseph J. Kieber. 2002. "Cytokinin Signaling in Arabidopsis." *The Plant Cell* 14 (suppl_1): S47–59. https://doi.org/10.1105/TPC.010444.

Ioio, Raffaele Dello, Kinu Nakamura, Laila Moubayidin, Serena Perilli, Masatoshi Taniguchi, Miyo T. Morita, Takashi Aoyama, Paolo Costantino, and Sabrina Sabatini. 2008. "A Genetic Framework for the Control of Cell Division and Differentiation in the Root Meristem." *Science* 322 (5906): 1380–84. https://doi.org/10.1126/SCIENCE.1164147.

Irfan, Muhammad, Tariq Aziz, Muhammad Aamer Maqsood, Hafiz Muhammad Bilal, Kadambot H. M. Siddique, and Minggang Xu. 2020. "Phosphorus (P) Use Efficiency in Rice Is Linked to Tissue-Specific Biomass and P Allocation Patterns." *Scientific Reports* 10 (1): 1–14. https://doi.org/10.1038/s41598-020-61147-3.

Jaillais, Yvon, Michael Hothorn, Youssef Belkhadir, Tsegaye Dabi, Zachary L. Nimchuk, Elliot M. Meyerowitz, and Joanne Chory. 2011. "Tyrosine Phosphorylation Controls Brassinosteroid Receptor Activation by Triggering Membrane Release of Its Kinase Inhibitor." *Genes & Development* 25 (3): 232–37. https://doi.org/10.1101/gad.2001911.

Jia, Hongfang, Songtao Zhang, Lizhi Wang, Yongxia Yang, Hongying Zhang, Hong Cui, Huifang Shao, and Guohua Xu. 2017. "OsPht1;8, a Phosphate Transporter, Is Involved in Auxin and Phosphate Starvation

Response in Rice." *Journal of Experimental Botany* 68 (18): 5057–68. https://doi.org/10.1093/JXB/ERX317.

Jia, Zhongtao, Ricardo F. H. Giehl, and Nicolaus von Wirén. 2022. "Nutrient–Hormone Relations: Driving Root Plasticity in Plants." *Molecular Plant* 15 (1): 86–103. https://doi.org/10.1016/J.MOLP.2021.12.004.

Jiang, Caifu, Xiuhua Gao, Lili Liao, Nicholas P. Harberd, and Xiangdong Fu. 2007. "Phosphate Starvation Root Architecture and Anthocyanin Accumulation Responses Are Modulated by the Gibberellin-DELLA Signaling Pathway in Arabidopsis." *Plant Physiology* 145 (4): 1460–70. https://doi.org/10.1104/pp.107.103788.

Kapulnik, Yoram, and Hinanit Koltai. 2016. "Fine-Tuning by Strigolactones of Root Response to Low Phosphate." *Journal of Integrative Plant Biology* 58 (3): 203–12. https://doi.org/10.1111/JIPB.12454.

Karimi, Sohail M., Matthias Freund, Brittney M. Wager, Michael Knoblauch, Jörg Fromm, Heike M. Mueller, Peter Ache, et al. 2021. "Under Salt Stress Guard Cells Rewire Ion Transport and Abscisic Acid Signaling." *New Phytologist* 231 (3): 1040–55. https://doi.org/10.1111/NPH.17376.

Katsir, Leron, Anthony L. Schilmiller, Paul E. Staswick, Yang He Sheng, and Gregg A. Howe. 2008. "COI1 Is a Critical Component of a Receptor for Jasmonate and the Bacterial Virulence Factor Coronatine." *Proceedings of the National Academy of Sciences of the United States of America* 105 (19): 7100–5. https://doi.org/10.1073/PNAS.0802332105/SUPPL_FILE/0802332105SI.PDF.

Kawa, Dorota, Magdalena M. Julkowska, Hector Montero Sommerfeld, Anneliek Ter Horst, Michel A. Haring, and Christa Testerink. 2016. "Phosphate-Dependent Root System Architecture Responses to Salt Stress." *Plant Physiology* 172 (2). doi:10.1104/pp.16.00712.

Kazan, Kemal, and John M. Manners. 2013. "MYc2: The Master in Action." *Molecular Plant* 6 (3): 686–703. https://doi.org/10.1093/mp/sss128.

Khan, Ghazanfar Abbas, Evangelia Vogiatzaki, Gaétan Glauser, and Yves Poirier. 2016. "Phosphate Deficiency Induces the Jasmonate Pathway and Enhances Resistance to Insect Herbivory." *Plant Physiology* 171 (1): 632. https://doi.org/10.1104/PP.16.00278.

Kiba, Takatoshi, Kentaro Takei, Mikiko Kojima, and Hitoshi Sakakibara. 2013. "Side-Chain Modification of Cytokinins Controls Shoot Growth in Arabidopsis." *Developmental Cell* 27 (4): 452–61. https://doi.org/10.1016/J.DEVCEL.2013.10.004.

Kim, Eun Ji, Se Hwa Lee, Chan Ho Park, So Hee Kim, Chuan Chih Hsu, Shouling Xu, Zhi Yong Wang, Seong Ki Kim, and Tae Wuk Kim. 2019. "Plant U-Box40 Mediates Degradation of the Brassinosteroid-Responsive Transcription Factor Bzr1 in Arabidopsis Roots." *Plant Cell* 31 (4): 791–808. https://doi.org/10.1105/tpc.18.00941.

Kim, Hye Ji, Jonathan P. Lynch, and Kathleen M. Brown. 2008. "Ethylene Insensitivity Impedes a Subset of Responses to Phosphorus Deficiency in Tomato and Petunia." *Plant, Cell & Environment* 31 (12): 1744–55. https://doi.org/10.1111/J.1365-3040.2008.01886.X.

Kohlen, Wouter, Tatsiana Charnikhova, Qing Liu, Ralph Bours, Malgorzata A. Domagalska, Sebastien Beguerie, Francel Verstappen, Ottoline Leyser, Harro Bouwmeester, and Carolien Ruyter-Spira. 2011. "Strigolactones Are Transported through the Xylem and Play a Key Role in Shoot Architectural Response to Phosphate Deficiency in Nonarbuscular Mycorrhizal Host Arabidopsis." *Plant Physiology* 155 (2): 974–87. https://doi.org/10.1104/PP.110.164640.

Kong, Yaze, Gang Wang, Xian Chen, Linying Li, Xueying Zhang, Sangtian Chen, Yuqing He, and Gaojie Hong. 2021. "OsPHR2 Modulates Phosphate Starvation-Induced OsMYC2 Signalling and Resistance to Xanthomonas Oryzae Pv. Oryzae." *Plant, Cell & Environment* 44 (10): 3432–44. https://doi.org/10.1111/PCE.14078.

Kumari, Priyanka, Loitongbam Lorinda Devi, Amresh Kumar, Ashutosh Pandey, Subodh Kumar Sinha, and Amar Pal Singh. 2022. "Differential Response of Rice Genotypes to Nitrogen Availability Is Associated with the Altered Nitrogen Metabolism and Ionomic Balance." *Environmental & Experimental Botany*, March: 104847. https://doi.org/10.1016/J.ENVEXPBOT.2022.104847.

Kuroha, T., H. Kato, T. Asami, S. Yoshida, H. Kamada, and S. Satoh. 2002. "A Trans-Zeatin Riboside in Root Xylem Sap Negatively Regulates Adventitious Root Formation on Cucumber Hypocotyls." *Journal of Experimental Botany* 53 (378): 2193–200. https://doi.org/10.1093/JXB/ERF077.

Lee, Hak-Soo, Yoon Kim, Giang Pham, Ju Won Kim, Ji-Hye Song, Yew Lee, Yong-Sic Hwang, Stanley J. Roux, and Soo-Hwan Kim. 2015. "Brassinazole Resistant 1 (BZR1)-Dependent Brassinosteroid Signalling Pathway Leads to Ectopic Activation of Quiescent Cell Division and Suppresses Columella Stem Cell Differentiation." *Journal of Experimental Botany* 66 (15): 4835–49. https://doi.org/10.1093/JXB/ERV316.

Lee, Sorcheng, Hui Cheng, Kathryn E. King, Weefuen Wang, Yawen He, Alamgir Hussain, Jane Lo, Nicholas P. Harberd, and Jinrong Peng. 2002. "Gibberellin Regulates Arabidopsis Seed Germination via RGL2, a

GAI/RGA-Like Gene Whose Expression Is Up-Regulated Following Imbibition." *Genes & Development* 16 (5): 646. https://doi.org/10.1101/GAD.969002.

Lei, Mingguang, Chuanmei Zhu, Yidan Liu, Athikkattuvalasu S. Karthikeyan, Ray A. Bressan, Kashchandra G. Raghothama, and Dong Liu. 2011. "Ethylene Signalling Is Involved in Regulation of Phosphate Starvation-Induced Gene Expression and Production of Acid Phosphatases and Anthocyanin in Arabidopsis." *The New Phytologist* 189 (4): 1084–95. https://doi.org/10.1111/J.1469-8137.2010.03555.X.

Li, Qian Feng, Chunming Wang, Lei Jiang, Shuo Li, Samuel S. M. Sun, and Jun Xian He. 2012. "An Interaction between BZR1 and DELLAs Mediates Direct Signaling Crosstalk between Brassinosteroids and Gibberellins in Arabidopsis." *Science Signaling* 5 (244). https://doi.org/10.1126/SCISIGNAL.2002908.

Li, Yan Su, Xiao Tao Mao, Qiu Ying Tian, Ling Hao Li, and Wen Hao Zhang. 2009. "Phosphorus Deficiency-Induced Reduction in Root Hydraulic Conductivity in Medicago falcata Is Associated with Ethylene Production." *Environmental & Experimental Botany* 67 (1): 172–77. https://doi.org/10.1016/J.ENVEXPBOT.2009.05.013.

Lin, Hong, Yan Yao De Li, Li Hua Jia, Jin Fang Tan, Zhi Hong Xu, Wen Ming Zheng, and Hong Wei Xue. 2020. "Phospholipase D-Derived Phosphatidic Acid Promotes Root Hair Development under Phosphorus Deficiency by Suppressing Vacuolar Degradation of PIN-FORMED2." *The New Phytologist* 226 (1): 142. https://doi.org/10.1111/NPH.16330.

Lin, Shu I., Su Fen Chiang, Wei Yi Lin, June Wei Chen, Ching Ying Tseng, Pei Chi Wu, and Tzyy Jen Chiou. 2008. "Regulatory Network of MicroRNA399 and PHO2 by Systemic Signaling." *Plant Physiology* 147 (2): 732–46. https://doi.org/10.1104/PP.108.116269.

Lin, Wei Yi, Huang Teng Kuei, Shang Jye Leong, and Tzyy Jen Chiou. 2014. "Long-Distance Call from Phosphate: Systemic Regulation of Phosphate Starvation Responses." *Journal of Experimental Botany* 65 (7): 1817–27. https://doi.org/10.1093/JXB/ERT431.

Liu, Junqi, Deborah A. Samac, Bruna Bucciarelli, Deborah L. Allan, and Carroll P. Vance. 2005. "Signaling of Phosphorus Deficiency-Induced Gene Expression in White Lupin Requires Sugar and Phloem Transport." *Plant Journal* 41 (2). doi:10.1111/j.1365-313X.2004.02289.x.

Liu, Dong. 2021. "Root Developmental Responses to Phosphorus Nutrition." *Journal of Integrative Plant Biology* 63 (6): 1065–90. https://doi.org/10.1111/JIPB.13090.

Liu, Fang, Xiao Jian Chang, Ye Ying, Wei Bo Xie, Ping Wu, and Xing Ming Lian. 2011. "Comprehensive Sequence and Whole-Life-Cycle Expression Profile Analysis of the Phosphate Transporter Gene Family in Rice." *Molecular Plant* 4 (6): 1105–22. https://doi.org/10.1093/MP/SSR058.

Liu, Q., G. Q. Zhou, F. Xu, X. L. Yan, H. Liao, and J. X. Wang. 2013. "The Involvement of Auxin in Root Architecture Plasticity in Arabidopsis Induced by Heterogeneous Phosphorus Availability." *Biologia Plantarum* 57 (4): 739–48. https://doi.org/10.1007/S10535-013-0327-Z.

Locascio, Antonella, Miguel A. Blázquez, and David Alabadí. 2013. "Genomic Analysis of DELLA Protein Activity." *Plant & Cell Physiology* 54 (8): 1229–37. https://doi.org/10.1093/PCP/PCT082.

López-Bucio, José, Alfredo Cruz-Ramírez, and Luis Herrera-Estrella. 2003. "The Role of Nutrient Availability in Regulating Root Architecture." *Current Opinion in Plant Biology*. https://doi.org/10.1016/S1369-5266(03)00035-9.

López-Bucio, José, Esmeralda Hernández-Abreu, Lenin Sánchez-Calderón, María Fernanda Nieto-Jacobo, June Simpson, and Luis Herrera-Estrella. 2002. "Phosphate Availability Alters Architecture and Causes Changes in Hormone Sensitivity in the Arabidopsis Root System." *Plant Physiology* 129 (1): 244–56. https://doi.org/10.1104/PP.010934.

López-Bucio, José, Esmeralda Hernández-Abreu, Lenin Sánchez-Calderón, Anahí Pérez-Torres, Rebekah A. Rampey, Bonnie Bartel, and Luis Herrera-Estrella. 2005. "An Auxin Transport Independent Pathway Is Involved in Phosphate Stress-Induced Root Architectural Alterations in Arabidopsis. Identification of BIG as a Mediator of Auxin in Pericycle Cell Activation." *Plant Physiology* 137 (2): 681. https://doi.org/10.1104/PP.104.049577.

López-Ráez, Juan Antonio, and Harro Bouwmeester. 2008. "Fine-Tuning Regulation of Strigolactone Biosynthesis under Phosphate Starvation." *Plant Signaling & Behavior* 3 (11): 963. https://doi.org/10.4161/PSB.6126.

López-Ráez, Juan Antonio, Tatsiana Charnikhova, Victoria Gómez-Roldán, Radoslava Matusova, Wouter Kohlen, Ric De Vos, France Verstappen, et al. 2008. "Tomato Strigolactones Are Derived from Carotenoids and Their Biosynthesis Is Promoted by Phosphate Starvation." *New Phytologist* 178 (4): 863–74. https://doi.org/10.1111/J.1469-8137.2008.02406.X.

Madmon, Ortal, Moran Mazuz, Puja Kumari, Anandamoy Dam, Aurel Ion, Einav Mayzlish-Gati, Eduard Belausov, et al. 2016. "Expression of MAX2 under SCARECROW Promoter Enhances the Strigolactone/

MAX2 Dependent Response of Arabidopsis Roots to Low-Phosphate Conditions." *Planta* 243 (6): 1419–27. https://doi.org/10.1007/s00425-016-2477-7.
Malamy, J. E., and K. S. Ryan. 2001. "Environmental Regulation of Lateral Root Initiation in Arabidopsis." *Plant Physiology* 127 (3): 899–909. https://doi.org/10.1104/pp.010406.
Mangano, Silvina, Silvina P. Denita-Juarez, Eliana Marzol, Cecilia Borassi, and José M. Estevez. 2018. "High Auxin and High Phosphate Impact on Rsl2 Expression and Ros-Homeostasis Linked to Root Hair Growth in Arabidopsis thaliana." *Frontiers in Plant Science* 9 (August): 1164. https://doi.org/10.3389/FPLS.2018.01164/BIBTEX.
Martín, Ana C., Juan Carlos Del Pozo, Joaquín Iglesias, Vicente Rubio, Roberto Solano, Alicia De La Peña, Antonio Leyva, and Javier Paz-Ares. 2000. "Influence of Cytokinins on the Expression of Phosphate Starvation Responsive Genes in Arabidopsis." *Plant Journal* 24 (5): 559–67. https://doi.org/10.1046/j.1365-313X.2000.00893.x.
Matsumoto-Kitano, Miho, Takami Kusumoto, Petr Tarkowski, Kaori Kinoshita-Tsujimura, Kateřina Václavíková, Kaori Miyawaki, and Tatsuo Kakimoto. 2008. "Cytokinins Are Central Regulators of Cambial Activity." *Proceedings of the National Academy of Sciences of the United States of America* 105 (50): 20027–31. https://doi.org/10.1073/PNAS.0805619105.
Mayzlish-Gati, Einav, Carolien De-Cuyper, Sofie Goormachtig, Tom Beeckman, Marnik Vuylsteke, Philip B. Brewer, Christine A. Beveridge, et al. 2012. "Strigolactones Are Involved in Root Response to Low Phosphate Conditions in Arabidopsis." *Plant Physiology* 160 (3): 1329–41. https://doi.org/10.1104/PP.112.202358.
Meng, Zhi Bin, Xue Di You, Dong Suo, Yun Long Chen, Caixian Tang, Jian Li Yang, and Shao Jian Zheng. 2013. "Root-Derived Auxin Contributes to the Phosphorus-Deficiency-Induced Cluster-Root Formation in White Lupin (Lupinus Albus)." *Physiologia Plantarum* 148 (4): 481–89. https://doi.org/10.1111/J.1399-3054.2012.01715.X.
Meyer, Gregor, Michael J. Bell, Casey L. Doolette, Gianluca Brunetti, Yaqi Zhang, Enzo Lombi, and Peter M. Kopittke. 2020. "Plant-Available Phosphorus in Highly Concentrated Fertilizer Bands: Effects of Soil Type, Phosphorus Form, and Coapplied Potassium." *Journal of Agricultural & Food Chemistry* 68 (29): 7571–80. https://doi.org/10.1021/ACS.JAFC.0C01287/SUPPL_FILE/JF0C01287_SI_001.PDF.
Miura, Kenji, Ana Rus, Altanbadralt Sharkhuu, Shuji Yokoi, Athikkattuvalasu S. Karthikeyan, Kashchandra G. Raghothama, Dongwon Baek, et al. 2005. "The Arabidopsis SUMO E3 Ligase SIZ1 Controls Phosphate Deficiency Responses." *Proceedings of the National Academy of Sciences of the United States of America* 102 (21): 7760–65. https://doi.org/10.1073/PNAS.0500778102.
Miyawaki, Kaori, Petr Tarkowski, Miho Matsumoto-Kitano, Tomohiko Kato, Shusei Sato, Danuse Tarkowska, Satoshi Tabata, Göran Sandberg, and Tatsuo Kakimoto. 2006. "Roles of Arabidopsis ATP/ADP Isopentenyltransferases and TRNA Isopentenyltransferases in Cytokinin Biosynthesis." *Proceedings of the National Academy of Sciences of the United States of America* 103 (44): 16598–603. https://doi.org/10.1073/PNAS.0603522103/SUPPL_FILE/03522FIG7.JPG.
Młodzińska, Ewa, and Magdalena Zboińska. 2016. "Phosphate Uptake and Allocation – A Closer Look at Arabidopsis thaliana L. and Oryza sativa L." *Frontiers in Plant Science* 7 (AUG2016). https://doi.org/10.3389/FPLS.2016.01198.
Mora-Macías, Javier, Jonathan Odilón Ojeda-Rivera, Dolores Gutiérrez-Alanís, Lenin Yong-Villalobos, Araceli Oropeza-Aburto, Javier Raya-González, Gabriel Jiménez-Domínguez, Gabriela Chávez-Calvillo, Rubén Rellán-Álvarez, and Luis Herrera-Estrell. 2017. "Malate-Dependent Fe Accumulation Is a Critical Checkpoint in the Root Developmental Response to Low Phosphate." *Proceedings of the National Academy of Sciences of the United States of America* 114 (17): E3563–72. https://doi.org/10.1073/pnas.1701952114.
Müller, J., Theresa Toev, Marcus Heisters, Janine Teller, Katie L. Moore, Gerd Hause, Dhurvas Chandrasekaran Dinesh, Katharina Bürstenbinder, and Steffen Abel. 2015. "Iron-Dependent Callose Deposition Adjusts Root Meristem Maintenance to Phosphate Availability." *Developmental Cell* 33 (2): 216–30. https://doi.org/10.1016/j.devcel.2015.02.007.
Nacry, Philippe, Geneviève Canivenc, Bertrand Muller, Abdelkrim Azmi, Harry Van Onckelen, Michel Rossignol, and Patrick Doumas. 2005. "A Role for Auxin Redistribution in the Responses of the Root System Architecture to Phosphate Starvation in Arabidopsis." *Plant Physiology* 138 (4): 2061. https://doi.org/10.1104/PP.105.060061.
Nagarajan, Vinay K., Ajay Jain, Michael D. Poling, Anthony J. Lewis, Kashchandra G. Raghothama, and Aaron P. Smith. 2011. "Arabidopsis Pht1;5 Mobilizes Phosphate between Source and Sink Organs and Influences the Interaction between Phosphate Homeostasis and Ethylene Signaling." *Plant Physiology* 156 (3): 1149–63. https://doi.org/10.1104/PP.111.174805.

Naumann, Christin, Marcus Heisters, Wolfgang Brandt, Philipp Janitza, Carolin Alfs, Nancy Tang, Alicia Toto Nienguesso, et al. 2022. "Bacterial-Type Ferroxidase Tunes Iron-Dependent Phosphate Sensing during Arabidopsis Root Development." *Current Biology*, April. https://doi.org/10.1016/J.CUB.2022.04.005.

Negi, Manisha, Raghavendrarao Sanagala, Vandna Rai, and Ajay Jain. 2016. "Deciphering Phosphate Deficiency-Mediated Temporal Effects on Different Root Traits in Rice Grown in a Modified Hydroponic System." *Frontiers in Plant Science* 7 (May). https://doi.org/10.3389/fpls.2016.00550.

Ng, Ley Moy, Karsten Melcher, Bin Tean Teh, and H. Eric Xu. 2014. "Abscisic Acid Perception and Signaling: Structural Mechanisms and Applications." *Acta Pharmacologica Sinica*, 35 (5):567–84. https://doi.org/10.1038/aps.2014.5.

Okushima, Yoko, Paul J. Overvoorde, Kazunari Arima, Jose M. Alonso, April Chan, Charlie Chang, Joseph R. Ecker, et al. 2005. "Functional Genomic Analysis of the Auxin Response Factor Gene Family Members in Arabidopsis thaliana: Unique and Overlapping Functions of ARF7 and ARF19." *The Plant Cell* 17 (2): 444–63. https://doi.org/10.1105/TPC.104.028316.

Pandey, Anshika, Loitongbam Lorinda Devi, and Amar Pal Singh. 2020. "Review: Emerging Roles of Brassinosteroid in Nutrient Foraging." *Plant Science*, March, 110474. https://doi.org/10.1016/j.plantsci.2020.110474.

Pandey, Bipin K., Lokesh Verma, Ankita Prusty, Ajit Pal Singh, Malcolm J. Bennett, Akhilesh K. Tyagi, Jitender Giri, and Poonam Mehra. 2021. "OsJAZ11 Regulates Phosphate Starvation Responses in Rice." *Planta* 254 (1): 1–16. https://doi.org/10.1007/S00425-021-03657-6/FIGURES/7.

Pant, Bikram Datt, Anja Buhtz, Julia Kehr, and Wolf Rüdiger Scheible. 2008. "MicroRNA399 Is a Long-Distance Signal for the Regulation of Plant Phosphate Homeostasis." *The Plant Journal : for Cell & Molecular Biology* 53 (5): 731–38. https://doi.org/10.1111/J.1365-313X.2007.03363.X.

Pantigoso, Hugo A., Jun Yuan, Yanhui He, Qinggang Guo, Charlie Vollmer, and Jorge M. Vivanco. 2020. "Role of Root Exudates on Assimilation of Phosphorus in Young and Old Arabidopsis thaliana Plants." *PLOS One* 15 (6). https://doi.org/10.1371/JOURNAL.PONE.0234216.

Pauwels, Laurens, and Alain Goossens. 2011. "The JAZ Proteins: A Crucial Interface in the Jasmonate Signaling Cascade." *The Plant Cell* 23 (9): 3089. https://doi.org/10.1105/TPC.111.089300.

Paz-Ares, Javier, Maria Isabel Puga, Monica Rojas-Triana, Iris Martinez-Hevia, Sergio Diaz, Cesar Poza-Carrión, Miguel Miñambres, and Antonio Leyva. 2022. "Plant Adaptation to Low Phosphorus Availability: Core Signaling, Crosstalks, and Applied Implications." *Molecular Plant* 15 (1): 104–24. https://doi.org/10.1016/J.MOLP.2021.12.005.

Pé, Benjamin, Mathilde Clé, Laurent Nussaume, and Thierry Desnos. 2011. "Root Developmental Adaptation to Phosphate Starvation: Better Safe Than Sorry." *Trends in Plant Science* 16 (8): 442–50. https://doi.org/10.1016/j.tplants.2011.05.006.

Péret, Benjamin, Mathilde Clément, Laurent Nussaume, and Thierry Desnos. 2011. "Root Developmental Adaptation to Phosphate Starvation: Better Safe Than Sorry." *Trends in Plant Science* 16 (8): 442–50. https://doi.org/10.1016/J.TPLANTS.2011.05.006.

Pérez-Torres, Claudia Anahí, José López-Bucio, Alfredo Cruz-Ramírez, Enrique Ibarra-Laclette, Sunethra Dharmasiri, Mark Estelle, and Luis Herrera-Estrella. 2008. "Phosphate Availability Alters Lateral Root Development in Arabidopsis by Modulating Auxin Sensitivity via a Mechanism Involving the TIR1 Auxin Receptor." *The Plant Cell* 20 (12): 3258. https://doi.org/10.1105/TPC.108.058719.

Pérez Torres, Claudia Anahí, José López Bucio, and Luis Herrera Estrella. 2009. "Low Phosphate Signaling Induces Changes in Cell Cycle Gene Expression by Increasing Auxin Sensitivity in the Arabidopsis Root System." *Plant Signaling & Behavior* 4 (8): 781–83. https://doi.org/10.1105/tpc.108.058719.

Planas-Riverola, Ainoa, Aditi Gupta, Isabel Betegoń-Putze, Nadja Bosch, Marta Ibanao, and Ana I. Cano-Delgado. 2019. "Brassinosteroid Signaling in Plant Development and Adaptation to Stress." *Development (Cambridge)*. Company of Biologists Ltd. https://doi.org/10.1242/dev.151894.

Poirier, Yves, and Marcel Bucher. 2002. "Phosphate Transport and Homeostasis in Arabidopsis." *The Arabidopsis Book / American Society of Plant Biologists* 1 (January): e0024. https://doi.org/10.1199/TAB.0024.

Qiu, Yuping, Ran Tao, Ying Feng, Zhina Xiao, Dan Zhang, Yang Peng, Xing Wen, Yichuan Wang, and Hongwei Guo. 2021. "EIN3 and RSL4 Interfere with an MYB–BHLH–WD40 Complex to Mediate Ethylene-Induced Ectopic Root Hair Formation in Arabidopsis." *Proceedings of the National Academy of Sciences of the United States of America* 118 (51). https://doi.org/10.1073/PNAS.2110004118/SUPPL_FILE/PNAS.2110004118.SAPP.PDF.

Rouached, Hatem, A. Bulak Arpat, and Yves Poirier. 2010. "Regulation of Phosphate Starvation Responses in Plants: Signaling Players and Cross-Talks." *Molecular Plant* 3 (2): 288–99. https://doi.org/10.1093/MP/SSP120.

Ruan, Wenyuan, Meina Guo, Lei Xu, Xueqing Wang, Hongyu Zhao, Junmin Wang, and Keke Yi. 2018. "An SPX-RLI1 Module Regulates Leaf Inclination in Response to Phosphate Availability in Rice." *The Plant Cell* 30 (4): 853–70. https://doi.org/10.1105/TPC.17.00738.

Russinova, Eugenia, Jan Willem Borst, Mark Kwaaitaal, Ana Caño-Delgado, Yanhai Yin, Joanne Chory, and Sacco C. De Vries. 2004. "Heterodimerization and Endocytosis of Arabidopsis Brassinosteroid Receptors BRI1 and AtSERK3 (BAK1)." *Plant Cell* 16 (12): 3216–29. https://doi.org/10.1105/tpc.104.025387.

Samuelson, Mariann E., and Carl Magnus Larsson. 1993. "Nitrate Regulation of Zeation Riboside Levels in Barley Roots: Effects of Inhibitors of N Assimilation and Comparison with Ammonium." *Plant Science* 93 (1–2): 77–84. https://doi.org/10.1016/0168-9452(93)90036-Y.

Sánchez-Calderón, Lenin, José López-Bucio, Alejandra Chacón-López, Alfredo Cruz-Ramírez, Fernanda Nieto-Jacobo, Joseph G. Dubrovsky, and Luis Herrera-Estrella. 2005. "Phosphate Starvation Induces a Determinate Developmental Program in the Roots of Arabidopsis thaliana." *Plant & Cell Physiology* 46 (1): 174–84. https://doi.org/10.1093/pcp/pci011.

Satbhai, Santosh B., Claudia Setzer, Florentina Freynschlag, Radka Slovak, Envel Kerdaffrec, and Wolfgang Busch. 2017. "Natural Allelic Variation of FRO2 Modulates Arabidopsis Root Growth under Iron Deficiency." *Nature Communications* 8 (1): 1–10. https://doi.org/10.1038/ncomms15603.

Shen, Chenjia, Suikang Wang, Saina Zhang, Yanxia Xu, Qian Qian, Yanhua Qi, and Jiang De An. 2013. "OsARF16, a Transcription Factor, Is Required for Auxin and Phosphate Starvation Response in Rice (Oryza sativa L.)." *Plant, Cell & Environment* 36 (3): 607–20. https://doi.org/10.1111/PCE.12001.

Shen, Chenjia, Runqing Yue, Tao Sun, Lei Zhang, Yanjun Yang, and Huizhong Wang. 2015. "OsARF16, a Transcription Factor Regulating Auxin Redistribution, Is Required for Iron Deficiency Response in Rice (Oryza sativa L.)." *Plant Science* 231 (February): 148–58. https://doi.org/10.1016/J.PLANTSCI.2014.12.003.

Shen, Chenjia, Runqing Yue, Yanjun Yang, Lei Zhang, Tao Sun, Shuanggui Tie, and Huizhong Wang. 2014. "OsARF16 Is Involved in Cytokinin-Mediated Inhibition of Phosphate Transport and Phosphate Signaling in Rice (Oryza sativa L.)." *PLOS One* 9 (11). https://doi.org/10.1371/JOURNAL.PONE.0112906.

Silva-Navas, Javier, Carlos M. Conesa, Angela Saez, Sara Navarro-Neila, Jose M. Garcia-Mina, Angel M. Zamarreño, Roberto Baigorri, Ranjan Swarup, and Juan C. Pozo. 2019. "Role of cis-Zeatin in Root Responses to Phosphate Starvation." *New Phytologist* 224 (1): 242–57. https://doi.org/10.1111/nph.16020.

Singh, Amar Pal, Yulia Fridman, Lilach Friedlander-Shani, Danuse Tarkowska, Miroslav Strnad, and Sigal Savaldi-Goldstein. 2014. "Activity of the Brassinosteroid Transcription Factors Brassinazole RESISTANT1 and Brassinosteroid INSENSITIVE1-Ethyl Methanesulfonate-SUPPRESSOR1/Brassinazole RESISTANT2 Blocks Developmental Reprogramming in Response to Low Phosphate Availability." *Plant Physiology* 166 (2): 578–88. https://doi.org/10.1104/pp.114.245019.

Singh, Amar Pal, Yulia Fridman, Neta Holland, Michal Ackerman-Lavert, Rani Zananiri, Yvon Jaillais, Arnon Henn, and Sigal Savaldi-Goldstein. 2018. "Interdependent Nutrient Availability and Steroid Hormone Signals Facilitate Root Growth Plasticity." *Developmental Cell* 46 (1): 59–72.e4. https://doi.org/10.1016/j.devcel.2018.06.002.

Singh, Amar Pal, and S. Savaldi-Goldstein. 2015. "Growth Control: Brassinosteroid Activity Gets Context." *Journal of Experimental Botany*. Oxford University Press. https://doi.org/10.1093/jxb/erv026.

Smith, Frank W., Stephen R. Mudge, Anne L. Rae, and Donna Glassop. 2003. "Phosphate Transport in Plants." *Plant & Soil* 248 (1): 71–83. https://doi.org/10.1023/A:1022376332180.

Song, Li, and Dong Liu. 2015. "Ethylene and Plant Responses to Phosphate Deficiency." *Frontiers in Plant Science* 6 (September). https://doi.org/10.3389/FPLS.2015.00796.

Song, Li, Yu Haopeng, Jinsong Dong, Ximing Che, Yuling Jiao, and Dong Liu. 2016. "The Molecular Mechanism of Ethylene-Mediated Root Hair Development Induced by Phosphate Starvation." *PLOS Genetics* 12 (7). https://doi.org/10.1371/JOURNAL.PGEN.1006194.

Sorefan, Karim, Jon Booker, Karine Haurogné, Magali Goussot, Katherine Bainbridge, Eloise Foo, Steven Chatfield, et al. 2003. "MAX4 and RMS1 Are Orthologous Dioxygenase-Like Genes That Regulate Shoot Branching in Arabidopsis and Pea." *Genes & Development* 17 (12): 1469–74. https://doi.org/10.1101/GAD.256603.

Stintzi, Annick, and John Browse. 2000. "The Arabidopsis Male-Sterile Mutant, Opr3, Lacks the 12-Oxophytodienoic Acid Reductase Required for Jasmonate Synthesis." *Proceedings of the National Academy of Sciences of the United States of America* 97 (19): 10625. https://doi.org/10.1073/PNAS.190264497.

Sun, Huwei, Xiaoli Guo, Fugui Xu, Daxia Wu, Xuhong Zhang, Manman Lou, Feifei Luo, Guohua Xu, and Yali Zhang. 2019. "Overexpression of OsPIN2 Regulates Root Growth and Formation in Response to

Phosphate Deficiency in Rice." *International Journal of Molecular Sciences* 20 (20). https://doi.org/10.3390/IJMS20205144.

Svistoonoff, Sergio, Audrey Creff, Matthieu Reymond, Cécile Sigoillot-Claude, Lilian Ricaud, Aline Blanchet, Laurent Nussaume, and Thierry Desnos. 2007. "Root Tip Contact with Low-Phosphate Media Reprograms Plant Root Architecture." *Nature Genetics* 39 (6): 792–96. https://doi.org/10.1038/ng2041.

Takei, Kentaro, Hitoshi Sakakibara, Mitsutaka Taniguchi, and Tatsuo Sugiyama. 2001. "Nitrogen-Dependent Accumulation of Cytokinins in Root and the Translocation to Leaf: Implication of Cytokinin Species That Induces Gene Expression of Maize Response Regulator." *Plant & Cell Physiology* 42 (1): 85–93. https://doi.org/10.1093/PCP/PCE009.

Thibaud, Marie Christine, Jean François Arrighi, Vincent Bayle, Serge Chiarenza, Audrey Creff, Regla Bustos, Javier Paz-Ares, Yves Poirier, and Laurent Nussaume. 2010. "Dissection of Local and Systemic Transcriptional Responses to Phosphate Starvation in Arabidopsis." *Plant Journal* 64 (5): 775–89. https://doi.org/10.1111/J.1365-313X.2010.04375.X.

Ticconi, C. A., C. A. Delatorre, and S. Abel. 2001. "Attenuation of Phosphate Starvation Responses by Phosphite in Arabidopsis." *Plant Physiology* 127 (3). doi:10.1104/pp.010396.

Ticconi, Carla A., Rocco D. Lucero, Siriwat Sakhonwasee, Aaron W. Adamson, Audrey Creff, Laurent Nussaume, Thierry Desnos, and Steffen Abel. 2009. "ER-Resident Proteins PDR2 and LPR1 Mediate the Developmental Response of Root Meristems to Phosphate Availability." *Proceedings of the National Academy of Sciences of the United States of America* 106 (33): 14174–79. https://doi.org/10.1073/PNAS.0901778106.

Umehara, Mikihisa, Atsushi Hanada, Hiroshi Magome, Noriko Takeda-Kamiya, and Shinjiro Yamaguchi. 2010. "Contribution of Strigolactones to the Inhibition of Tiller Bud Outgrowth under Phosphate Deficiency in Rice." *Plant & Cell Physiology* 51 (7): 1118–26. https://doi.org/10.1093/PCP/PCQ084.

Umehara, Mikihisa, Atsushi Hanada, Satoko Yoshida, Kohki Akiyama, Tomotsugu Arite, Noriko Takeda-Kamiya, Hiroshi Magome, et al. 2008. "Inhibition of Shoot Branching by New Terpenoid Plant Hormones." *Nature* (455) (7210): 195–200. https://doi.org/10.1038/nature07272.

Vijayan, Perumal, Jay Shockey, C. André Lévesque, R. James Cook, and John Browse. 1998. "A Role for Jasmonate in Pathogen Defense of Arabidopsis." *Proceedings of the National Academy of Sciences of the United States of America* 95 (12): 7209–14. https://doi.org/10.1073/PNAS.95.12.7209/ASSET/F4645A3E-84E9-40BD-ACE5-2F1EF8AE1D0A/ASSETS/GRAPHIC/PQ1181139004.JPEG.

Vinod, K. K., and Sigrid Heuer. 2012. "Approaches towards Nitrogen- and Phosphorus-Efficient Rice." *AoB Plants* 2012 (0). https://doi.org/10.1093/AOBPLA/PLS028.

Wang, Kevin L. C., Hai Li, and Joseph R. Ecker. 2002. "Ethylene Biosynthesis and Signaling Networks." *The Plant Cell* 14 (suppl_1): S131–51. https://doi.org/10.1105/TPC.001768.

Wang, Qian, Yu Feifei, and Qi Xie. 2020. "Balancing Growth and Adaptation to Stress: Crosstalk between Brassinosteroid and Abscisic Acid Signaling." *Plant, Cell & Environment* 43 (10): 2325–35. https://doi.org/10.1111/PCE.13846.

Wang, Xiaoyue, Zhen Wang, Zai Zheng, Jinsong Dong, Li Song, Liqian Sui, Laurent Nussaume, Thierry Desnos, and Dong Liu. 2019. "Genetic Dissection of Fe-Dependent Signaling in Root Developmental Responses to Phosphate Deficiency." *Plant Physiology* 179 (1): 300–16. https://doi.org/10.1104/pp.18.00907.

Wang, Xuming, Keke Yi, Yong Tao, Fang Wang, Zhongchang Wu, Dean Jiang, Xin Chen, Lihuang Zhu, and Ping Wu. 2006. "Cytokinin Represses Phosphate-Starvation Response through Increasing of Intracellular Phosphate Level." *Plant, Cell & Environment* 29 (10): 1924–35. https://doi.org/10.1111/J.1365-3040.2006.01568.X.

Wang, Yu, Jia Jian Cao, Kai Xin Wang, Xiao Jian Xia, Kai Shi, Yan Hong Zhou, Jing Quan Yu, and Jie Zhoua. 2019. "BZR1 Mediates Brassinosteroid-Induced Autophagy and Nitrogen Starvation in Tomato." *Plant Physiology* 179 (2): 671–85. https://doi.org/10.1104/pp.18.01028.

Wang, Zhengrui, A. B. M.Moshiur Rahman, Guoying Wang, Uwe Ludewig, Jianbo Shen, and Günter Neumann. 2015. "Hormonal Interactions during Cluster-Root Development in Phosphate-Deficient White Lupin (Lupinus Albus L.)." *Journal of Plant Physiology* 177 (April): 74–82. https://doi.org/10.1016/J.JPLPH.2014.10.022.

Wang, Zhi Yong, Takeshi Nakano, Joshua Gendron, Junxian He, Meng Chen, Dionne Vafeados, Yanli Yang, et al. 2002. "Nuclear-Localized BZR1 Mediates Brassinosteroid-Induced Growth and Feedback Suppression of Brassinosteroid Biosynthesis." *Developmental Cell* 2 (4): 505–13. https://doi.org/10.1016/S1534-5807(02)00153-3.

Wang, Zhi Yong, Hideharu Seto, Shozo Fujioka, Shigeo Yoshida, and Joanne Chory. 2001. "BRI1 Is a Critical Component of a Plasma-Membrane Receptor for Plant Steroids." *Nature* 410 (6826): 380–83. https://doi.org/10.1038/35066597.

Xiao, Fei, Yiyi Zhang, Shuangshuang Zhao, and Huapeng Zhou. 2021. "MYB30 and ETHYLENE INSENEITIVE3 Antagonistically Regulate Root Hair Growth and Phosphorus Uptake under Phosphate Deficiency in Arabidopsis." 16 (7). https://doi.org/10.1080/15592324.2021.1913310.

Xie, Dao Xin, Bart F. Feys, Sarah James, Manuela Nieto-Rostro, and John G. Turner. 1998. "COI1: An Arabidopsis Gene Required for Jasmonate-Regulated Defense and Fertility." *Science* 280 (5366): 1091–94. https://doi.org/10.1126/SCIENCE.280.5366.1091/ASSET/F0BFC4B3-8E2A-4352-B1A2-A7C6AF673824/ASSETS/GRAPHIC/SE2086511003.JPEG.

Xie, Ye, Huijuan Tan, Zhaoxue Ma, and Jirong Huang. 2016. "DELLA Proteins Promote Anthocyanin Biosynthesis via Sequestering MYBL2 and JAZ Suppressors of the MYB/BHLH/WD40 Complex in Arabidopsis thaliana." *Molecular Plant* 9 (5): 711–21. https://doi.org/10.1016/J.MOLP.2016.01.014.

Xu, Juan, and Shuqun Zhang. 2015. "Ethylene Biosynthesis and Regulation in Plants." *Ethylene in Plants*: 1–25. https://doi.org/10.1007/978-94-017-9484-8_1.

Yan, Huimin, Yale Wang, Bo Chen, Weijie Wang, Hongzheng Sun, Huwei Sun, Junzhou Li, and Quanzhi Zhao. 2022. "OsCKX2 Regulates Phosphate Deficiency Tolerance by Modulating Cytokinin in Rice." *Plant Science : An International Journal of Experimental Plant Biology* 319 (June). https://doi.org/10.1016/J.PLANTSCI.2022.111257.

Yan, Yuanxin, Shawn Christensen, Tom Isakeit, Jürgen Engelberth, Robert Meeley, Allison Hayward, R. J. Neil Emery, and Michael V. Kolomiets. 2012. "Disruption of OPR7 and OPR8 Reveals the Versatile Functions of Jasmonic Acid in Maize Development and Defense." *The Plant Cell* 24 (4): 1420–36. https://doi.org/10.1105/TPC.111.094151.

Ye, Ying, Jing Yuan, Xiaojian Chang, Meng Yang, Lejing Zhang, Kai Lu, and Xingming Lian. 2015. "The Phosphate Transporter Gene OsPht1;4 Is Involved in Phosphate Homeostasis in Rice." *PLOS ONE* 10 (5): e0126186. https://doi.org/10.1371/JOURNAL.PONE.0126186.

Yin, Yanhai, Zhi Yong Wang, Santiago Mora-Garcia, Jianming Li, Shigeo Yoshida, Tadao Asami, and Joanne Chory. 2002. "BES1 Accumulates in the Nucleus in Response to Brassinosteroids to Regulate Gene Expression and Promote Stem Elongation." *Cell* 109 (2): 181–91. https://doi.org/10.1016/S0092-8674(02)00721-3.

Yoneyama, Kaori, Koichi Yoneyama, Yasutomo Takeuchi, and Hitoshi Sekimoto. 2007. "Phosphorus Deficiency in Red Clover Promotes Exudation of Orobanchol, the Signal for Mycorrhizal Symbionts and Germination Stimulant for Root Parasites." *Planta* 225 (4): 1031–38. https://doi.org/10.1007/S00425-006-0410-1.

Zhang, Yu, Ting-Ting Li, Lin-Feng Wang, Jia-Xing Guo, Kai-Kai Lu, Ru-Feng Song, Jia-Xin Zuo, Hui-Hui Chen, and Wen-Cheng Liu. 2022. "Abscisic Acid Facilitates Phosphate Acquisition through the Transcription Factor <scp>ABA INSENSITIVE5</Scp> in Arabidopsis." *The Plant Journal*, May. https://doi.org/10.1111/TPJ.15791.

Zhao, Hongyun, Kaili Nie, Huapeng Zhou, Xiaojing Yan, Qidi Zhan, Yuan Zheng, and Chun Peng Song. 2020. "ABI5 Modulates Seed Germination via Feedback Regulation of the Expression of the PYR/PYL/RCAR ABA Receptor Genes." *The New Phytologist* 228 (2): 596–608. https://doi.org/10.1111/NPH.16713.

Zheng, Hongyan, Xiaoying Pan, Yuxia Deng, Huamao Wu, Pei Liu, and Xuexian Li. 2016. "AtOPR3 Specifically Inhibits Primary Root Growth in Arabidopsis under Phosphate Deficiency." *Scientific Reports* 6 (1): 1–11. https://doi.org/10.1038/srep24778.

Zou, Linyuan, Minghao Qu, Longjun Zeng, and Guosheng Xiong. 2020. "The Molecular Basis of the Interaction between Brassinosteroid Induced and Phosphorous Deficiency Induced Leaf Inclination in Rice." *Plant Growth Regulation* 91 (2): 263–76. https://doi.org/10.1007/S10725-020-00604-5.

6 The Pivotal Role of Phosphate in Shaping Beneficial Plant-Microbe Interactions

Arianna Capparotto and Marco Giovannetti

6.1 TRICKS AND TIPS OF A LONG-LASTING INTERACTION

Arbuscular mycorrhizal (AM) symbiosis can contribute to the majority of phosphate needed by the plant (Paries and Gutjhar 2023; Giovannetti et al. 2017). This is possible thanks to physical and biochemical fungal properties: on the one hand AM fungal mycelia can explore soil niches better than plant roots, extending further than the phosphate depletion area; on the other hand, it was suggested that AM fungi are also able to have access to soil organic P throughout the deployment of acid phosphatases (Feng et al. 2003; Shibata and Yano 2003; Sato et al. 2015). Anyhow, since it is known that AM fungi can host both endosymbionts (Salvioli di Fossalunga et al. 2017) and a community of conserved hyphal associated-bacteria (Emmett, Lévesque-Tremblay, and Harrison 2021) shaped by fungal exudates (Zhang, Feng, and Declerck 2018), it is tempting to speculate that they could have a role in mediating or enhancing phosphate mineralisation (Wang et al. 2022).

Fungi have a wide range of phosphate transporters that allow them to absorb soil phosphate (Xie et al. 2016; Benedetto et al. 2005) from extraradical mycelia and translocated it in the form of polyphosphates until reaching the intraradical mycelia. Here, the polyphosphate granules are cleaved, creating a large pool of negative charges that are counterbalanced by a simultaneous uptake of K^+, Ca^{2+}, Mg^{2+}, and Na^+ (Kikuchi et al. 2014). Once the phosphate reaches the fungal arbuscule, it is released into the periarbuscular membrane space with machinery that is yet to be discovered. Interestingly, fungal phosphate transporters have been shown to be expressed both in the extraradical and in the intraradical mycelium, suggesting a possible role in the reabsorption of phosphate (Fiorilli, Lanfranco, and Bonfante 2013). The water flow, mediated by the action of aquaporins, or an indirect role played by aquaporins, could represent one of the pivotal mechanisms allowing phosphate transport along the fungal mycelium: indeed, knockdown of a *Rhizophagus clarus* aquaglyceroprotein impaired polyphosphate translocation (Kikuchi et al. 2016), and recent experimental evidence demonstrated that water transported by arbuscular mycorrhizal fungi (AMF) accounted for 34.6% of the water transpired by host plants (within the interaction between *Rhizophagus intraradices* and *Avena barbata*) (Kakouridis et al. 2022).

If the lack of genetic tools has slowed our understanding of the molecular mechanisms driving the mycorrhizal phosphate flow from the fungal side, the crucial components of the plant phosphate uptake machinery are well described. Plants that have lost the capacity to form AM symbiosis have also lost a copy of a phosphate transporter that is crucial for the uptake of phosphate from the periarbuscular space. This genetic evidence is well supported by several molecular and physiological studies on different plant species (Rausch et al. 2001; Harrison, Dewbre, and Liu 2002; Paszkowski et al. 2002; Nagy et al. 2005; Volpe et al. 2016; Sawers et al. 2017) demonstrating that AM-specific PHT1 phosphate transporters are necessary for functional arbuscules. The phosphate transporter not only mediates the trafficking of the substrate but also shows a role, either direct or indirect, in regulating the correct formation and hyphal branching of the arbuscule. Thus, PHT1 phosphate transporters have been suggested to act as transceptors, similar to what happens in PHO1 yeast proteins (Popova et al. 2010). As transceptors, the sole binding of the protein to the substrate can trigger a signalling cascade in the cytoplasm. A similar action has been suggested for OsPT13 in rice: indeed, the protein is not mediating any phosphate transfer, but *ospt13* shows more degenerated arbuscules than the wild type (Yang et al. 2012). In *Lotus japonicus*, AM-specific phosphate transporters show a small accumulation of transcripts in the root tips, depending on the external phosphate concentration, and possibly regulate root system architecture (Volpe et al. 2016). Interestingly, phosphate sensing is intertwined with other nutrient sensing, such as ammonium, and external nutrient availability plays a key role in defining which element is the driving force that regulates the arbuscular lifetime. This could be interpreted as a faceting of Liebig's law of the minimum: when nitrogen is the limiting growth factor, Medicago *pt4* plants show functional and fully develop arbuscules (Javot et al. 2011; Breuillin-Sessoms et al. 2015). Altogether, this data confirms that local phosphate levels have a great impact on the development and the functioning of AM symbiosis, but this represents only one side of the actions played by phosphate within this relationship.

In addition, phosphate has also a strong and well-known impact on the symbiosis throughout signalling events at the plant systemic level: higher external phosphate levels inhibit AM colonisation in a dose-dependent manner (Breuillin et al. 2010; Balzergue et al. 2011). This inhibition has been demonstrated to be systemic by split-root experiments, in which only one-half of the root system is exposed to high Pi concentration, but AM colonisation is impaired on the whole root system (Branscheid et al. 2010). Different Pi-starvation signals have been investigated in the last decades, and a strong correlation between phosphate levels and miR399 or strigolactone levels was shown. Nevertheless, taking them singularly, no significant effect was demonstrated on the systemic inhibition of AM colonisation by high external phosphate: neither miR399 overexpression (Branscheid et al. 2010) nor synthetic strigolactone application (Breuillin et al. 2010; Balzergue et al. 2011) had a significant effect in different mycorrhizal plant systems.

Interesting further clues are given by the studies on PHR1: this protein is a member of the MYB transcription factor family and one of the master regulators of plant transcriptional responses to low phosphate (Rubio et al. 2001). Its sequence and mode of action are conserved in different plant species where PHR1 controls the expression of the majority of genes responsive to phosphate starvation (Morcuende et al. 2007; Bustos et al. 2010; Pant et al. 2015). Surprisingly, PHR1 transcript levels are not modulated by phosphate levels, but its activity is modulated by the SPX protein that sequesters it when cellular phosphate concentration is high. Recently, a pivotal role for PHR and SPX proteins in regulating a transcriptional cascade controlling AM symbiosis has been shown for rice, *Medicago*, *Lotus*, and tomato (Shi et al. 2021; Wang et al. 2021; Das et al. 2022; Liao et al. 2022), confirming the crucial role for the plant phosphate homeostasis in mediating AM. It remains unclear what the first metabolic changes mediating plant phosphate sensing are. One possibility is given by the role of inositol pyrophosphates (Gupta and Laxman 2021) as key molecules buffering cellular responses to phosphate (Wild et al. 2016; Jung et al. 2018). PP-InsPs are a family of signalling molecules characterised by the combinatorial placement of phosphate groups around the inositol six-carbon ring (Shears 2015). Plant cells with higher phosphate concentrations accumulate higher levels of inositol pyrophosphates that can selectively mediate the binding of SPX proteins to PHR, preventing the transcription of Pi-starvation response (PSR) genes (Jung et al. 2018). Consistently, mutants unable to synthesise PP-InsPs precursor show a constitutive PSR, despite the high Pi concentration (Jung et al. 2018). Understanding the first metabolic changes regulating mycorrhizal inhibition at high phosphate levels will possibly pave the way for future agricultural applications, keeping in mind the extent of natural variation of plant responses to phosphate (Giovannetti et al. 2019) and AM fungi (Lehnert et al. 2017; Martín-Robles et al. 2018).

6.2 NO ONE CAN WHISTLE A SYMPHONY: INTEGRATING A PROKARYOTES-BASED SOLUTION TO FACE NUTRITIONAL CHALLENGES

Fungi do not act as isolated entities in providing plant resilience under stress conditions; rather they interact with bacteria and archaea constituting an holobiontic system that should be considered in its integrity (Vandenkoornhuyse et al. 2015). In the soil, bacteria provide several ecosystem services such as the regulation of nutrient availability, the decomposition of organic matter, the degradation of pollutants, and the turnover of important inorganic molecules by taking part in biogeochemical cycles. In addition, they establish a mutual interplay with the resident living organisms including plants and fungi, both supporting their development and nutrition, and providing the first line of defence against pathogens (Karimi et al. 2018; Fuke et al. 2021; Coban, De Deyn, and van der Ploeg 2022). Due to their peculiar involvement in preserving soil structure and quality, microbes have been recognised as one of the indicators for the assessment of soil health (Fierer, Wood, and Bueno de Mesquita 2021; Timmis and Ramos 2021). In this perspective, soil biodiversity has been included by the European Union in their 2022 soil deal as one of the metrics to monitor soil health.

Although the distribution of prokaryotes in the soil is not homogeneous, it could be considered as two different groups: the bulk soil community and the rhizosphere community (Bulgarelli et al. 2013; Lidbury, Raguideau et al. 2022). The bulk component is the part of the soil that is not influenced by the plant root apparatus. The resident bacteria of this community are shaped by phenomena of dispersal and evolution, while its abundance is modulated by events of ecological drift and selection (Hanson et al. 2012; Fitzpatrick et al. 2020; Gupta et al. 2021). Even if environmental factors and geographical variation represent the main driving forces acting in determining the microbiome composition of the bulk soil, antagonistic and beneficial interactions between microorganisms can alter its structure (Fitzpatrick et al. 2020; Zhou et al. 2022). According to Trivedi et al. (2020), the bulk soil can be referred to as the microbes' "seed bank", a mixture of bacteria and archaea with different functions that serve as reservoirs for the assembly of the rhizosphere community.

On the other hand, the rhizosphere is the part of the soil located in proximity to the root apparatus and is directly influenced by the metabolic, biochemical, and respiratory activity of plants (Bulgarelli et al. 2013). Given the myriad of microorganisms it hosts and its involvement in the interplay between the soil and plant roots, the rhizosphere is considered to be one of the most dynamic interfaces on Earth (Philippot et al. 2013; Lidbury, Raguideau, et al. 2022). Even if soil features and environmental factors are still the main drivers of bacterial community assembly in this soil layer, plants however exert a significant selective force in a species-dependent manner and, to a less extent, in a genotype-dependent manner (Walters et al. 2018; Trivedi et al. 2020). The main plant pressure that acts in selecting the bacteria of the bulk soil to become part of the rhizosphere is rhizodeposition, considered as the secretion of several compounds (such as sugars, organic acid ions, organic carbon molecules, and nitrogen) by the plant root apparatus (Dennis, Miller, and Hirsch 2010; Pausch and Kuzyakov 2018). In fact, the accumulation of these substances in the proximity of the roots has been correlated with an accumulation of organotrophic bacteria, in particular those belonging to the genera Proteobacteria – a large fraction of which is known to be involved in the utilisation of root-derived carbon substrates (Bulgarelli et al. 2013; Philippot et al. 2013; Lidbury, Raguideau, et al. 2022; Ling, Wang, and Kuzyakov 2022). Despite the existence of a general pattern of rhizodeposit profiles, their amount and composition can be affected by several factors and can vary in time and space, and according to different biotic and abiotic stresses acting on plants (Jones, Hodge, and Kuzyakov 2004; Dennis, Miller, and Hirsch 2010; Cotton et al. 2019). In turn, root exudate modifications of stressed plants can select bulk soil bacteria for some advantageous functional traits, reshaping the rhizosphere community's composition and finally improving plant fitness (Yan et al. 2017; Matthews et al. 2019). The model at the base of this mechanism has been termed the "cry-for-help" hypothesis because such a change in the secretory activity under stress conditions can be considered as the plant's mayday signal or as its effort to call for backup (Rolfe, Griffiths, and Ton 2019).

6.3 THE CRY FOR PHOSPHATE

A practical example of the cry-for-help model occurs under phosphate deficiency, one of the main conditions limiting plant growth and development. Despite the large amount of phosphorus contained in the soil, only a small fraction (less than 20% of the total) is present as mineral phosphate available to plants, while the rest is in the form of unavailable organic phosphate (Schachtman, Reid, and Ayling 1998; Vance, Uhde-Stone, and Allan 2003). Several studies have demonstrated how plants respond to P deficiency by releasing different types and amounts of exudates at different growth stages, modulating the microbiome assembly in the rhizosphere (Pantigoso, Manter, and Vivanco 2020; Castagno et al. 2021; Amy et al. 2022). The nature of the compounds that have been shown to affect the soil microbiome is diverse and is linked either to plant hormone metabolism or to nutrient cycling. For example, under P starvation conditions, an increase in gamma-aminobutyric acid, C-containing compounds (malate and citrate), and strigolactones has been detected (Carvalhais et al. 2011; Nasir et al. 2019). A study conducted on rice demonstrated the involvement of strigolactones in shaping the rhizosphere community. In particular, compared to wild-type plants, both rice mutants lacking the gene for strigolactones biosynthesis *DWARF 17* (d17) and receptor *DWARF 14* (d14) are impaired in the ability to accumulate beneficial bacterial strains in their rhizosphere (Nasir et al. 2019). Another compound with a crucial effect on rhizospheric microbes is citrate: its exudation has been correlated with an increased expression of the LHA1 PM H^+-ATPase gene as well as with the consequent expression of H^+-ATPase protein, which is involved in the proton extrusion and, therefore, in the acidification of the rhizospheric soil (Tomasi et al. 2009; Lambers, Clements, and Nelson 2013). In turn, evidence has shown that citrate accumulation in the roots' tissues occurs as a consequence of the citrate catabolism down-regulation (Kania et al. 2003). Another example is coumarin scopoletin, released under conditions of both phosphorous and iron deficiency by the activity of MYB72 and the β-glucosidase BGLU42 (Stringlis et al. 2018). This metabolite has an antimicrobial activity that favours the selection of beneficial microorganisms by the plant. Finally, plant exudates stimulate the expression of bacterial genes that encode motility and adhesion, affecting the colonisation of the plant's internal compartments (Coutinho et al. 2015; Trivedi et al. 2020).

6.4 RHIZOBACTERIA AT WORK

The main strategy applied in agriculture to face phosphate deficiency in the soil relies on the use of phosphate-based fertilisers. However, it has been observed that the majority of phosphate applied in the fields is lost before consumption, causing severe problems in terms of microbial biodiversity loss in fields and eutrophication events in the water (Kang et al. 2011; Weihrauch and Opp 2018). Given the ability of some soil microorganisms to solubilise and mineralise the organic insoluble phosphorus, bacteria-based solutions have been pointed out as possible eco-friendly candidates for more sustainable food production technology (Nassal et al. 2018; Khan 2022). Bacteria that are able to convert organic phosphorus into an accessible form for the plant have been termed phosphobacteria (PSB). They belong to a larger class of plant growth-promoting rhizobacteria (PGPR) that are known to increase plant growth by producing phytostimulators and plant growth regulators such as indole-acetic acid (IAA), gibberellic acid, cytokinins, and ethylene (Singh, Parmar, and Kuhad 2011; Kenneth et al. 2019). Among PSB, *Bacillus* and *Pseudomonas* are the genera containing the largest number of species reported to be phosphate solubilisers (Fierer, Wood, and Bueno de Mesquita 2021; Timmis and Ramos 2021). However, this classification also includes other genera, such as *Burkholderia, Rhizobium, Azotobacter, Enterobacter, Erwinia, Kushneria, Paenibacillus, Ralstonia, Rhodococcus, Serratia, Bradyrhizobium, Salmonella, Sinomonas,* and *Thiobacillus*

(Alori, Glick, and Babalola 2017; Castagno et al. 2021). PSB solubilises phosphate via different mechanisms. The first one relies on the production of low-molecular-weight organic acids such as gluconic, acetic, formic, lactic, glycolic, malic, and citric acids. These can decrease the environmental pH from 7.0 to 2.0, favouring the transformation of PO_4^{3-} into its accessible form to plants, HPO_4^{2-} and $H_2PO_4^-$ (Suleman et al. 2018; Prabhu et al. 2019; Castagno et al. 2021). Among organic acids, PSB mainly produce gluconic acid thanks to the presence of a membrane-bound enzyme, the glucose dehydrogenase, which oxidates glucose (Deppenmeier, Hoffmeister, and Prust 2002; Rasul et al. 2019). A study conducted to understand which plant-released carbon source has the best effect on P solubilising activity remarkably revealed that P mineralisation efficiency was highest when glucose was supplied to PSB in a liquid medium (Suleman et al. 2018). In addition to this capacity, PSB can contribute to phosphate solubilisation throughout chelation. Some organic acids such as 2-ketogluconic, humic, and fulvic acids are able to chelate cations in the P surface through their hydroxyl and carboxyl groups and to convert them into soluble molecules (Singh, Parmar, and Kuhad 2011; Nadiéline et al. 2019). Other less impacting mechanisms include the production of inorganic acids able to react with P-containing molecules, favouring their mineralisation and the extrusion of H^+ ions produced during $NH4^+$ assimilation (Prabhu et al. 2019; Castagno et al. 2021). Furthermore, conversely to plants, several PSB can produce the so-called phosphohydrolases or phytases, a class of enzymes involved in the mineralisation of phytate (*myo*-inositol phosphate), one of the most common forms of unavailable phosphate present in the soil (Singh et al. 2020; Lidbury, Scanlan, et al. 2022). Phosphatases can be periplasmic or outer membrane-bound enzymes and are involved in the cleavage of Pi moiety upon phosphate deficiency, while at standard Pi concentrations, their action is inhibited (Lidbury et al. 2016; Lidbury, Scanlan, et al. 2022). This reciprocal communication between rhizobacteria and plants is at the base of a mutual relationship on which both parts rely. On the one hand, plants select the bacteria that they need in order to face particular challenges by secreting specific exudates and, on the other hand, bacteria use those chemical compounds for energy and biomass production (Haichar et al. 2008; Bulgarelli et al. 2013). Altogether, the underlying mechanism at the base of the plant–phosphobacteria interaction can be better exploited as a possible biofertiliser to improve the phosphorous uptake by plants.

6.5 HOME SWEET HOME: THE PI-MEDIATED BACTERIAL ACCOMMODATION INSIDE THE HOST

Just as for the rhizosphere community, the endosphere compartments can be also colonised by bacteria in a host-genotype-dependent manner (Trivedi et al. 2020; Brachi et al. 2022). Endophytic bacteria are referred to as microorganisms living in the plant's internal organs without causing them any harm. Their role here is to supply nutrients such as phosphorous, and, for this reason, they are specifically recruited by the plant (Compant et al. 2021; Dudeja et al. 2021). As described above, rhizodeposition is a crucial process used by plants to select particular types of microorganisms in the soil. In the last decade, many studies have started to elucidate the molecular connection between the ability of the host to perceive phosphate levels and the downstream plant exudation activity (Ziegler et al. 2016; Stringlis et al. 2018; Chutia, Abel, and Ziegler 2019).

It is well known that plants possess an adaptive PSR mechanism to cope with phosphate deficiency in the surrounding environment. With regard to this, light has been shed on the role of phosphate response (PHR) proteins in the modulation of the transcriptomic and metabolic changes induced by phosphate status alterations. PHR proteins have been demonstrated to bind to a specific region, the P1BS motif, located upstream of the promoter of several phosphate starvation-induced (PSI) genes (Rubio et al. 2001; Kiers, Hutton, and Denison 2007).

New pieces of evidence suggest that PSR has a role in the alteration of plant exudate profiles, which in turn culminate with the rearrangement of rhizospheric bacterial communities (Chutia, Abel, and Ziegler 2019). Among the most widely studied plant exudates, there is the class of coumarins, which are derived from the phenylpropanoid pathway. Recent studies demonstrated that, according to the different Fe levels, PHR1 is involved in the Pi-deficiency-induced changes in coumarin biosynthesis. For example, *phr1* mutants grown on low Pi/high Fe media have shown an accumulation of coumarin precursors, indicating that in the wild-type plants, PHR1 has a role in the down-regulation of coumarin biosynthesis under Pi deficiency conditions (Ziegler et al. 2016; Chutia, Abel, and Ziegler 2019).

Thus, thanks to the mechanism described above, bacteria are recruited in the rhizosphere, where they can be additionally selected to be hosted in the internal plant tissues (Figure 6.1). Several

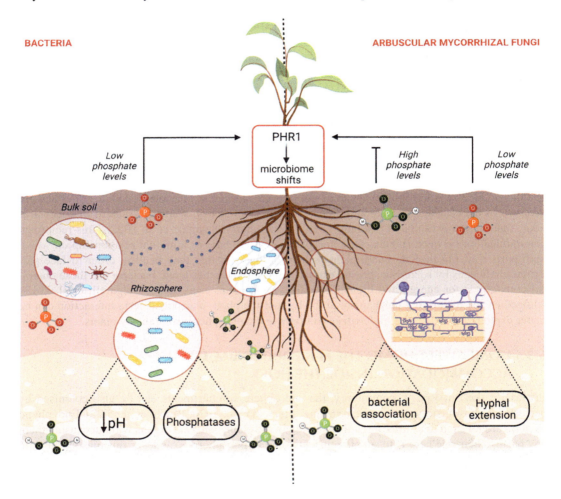

FIGURE 6.1 Phosphate deficiency-induced microbiome shifts. The plant-associated microbiome is a dynamic entity that varies in response to environmental changes and plant nutritional status. Given the small proportion of available phosphate in the soil, plants evolved a sophisticated mechanism to both sense the Pi levels and attract other microorganisms with the ability to solubilise this nutrient and increase its uptake. Low levels of $H_2PO_4^-$ (in green) trigger the activation of the PHR1 master regulator, which activates phosphate starvation-induced (PSI) genes. Among those, some are involved in the modulation of the root secretory activity, which in turn plays a role in the selection of rhizosphere bacteria that improve plant fitness. Furthermore, PHR protein activation by the change in phosphate homeostasis allows the colonisation by AM fungi, which form a tight hyphal network crucial to explore new soil niches. In the soil, microorganisms decrease the environmental pH and release acid phosphatase enzymes, favouring the conversion of the unavailable PO_4^{3-} (in red) moiety in its available forms, HPO_4^{2-} and $H_2PO_4^-$ (in green). On the other hand, when the nutrient concentration in the soil is high, phosphate inhibits the PHR1 action, and, as a consequence, it impairs AM colonisation and, indirectly, alters the soil microbial community.

studies have highlighted the involvement of PHRs proteins in the Pi-dependent root endophytes accommodation (Fabiańska, Sosa-Lopez, and Bucher 2019; Thalineau et al. 2020), denoting the connection between phosphate homeostasis and plant immune system in modulating this fine-tuning. A recent study demonstrated that *Arabidopsis* PSR mutants assemble an atypical bacterial community in the root compartments of phosphate-replete soil, following the colonisation with a synthetic bacterial community (SynCom) (Castrillo et al. 2017). In general, phosphate-utiliser bacteria compete for the phosphate uptake in the soil, leading to the activation of PSR (Richardson et al. 2009; Castrillo et al. 2017). However, PSR impairment has been accompanied by transcriptional changes in plant defence mechanisms, culminating in enhanced immune activation and vice versa (Castrillo et al. 2017; Dindas et al. 2022). Therefore, SynCom enhances the activity of PHR1, which directly regulates a set of plant immune system genes inextricably connected to the assembly of the root microbiome (Castrillo et al. 2017). In particular, PHR1 negatively regulates some plant immune system genes, which increase the plant's susceptibility to pathogens attacks but at the same time facilitate the entrance of beneficial microbes (Castrillo et al. 2017; Liu et al. 2018; Finkel et al. 2019). A similar publication reported the effects of PSR gene impairment on the bacterial composition of the plant microbiota, independently of the P levels, which showed a strong dependency of bacterial microbiota on the soil bacterial community composition, in contrast to the fungal microbiota, meaning that plants are much more selective with fungi than with bacteria (Finkel et al. 2019). Once inside the plant, the host maintains a stable metabolite availability to alleviate the selective pressures on endophytes while assuring their continuous biosynthetic activity over time (Cole et al. 2017; Liu et al. 2018). However, the experiments on bacterial and fungal community assembly under P-deficient conditions suffer from low reproducibility, therefore more deep insight into this topic needs to be achieved (Schloss 2018). Additionally, further investigations should be performed to fully elucidate the link between the PSR and the genetic component regulating rhizodeposition.

6.6 THE WHOLE IS GREATER THAN THE SUM OF THE PARTS: DEVELOPMENT OF SYNTHETIC MICROBIAL COMMUNITIES (SYNCOM) TO INCREASE PI MINERALISATION AND UPTAKE BY THE PLANT

In the last few years, advances in sequencing technologies and omics approaches have increased the microbiome community resolution in the soil and in different plant organs (Mitter et al. 2017; Finkel et al. 2017; Hartman et al. 2017), but a full picture of the tight network of interactions taking place in the belowground still need to be achieved (Zhuang et al. 2021). Thus, despite the increasing knowledge of natural microbial communities, their intrinsic complexity and variability have made them a difficult system to study and define (Großkopf and Soyer 2014). To overcome the complexity found in nature, a reductionist approach based on developing synthetic microbial communities has been implemented in agricultural practices (Großkopf and Soyer 2014; Vorholt et al. 2017). In this scenario, several types of research have focused on the isolation and characterisation of PSB strains to further study their effects on plant growth promotion and phosphate solubilisation, alone or in combination with other microorganisms (Kaur and Sudhakara Reddy 2014; Prabhu et al. 2019). SynCom can be divided into low (counting between 1 and 10 members) and high-complexity communities (containing > 100 microorganisms) (Vorholt et al. 2017). Several approaches have been applied over the years to formulate SynCom with the best trade-off between their plant growth-promoting effects and their survival rate when inoculated in the plant or in the soil (Khan 2022). Top-down and bottom-up methods have been implemented to dissect the natural microbial community and identify the best candidates to use in agriculture (Vorholt et al. 2017; Khan 2022). Among those, cultivation-dependent approaches and sequencing technologies are in general applied to determine bacterial strains that increase plant fitness under different biotic or abiotic stresses (Bai et al. 2015). The selected bacterial isolates are then grouped in collections to further screen their combinatorial effect, according to a bottom-up procedure. Tests of this type include binary microbe-microbe interaction assays, binary or complex plant-microbe interaction assays, and functional genomic approaches (Baldan et al. 2015; Samad et al. 2017; Vorholt et al. 2017).

Practical applications of PSB-based SynCom have already demonstrated their positive effects on plant growth and nutrition. For example, the inoculation of two PSB strains in the soil of maize and wheat crops at different agroclimatic locations favoured an improvement in crop productivity and soil fertility, and an increase in grain yield (Kaur and Sudhakara Reddy 2014). However, this beneficial effect was more evident when the treatment was coupled with a rocky phosphate supplement, probably because of the better P utilisation from the pool of nutrients (Kaur and Sudhakara Reddy 2014). PSB can also improve the germination and vigour of wheat seedlings when compared to the non-inoculated ones, and grain yield increased to a larger extent in pots (38.5%) compared to the field condition (17–18%), demonstrating the efficiency gaps that still limit the use of PSB in less controlled environments (Suleman et al. 2018). In a similar experiment, PSB significantly neutralised the plant growth inhibitory effects of soil liming produced as a side effect of calcification and salinisation processes, pointing out their possible application in a large range of conditions (Adnan et al. 2020). Additionally, the use of these biofertilisers significantly increased the shoot height, the shoot dry weight, the pigment content, the P uptake, and the antioxidant activity of inoculated rice (*Oryza sativa L.*) plants. Nevertheless, the maximum level of increase of those parameters was achieved when the treatment was performed in combination with 50% less chemical fertiliser (Rawat, Shankhdhar, and Shankhdhar 2021). Plant growth-promoting effects, increased P shoot accumulation, and increased P mineralisation activity have also been documented in tomatoes after the inoculation of the plants with two different *Pseudomonas* strains, HT-RU47 and RU47, known to play a role in Pi nutrition (Nassal et al. 2018). Overall, PSB represent an encouraging sustainable tool to increase crop yield and health, especially when tested in combination with reduced doses of common-used fertilisers.

Altogether, because rock phosphate deposits are limited, it is crucial to take into account the complexity of plant–fungal–bacterial interactions as a source for improving plant phosphate use efficiency and to be aware that cocktails of more than one beneficial microbe provide more functional benefits to the plant than a single-strain inoculant (Kaminsky et al. 2019). Fungi and fungal hyphae work as a bridge between plants and bacteria, and, through their extraradical hyphae, they secrete C-based metabolites, selecting microbes with which to cooperate and improve their capacity of exchanging mineral nutrients with plant carbohydrates. In this scenario, PSB activity plays a fundamental role in enhancing the mineralisation of organic P and, together with AM fungi, could allow to grow healthier plants with a reduced environmental impact (Zhang et al. 2016; Zhang, Feng, and Declerck 2018; Giovannetti et al., 2023). A common effort to disentangle this complex matrix is needed to improve the sustainability of the agricultural system without harnessing soil health.

6.7 ACKNOWLEDGEMENTS

Figure 6.1 was created with Biorender.com. This work was supported by grants from the University of Padova, Italy (PRID prot. BIRD214519), from the Italian Ministry of University and Research (Borse PON "Ricerca e innovazione" 2014–2020), and from the European Union – NextGenerationEU (2021 STARS Grants@Unipd programme P-NICHE).

REFERENCES

Adnan, Muhammad, Shah Fahad, Muhammad Zamin, Shahen Shah, Ishaq Ahmad Mian, Subhan Danish, Muhammad, Zafar-ul-Hye, et al. 2020. 'Coupling Phosphate-Solubilizing Bacteria with Phosphorus Supplements Improve Maize Phosphorus Acquisition and Growth under Lime Induced Salinity Stress'. *Plants* 9(7): 900. doi:10.3390/plants9070900.
Alori, Elizabeth T., Bernard R. Glick, and Olubukola O. Babalola. 2017. 'Microbial Phosphorus Solubilization and Its Potential for Use in Sustainable Agriculture'. *Frontiers in Microbiology* 8. https://www.frontiersin.org/articles/10.3389/fmicb.2017.00971.

Amy, Charlotte, Jean-Christophe Avice, Karine Laval, and Mélanie Bressan. 2022. 'Are Native Phosphate Solubilizing Bacteria a Relevant Alternative to Mineral Fertilizations for Crops? Part I. When Rhizobacteria Meet Plant P Requirements'. *Rhizosphere* 21: 100476. doi:10.1016/j.rhisph.2022.100476.

Bai, Yang, Daniel B. Müller, Girish Srinivas, Ruben Garrido-Oter, Eva Potthoff, Matthias Rott, Nina Dombrowski, et al. 2015. 'Functional Overlap of the Arabidopsis Leaf and Root Microbiota'. *Nature* 528(7582): 364–69. doi:10.1038/nature16192.

Baldan, Enrico, Sebastiano Nigris, Chiara Romualdi, Stefano D'Alessandro, Anna Clocchiatti, Michela Zottini, Piergiorgio Stevanato, Andrea Squartini, and Barbara Baldan. 2015. 'Beneficial Bacteria Isolated from Grapevine Inner Tissues Shape Arabidopsis thaliana Roots'. *PLOS One* 10(10). e0140252. doi:10.1371/journal.pone.0140252.

Balzergue, Coline, Virginie Puech-Pagès, Guillaume Bécard, and Soizic F. Rochange. 2011. 'The Regulation of Arbuscular Mycorrhizal Symbiosis by Phosphate in Pea Involves Early and Systemic Signalling Events'. *Journal of Experimental Botany* 62(3): 1049–60. doi:10.1093/jxb/erq335.

Benedetto, Alessandra, Franco Magurno, Paola Bonfante, and Luisa Lanfranco. 2005. 'Expression Profiles of a Phosphate Transporter Gene (GmosPT) from the Endomycorrhizal Fungus Glomus Mosseae'. *Mycorrhiza* 15(8): 620–27. doi:10.1007/s00572-005-0006-9.

Brachi, Benjamin, Daniele Filiault, Hannah Whitehurst, Paul Darme, Pierre Le Gars, Marine Le Mentec, Timothy C. Morton, et al. 2022. 'Plant Genetic Effects on Microbial Hubs Impact Host Fitness in Repeated Field Trials'. *Proceedings of the National Academy of Sciences* 119(30): e2201285119. doi:10.1073/pnas.2201285119.

Branscheid, Anja, Daniela Sieh, Bikram Datt Pant, Patrick May, Emanuel A. Devers, Anders Elkrog, Leif Schauser, Wolf-Rüdiger Scheible, and Franziska Krajinski. 2010. 'Expression Pattern Suggests a Role of MiR399 in the Regulation of the Cellular Response to Local Pi Increase During Arbuscular Mycorrhizal Symbiosis'. *Molecular Plant–Microbe Interactions®* 23(7): 915–26. doi:10.1094/MPMI-23-7-0915.

Breuillin, Florence, Jonathan Schramm, Mohammad Hajirezaei, Amir Ahkami, Patrick Favre, Uwe Druege, Bettina Hause, et al. 2010. 'Phosphate Systemically Inhibits Development of Arbuscular Mycorrhiza in Petunia Hybrida and Represses Genes Involved in Mycorrhizal Functioning'. *The Plant Journal* 64(6): 1002–17. doi:10.1111/j.1365-313X.2010.04385.x.

Breuillin-Sessoms, Florence, Daniela S. Floss, S. Karen Gomez, Nathan Pumplin, Yi Ding, Veronique Levesque-Tremblay, Roslyn D. Noar, et al. 2015. 'Suppression of Arbuscule Degeneration in Medicago truncatula Phosphate Transporter4 Mutants Is Dependent on the Ammonium Transporter 2 Family Protein AMT2;3'. *The Plant Cell* 27(4): 1352–66. doi:10.1105/tpc.114.131144.

Bulgarelli, Davide, Klaus Schlaeppi, Stijn Spaepen, Emiel Ver Loren van Themaat, and Paul Schulze-Lefert. 2013. 'Structure and Functions of the Bacterial Microbiota of Plants'. *Annual Review of Plant Biology* 64(1): 807–38. doi:10.1146/annurev-arplant-050312-120106.

Bustos, Regla, Gabriel Castrillo, Francisco Linhares, María Isabel Puga, Vicente Rubio, Julian Pérez-Pérez, Roberto Solano, Antonio Leyva, and Javier Paz-Ares. 2010. 'A Central Regulatory System Largely Controls Transcriptional Activation and Repression Responses to Phosphate Starvation in Arabidopsis'. *PLOS Genetics* 6(9): e1001102. doi:10.1371/journal.pgen.1001102.

Carvalhais, Lilia C., Paul G. Dennis, Dmitri Fedoseyenko, Mohammad-Reza Hajirezaei, Rainer Borriss, and Nicolaus von Wirén. 2011. 'Root Exudation of Sugars, Amino Acids, and Organic Acids by Maize as Affected by Nitrogen, Phosphorus, Potassium, and Iron Deficiency'. *Journal of Plant Nutrition and Soil Science* 174(1): 3–11. doi:10.1002/jpln.201000085.

Castagno, Luis N., Analía I. Sannazzaro, María E. Gonzalez, Fernando L. Pieckenstain, and María J. Estrella. 2021. 'Phosphobacteria as Key Actors to Overcome Phosphorus Deficiency in Plants'. *Annals of Applied Biology* 178(2): 256–67. doi:10.1111/aab.12673.

Castrillo, Gabriel, Paulo J. Pereira Lima Teixeira, Sur Herrera Paredes, Theresa F. Law, Laura de Lorenzo, Meghan E. Feltcher, Omri M. Finkel, et al. 2017. 'Root Microbiota Drive Direct Integration of Phosphate Stress and Immunity'. *Nature* 543(7646): 513–18. doi:10.1038/nature21417.

Chutia, Ranju, Steffen Abel, and Jörg Ziegler. 2019. 'Iron and Phosphate Deficiency Regulators Concertedly Control Coumarin Profiles in Arabidopsis thaliana Roots During Iron, Phosphate, and Combined Deficiencies'. *Frontiers in Plant Science* 10. https://www.frontiersin.org/articles/10.3389/fpls.2019.00113.

Coban, Oksana, Gerlinde B. De Deyn, and Martine van der Ploeg. 2022. 'Soil Microbiota as Game-Changers in Restoration of Degraded Lands'. *Science* 375(6584): abe0725. doi:10.1126/science.abe0725.

Cole, Benjamin J., Meghan E. Feltcher, Robert J. Waters, Kelly M. Wetmore, Tatiana S. Mucyn, Elizabeth M. Ryan, Gaoyan Wang, et al. 2017. 'Genome-Wide Identification of Bacterial Plant Colonization Genes'. Edited by Xinnian Dong. *PLOS Biology* 15(9): e2002860. doi:10.1371/journal.pbio.2002860.

Compant, Stéphane, Marine C. Cambon, Corinne Vacher, Birgit Mitter, Abdul Samad, and Angela Sessitsch. 2021. 'The Plant Endosphere World – Bacterial Life within Plants'. *Environmental Microbiology* 23(4): 1812–29. doi:10.1111/1462-2920.15240.

Cotton, T. E. Anne, Pierre Pétriacq, Duncan D. Cameron, Moaed Al Meselmani, Roland Schwarzenbacher, Stephen A. Rolfe, and Jurriaan Ton. 2019. 'Metabolic Regulation of the Maize Rhizobiome by Benzoxazinoids'. *The ISME Journal* 13(7): 1647–58. doi:10.1038/s41396-019-0375-2.

Coutinho, Bruna G., Danilo Licastro, Lucia Mendonça-Previato, Miguel Cámara, and Vittorio Venturi. 2015. 'Plant-Influenced Gene Expression in the Rice Endophyte *Burkholderia kururiensis* M130'. *Molecular Plant–Microbe Interactions®* 28(1): 10–21. doi:10.1094/MPMI-07-14-0225-R.

Das, Debatosh, Michael Paries, Karen Hobecker, Michael Gigl, Corinna Dawid, Hon-Ming Lam, Jianhua Zhang, Moxian Chen, and Caroline Gutjahr. 2022. 'Phosphate Starvation Response Transcription Factors Enable Arbuscular Mycorrhiza Symbiosis'. *Nature Communications* 13(1): 477. doi:10.1038/s41467-022-27976-8.

Dennis, Paul G., Anthony J. Miller, and Penny R. Hirsch. 2010. 'Are Root Exudates More Important than Other Sources of Rhizodeposits in Structuring Rhizosphere Bacterial Communities?: Root Exudates and Rhizosphere Bacteria'. *FEMS Microbiology Ecology* 72(3): 313–27. doi:10.1111/j.1574-6941.2010.00860.x.

Deppenmeier, Uwe, Marc Hoffmeister, and Christina Prust. 2002. 'Biochemistry and Biotechnological Applications of Gluconobacter Strains'. *Applied Microbiology and Biotechnology* 60(3): 233–42. doi:10.1007/s00253-002-1114-5.

Dindas, Julian, Thomas A. DeFalco, Yu Gang, Lu Zhang, Pascale David, Marta Bjornson, Marie-Christine Thibaud, et al. 2022. 'Direct Inhibition of Phosphate Transport by Immune Signaling in Arabidopsis'. *Current Biology* 32(2): 488–495.e5. doi:10.1016/j.cub.2021.11.063.

Dudeja, Surjit S., Pooja Suneja-Madan, Paul Minakshi, Rajat Maheswari, and Erika Kothe. 2021. 'Bacterial Endophytes: Molecular Interactions with Their Hosts'. *Journal of Basic Microbiology* 61(6): 475–505. doi:10.1002/jobm.202000657.

Emmett, Bryan D., Véronique Lévesque-Tremblay, and Maria J. Harrison. 2021. 'Conserved and Reproducible Bacterial Communities Associate with Extraradical Hyphae of Arbuscular Mycorrhizal Fungi'. *The ISME Journal* 15(8): 2276–88. doi:10.1038/s41396-021-00920-2.

Fabiańska, Izabela, Esperanza Sosa-Lopez, and Marcel Bucher. 2019. 'The Role of Nutrient Balance in Shaping Plant Root-Fungal Interactions: Facts and Speculation'. *Current Opinion in Microbiology*. Environmental Microbiology 49: 90–6. doi:10.1016/j.mib.2019.10.004.

Feng, Gu, Yongchun C. Song, Xiaolin L. Li, and Peter Christie. 2003. 'Contribution of Arbuscular Mycorrhizal Fungi to Utilization of Organic Sources of Phosphorus by Red Clover in a Calcareous Soil'. *Applied Soil Ecology* 22(2): 139–48. doi:10.1016/S0929-1393(02)00133-6.

Fierer, Noah, Stephen A. Wood, and Clifton P. Bueno de Mesquita. 2021. 'How Microbes Can, and Cannot, Be Used to Assess Soil Health'. *Soil Biology and Biochemistry* 153: 108111. doi:10.1016/j.soilbio.2020.108111.

Finkel, Omri M., Gabriel Castrillo, Sur Herrera Paredes, Isai Salas González, and Jeffery L. Dangl. 2017. 'Understanding and Exploiting Plant Beneficial Microbes'. *Current Opinion in Plant Biology* 38: 155–63. doi:10.1016/j.pbi.2017.04.018.

Finkel, Omri M., Isai Salas-González, Gabriel Castrillo, Stijn Spaepen, Theresa F. Law, Paulo J. Pereira Lima Teixeira, Corbin D. Jones, and Jeffery L. Dangl. 2019. 'The Effects of Soil Phosphorus Content on Plant Microbiota Are Driven by the Plant Phosphate Starvation Response'. *PLOS Biology* 17(11): e3000534. doi:10.1371/journal.pbio.3000534.

Fiorilli, Valentina, Luisa Lanfranco, and Paola Bonfante. 2013. 'The Expression of GintPT, the Phosphate Transporter of Rhizophagus irregularis, Depends on the Symbiotic Status and Phosphate Availability'. *Planta* 237(5): 1267–77.

Fitzpatrick, Connor R., Isai Salas-González, Jonathan M. Conway, Omri M. Finkel, Sarah Gilbert, Russ Dor, Paulo José Pereira Lima Teixeira, and Jeffery L. Dangl. 2020. 'The Plant Microbiome: From Ecology to Reductionism and Beyond'. *Annual Review of Microbiology* 74(1): 81–100. doi:10.1146/annurev-micro-022620-014327.

Fuke, Priya, T. Mohan Manu, Manish Kumar, Ankush D. Sawarkar, Ashok Pandey, and Lal Singh. 2021. 'Role of Microbial Diversity to Influence the Growth and Environmental Remediation Capacity of Bamboo: A Review'. *Industrial Crops and Products* 167: 113567. doi:10.1016/j.indcrop.2021.113567.

Giovannetti, Marco, Veronica Volpe, Alessandra Salvioli, and Paola Bonfante. 2017. 'Chapter 7 - Fungal and Plant Tools for the Uptake of Nutrients in Arbuscular Mycorrhizas: A Molecular View'. In *Mycorrhizal Mediation of Soil*, edited by Nancy Collins Johnson, Catherine Gehring, and Jan Jansa, 107–28. Elsevier. doi:10.1016/B978-0-12-804312-7.00007-3.

Giovannetti, Marco, Christian Göschl, Christof Dietzen, Stig U. Andersen, Stanislav Kopriva, and Wolfgang Busch. 2019. 'Identification of Novel Genes Involved in Phosphate Accumulation in Lotus japonicus through Genome Wide Association Mapping of Root System Architecture and Anion Content'. *PLOS Genetics* 15(12): e1008126. doi:10.1371/journal.pgen.1008126.

Giovannetti, Marco, Alessandra Salvioli di Fossalunga, Ioannis A. Stringlis, Silvia Proietti and Valentina Fiorilli (2023). 'Unearthing soil-plant-microbiota crosstalk: Looking back to move forward'. *Frontiers in Plant Science* 13: 1082752. doi: 10.3389/fpls.2022.1082752.

Großkopf, Tobias, and Orkun S. Soyer. 2014. 'Synthetic Microbial Communities'. *Current Opinion in Microbiology* 18(100): 72–7. doi:10.1016/j.mib.2014.02.002.

Gupta, Ritu, and Sunil Laxman. 2021. 'Cycles, Sources, and Sinks: Conceptualizing How Phosphate Balance Modulates Carbon Flux Using Yeast Metabolic Networks'. *ELife* 10: e63341. doi:10.7554/eLife.63341.

Gupta, Rupali, Gautam Anand, Rajeeva Gaur, and Dinesh Yadav. 2021. 'Plant–Microbiome Interactions for Sustainable Agriculture: A Review'. *Physiology and Molecular Biology of Plants* 27(1): 165–79. doi:10.1007/s12298-021-00927-1.

el Zahar, Feth, Christine Marol, Odile Berge, Juan I. Rangel-Castro, James I. Prosser, Jérôme Balesdent, Thierry Heulin, and Wafa Achouak. 2008. 'Plant Host Habitat and Root Exudates Shape Soil Bacterial Community Structure'. *The ISME Journal* 2(12): 1221–30. doi:10.1038/ismej.2008.80.

Hanson, China A., Jed A. Fuhrman, M. Claire Horner-Devine, and Jennifer B. H. Martiny. 2012. 'Beyond Biogeographic Patterns: Processes Shaping the Microbial Landscape'. *Nature Reviews in Microbiology* 10(7): 497–506. doi:10.1038/nrmicro2795.

Harrison, Maria J., Gary R. Dewbre, and Jinyuan Liu. 2002. 'A Phosphate Transporter from Medicago truncatula Involved in the Acquisition of Phosphate Released by Arbuscular Mycorrhizal Fungi'. *The Plant Cell* 14(10): 2413–29. doi:10.1105/tpc.004861.

Hartman, Kyle, Marcel G.A. van der Heijden, Valexia Roussely-Provent, Jean-Claude Walser, and Klaus Schlaeppi. 2017. 'Deciphering Composition and Function of the Root Microbiome of a Legume Plant'. *Microbiome* 5(1): 2. doi:10.1186/s40168-016-0220-z.

Javot, Hélène, Varma R. Penmetsa, Florence Breuillin, Kishor K. Bhattarai, Roslyn D. Noar, Karen S. Gomez, Quan Zhang, Douglas R. Cook, and Maria J. Harrison. 2011. 'Medicago truncatula Mtpt4 Mutants Reveal a Role for Nitrogen in the Regulation of Arbuscule Degeneration in Arbuscular Mycorrhizal Symbiosis'. *The Plant Journal* 68(6): 954–65. doi:10.1111/j.1365-313X.2011.04746.x.

Jones, David L., Angela Hodge, and Yakov Kuzyakov. 2004. 'Plant and Mycorrhizal Regulation of Rhizodeposition'. *New Phytologist* 163(3): 459–80. doi:10.1111/j.1469-8137.2004.01130.x.

Jung, Ji-Yul, Martina K. Ried, Michael Hothorn, and Yves Poirier. 2018. 'Control of Plant Phosphate Homeostasis by Inositol Pyrophosphates and the SPX Domain'. *Current Opinion in Biotechnology* 49: 156–62. doi:10.1016/j.copbio.2017.08.012.

Kakouridis, Anne, John A. Hagen, Megan P. Kan, Stefania Mambelli, Lewis J. Feldman, Donald J. Herman, Peter K. Weber, Jennifer Pett-Ridge, and Mary K. Firestone. 2022. 'Routes to Roots: Direct Evidence of Water Transport by Arbuscular Mycorrhizal Fungi to Host Plants'. *New Phytologist* n/a(n/a). doi:10.1111/nph.18281.

Kaminsky, Laura M., Ryan V. Trexler, Rondy J. Malik, Kevin L. Hockett, and Terrence H. Bell. 2019. 'The Inherent Conflicts in Developing Soil Microbial Inoculants'. *Trends in Biotechnology* 37(2): 140–51. doi:10.1016/j.tibtech.2018.11.011.

Kang, Jihoon, Aziz Amoozegar, Dean Hesterberg, and Deanna L. Osmond. 2011. 'Phosphorus Leaching in a Sandy Soil as Affected by Organic and Inorganic Fertilizer Sources'. *Geoderma* 161(3–4): 194–201. doi:10.1016/j.geoderma.2010.12.019.

Kania, Angelika, Nicolas Langlade, Enrico Martinoia, and Günter Neumann. 2003. 'Phosphorus Deficiency-Induced Modifications in Citrate Catabolism and in Cytosolic PH as Related to Citrate Exudation in Cluster Roots of White Lupin'. *Plant and Soil* 248(1): 117–27. doi:10.1023/A:1022371115788.

Karimi, Battle, Sébastien Terrat, Samuel Dequiedt, Nicolas P. A. Saby, Walid Horrigue, Mélanie Lelièvre, Virginie Nowak, et al. 2018. 'Biogeography of Soil Bacteria and Archaea across France'. *Science Advances* 4(7): eaat1808. doi:10.1126/sciadv.aat1808.

Kaur, Gurdeep, and Mondem Sudhakara Reddy. 2014. 'Influence of P-Solubilizing Bacteria on Crop Yield and Soil Fertility at Multilocational Sites'. *European Journal of Soil Biology* 61: 35–40. doi:10.1016/j.ejsobi.2013.12.009.

Kenneth, Odoh Chuks, Eze Chibuzor Nwadibe, Akpi Uchenna Kalu, and Unah Victor Unah. 2019. 'Plant Growth Promoting Rhizobacteria (PGPR): A Novel Agent for Sustainable Food Production'. *American Journal of Agricultural and Biological Sciences* 14(1): 35–54. doi:10.3844/ajabssp.2019.35.54.

Khan, Shams Tabrez. 2022. 'Consortia-Based Microbial Inoculants for Sustaining Agricultural Activities'. *Applied Soil Ecology* 176: 104503. doi:10.1016/j.apsoil.2022.104503.

Kiers, E. Toby, Mark G. Hutton, and Ford R. Denison. 2007. 'Human Selection and the Relaxation of Legume Defences against Ineffective Rhizobia'. *Proceedings of the Royal Society B: Biological Sciences* 274(1629): 3119–26. doi:10.1098/rspb.2007.1187.

Kikuchi, Yusuke, Nowaki Hijikata, Ryo Ohtomo, Yoshihiro Handa, Masayoshi Kawaguchi, Katsuharu Saito, Chikara Masuta, and Tatsuhiro Ezawa. 2016. 'Aquaporin-Mediated Long-Distance Polyphosphate Translocation Directed towards the Host in Arbuscular Mycorrhizal Symbiosis: Application of Virus-Induced Gene Silencing'. *New Phytologist* 211(4): 1202–8. doi:10.1111/nph.14016.

Kikuchi, Yusuke, Nowaki Hijikata, Kaede Yokoyama, Ryo Ohtomo, Yoshihiro Handa, Masayoshi Kawaguchi, Katsuharu Saito, and Tatsuhiro Ezawa. 2014. 'Polyphosphate Accumulation Is Driven by Transcriptome Alterations That Lead to Near-Synchronous and Near-Equivalent Uptake of Inorganic Cations in an Arbuscular Mycorrhizal Fungus'. *New Phytologist* 204(3): 638–49. doi:10.1111/nph.12937.

Lambers, Hans, Jon C. Clements, and Matthew N. Nelson. 2013. 'How a Phosphorus-Acquisition Strategy Based on Carboxylate Exudation Powers the Success and Agronomic Potential of Lupines (Lupinus, Fabaceae)'. *American Journal of Botany* 100(2): 263–88. doi:10.3732/ajb.1200474.

Lehnert, Heike, Albrecht Serfling, Matthias Enders, Wolfgang Friedt, and Frank Ordon. 2017. 'Genetics of Mycorrhizal Symbiosis in Winter Wheat (Triticum aestivum)'. *New Phytologist* 215(2): 779–91. doi:10.1111/nph.14595.

Liao, Dehua, Chao Sun, Haiyan Liang, Yang Wang, Xinxin Bian, Chaoqun Dong, Xufang Niu, et al. 2022. 'SlSPX1-SlPHR Complexes Mediate the Suppression of Arbuscular Mycorrhizal Symbiosis by Phosphate Repletion in Tomato'. *The Plant Cell*: koac212. doi:10.1093/plcell/koac212.

Lidbury, Ian D. E. A., Andrew R. J. Murphy, David J. Scanlan, Gary D. Bending, Alexandra M. E. Jones, Jonathan D. Moore, Andrew Goodall, John P. Hammond, and Elizabeth M. H. Wellington. 2016. 'Comparative Genomic, Proteomic and Exoproteomic Analyses of Three *Pseudomonas* Strains Reveals Novel Insights into the Phosphorus Scavenging Capabilities of Soil Bacteria'. *Environmental Microbiology* 18(10): 3535–49. doi:10.1111/1462-2920.13390.

Lidbury, Ian D. E. A., Sebastien Raguideau, Chiara Borsetto, Andrew R. J. Murphy, Andrew Bottrill, Senlin Liu, Richard Stark, et al. 2022. 'Stimulation of Distinct Rhizosphere Bacteria Drives Phosphorus and Nitrogen Mineralization in Oilseed Rape under Field Conditions'. *Msystems*: e00025-22. doi:10.1128/msystems.00025-22.

Lidbury, Ian D. E. A., David J. Scanlan, Andrew R. J. Murphy, Joseph A. Christie-Oleza, Maria M. Aguilo-Ferretjans, Andrew Hitchcock, and Tim J. Daniell. 2022. 'A Widely Distributed Phosphate-Insensitive Phosphatase Presents a Route for Rapid Organophosphorus Remineralization in the Biosphere'. *Proceedings of the National Academy of Sciences* 119(5): e2118122119. doi:10.1073/pnas.2118122119.

Ling, Ning, Tingting Wang, and Yakov Kuzyakov. 2022. 'Rhizosphere Bacteriome Structure and Functions'. *Nature Communications* 13(1): 836. doi:10.1038/s41467-022-28448-9.

Liu, Zhexian, Polina Beskrovnaya, Ryan A. Melnyk, Sarzana S. Hossain, Sophie Khorasani, Lucy R. O'Sullivan, Christina L. Wiesmann, Jen Bush, Joël D. Richard, and Cara H. Haney. 2018. 'A Genome-Wide Screen Identifies Genes in Rhizosphere-Associated Pseudomonas Required to Evade Plant Defenses'. 9(6): 17.

Martín-Robles, Nieves, Anika Lehmann, Erica Seco, Ricardo Aroca, Matthias C. Rillig, and Rubén Milla. 2018. 'Impacts of Domestication on the Arbuscular Mycorrhizal Symbiosis of 27 Crop Species'. *New Phytologist* 218(1): 322–34. doi:10.1111/nph.14962.

Matthews, Andrew, Sarah Pierce, Helen Hipperson, and Ben Raymond. 2019. 'Rhizobacterial Community Assembly Patterns Vary Between Crop Species'. *Frontiers in Microbiology* 10: 581. doi:10.3389/fmicb.2019.00581.

Mitter, Birgit, Nikolaus Pfaffenbichler, Richard Flavell, Stéphane Compant, Livio Antonielli, Alexandra Petric, Teresa Berninger, et al. 2017. 'A New Approach to Modify Plant Microbiomes and Traits by Introducing Beneficial Bacteria at Flowering into Progeny Seeds'. *Frontiers in Microbiology* 8. doi:10.3389/fmicb.2017.00011.

Morcuende, Rosa, Rajendra Bari, Yves Gibon, Wenming Zheng, Bikram Datt Pant, Oliver Bläsing, Björn Usadel, et al. 2007. 'Genome-Wide Reprogramming of Metabolism and Regulatory Networks of Arabidopsis in Response to Phosphorus'. *Plant, Cell and Environment* 30(1): 85–112. doi:10.1111/j.1365-3040.2006.01608.x.

Nadiéline Valentin, Christian, Mame Farma Ndiaye, Saliou Fall, Tatiana Krasova, Antoine Le Quéré, and Diégane Diouf. 2019 'Isolation and Characterization of Potential Phosphate Solubilizing Bacteria in Two Regions of Senegal'. *African Journal of Microbiology Research* 10.

Nagy, Réka, Vladimir Karandashov, Véronique Chague, Katsiaryna Kalinkevich, Guohua Xu M'Barek Tamasloukht, Iver Jakobsen, Avraham A. Levy, Nikolaus Amrhein, and Marcel Bucher. 2005. 'The Characterization of Novel Mycorrhiza-Specific Phosphate Transporters from Lycopersicon esculentum and Solanum tuberosum Uncovers Functional Redundancy in Symbiotic Phosphate Transport in Solanaceous Species'. *The Plant Journal: For Cell and Molecular Biology* 42(2): 236–50. doi:10.1111/j.1365-313X.2005.02364.x.

Nasir, Fahad, Shaohua Shi, Lei Tian, Chunling Chang, Lina Ma, Xiujun Li, Yingzhi Gao, and Chunjie Tian. 2019. 'Strigolactones Shape the Rhizomicrobiome in Rice (Oryza sativa)'. *Plant Science* 286: 118–33. doi:10.1016/j.plantsci.2019.05.016.

Nassal, Dinah, Marie Spohn, Namis Eltlbany, Samuel Jacquiod, Kornelia Smalla, Sven Marhan, and Ellen Kandeler. 2018. 'Effects of Phosphorus-Mobilizing Bacteria on Tomato Growth and Soil Microbial Activity'. *Plant and Soil* 427(1–2): 17–37. doi:10.1007/s11104-017-3528-y.

Pant, Bikram-Datt, Pooja Pant, Alexander Erban, David Huhman, Joachim Kopka, and Wolf-Rüdiger Scheible. 2015. 'Identification of Primary and Secondary Metabolites with Phosphorus Status-Dependent Abundance in Arabidopsis, and of the Transcription Factor PHR1 as a Major Regulator of Metabolic Changes during Phosphorus Limitation'. *Plant, Cell and Environment* 38(1): 172–87. doi:10.1111/pce.12378.

Pantigoso, Hugo A., Daniel K. Manter, and Jorge M. Vivanco. 2020. 'Differential Effects of Phosphorus Fertilization on Plant Uptake and Rhizosphere Microbiome of Cultivated and Non-cultivated Potatoes'. *Microbial Ecology* 80(1): 169–80. doi:10.1007/s00248-020-01486-w.

Paries Michael, and Caroline Gutjahr. 2023. The good, the bad, and the phosphate: regulation of beneficial and detrimental plant-microbe interactions by the plant phosphate status. *New Phytologist* 239(1): 29–46.

Paszkowski, Uta, Scott Kroken, Christophe Roux, and Steven P. Briggs. 2002. 'Rice Phosphate Transporters Include an Evolutionarily Divergent Gene Specifically Activated in Arbuscular Mycorrhizal Symbiosis'. *Proceedings of the National Academy of Sciences* 99(20): 13324–29. doi:10.1073/pnas.202474599.

Pausch, Johanna, and Yakov Kuzyakov. 2018. 'Carbon Input by Roots into the Soil: Quantification of Rhizodeposition from Root to Ecosystem Scale'. *Global Change Biology* 24(1): 1–12. doi:10.1111/gcb.13850.

Philippot, Laurent, Jos M. Raaijmakers, Philippe Lemanceau, and Wim H. van der Putten. 2013. 'Going Back to the Roots: The Microbial Ecology of the Rhizosphere'. *Nature Reviews Microbiology* 11(11): 789–99. doi:10.1038/nrmicro3109.

Popova, Yulia, Palvannan Thayumanavan, Elena Lonati, Margarida Agrochão, and Johan M. Thevelein. 2010. 'Transport and Signaling through the Phosphate-Binding Site of the Yeast Pho84 Phosphate Transceptor'. *Proceedings of the National Academy of Sciences of the United States of America* 107(7): 2890–95. doi:10.1073/pnas.0906546107.

Prabhu, Neha, Sunita Borkar, and Sandeep Garg. 2019. 'Phosphate Solubilization by Microorganisms: Overview, Mechanisms, Applications and Advances'. In *Advances in Biological Science Research: A Practical Approach*, edited by S. N. Meena and M. M. Naik, 161–76. Amsterdam, the Netherlands: Elsevier.

Rasul, Maria, Sumera Yasmin, Muhammad Suleman, Ahmad Zaheer, Thomas Reitz, Mika T. Tarkka, Ejazul Islam, and Muhammad Sajjad Mirza. 2019. 'Glucose Dehydrogenase Gene Containing Phosphobacteria for Biofortification of Phosphorus with Growth Promotion of Rice'. *Microbiological Research* 223–225: 1–12. doi:10.1016/j.micres.2019.03.004.

Rausch, Christine, Pierre Daram, Silvia Brunner, Jan Jansa, Maryse Laloi, Georg Leggewie, Nikolaus Amrhein, and Marcel Bucher. 2001. 'A Phosphate Transporter Expressed in Arbuscule-Containing Cells in Potato'. *Nature* 414(6862): 462–65. doi:10.1038/35106601.

Rawat, Pratibha, Deepti Shankhdhar, and Shailesh C. Shankhdhar. 2021. 'Synergistic Impact of Phosphate Solubilizing Bacteria and Phosphorus Rates on Growth, Antioxidative Defense System, and Yield Characteristics of Upland Rice (Oryza sativa L.)'. *Journal of Plant Growth Regulation*. doi:10.1007/s00344-021-10458-4.

Richardson, Alan E., José-Miguel Barea, Ann M. McNeill, and Claire Prigent-Combaret. 2009. 'Acquisition of Phosphorus and Nitrogen in the Rhizosphere and Plant Growth Promotion by Microorganisms'. *Plant and Soil* 321(1–2): 305–39. doi:10.1007/s11104-009-9895-2.

Rolfe, Stephen A., Joseph Griffiths, and Jurriaan Ton. 2019. 'Crying Out for Help with Root Exudates: Adaptive Mechanisms by Which Stressed Plants Assemble Health-Promoting Soil Microbiomes'. *Current Opinion in Microbiology* 49: 73–82. doi:10.1016/j.mib.2019.10.003.

Rubio, Vicente, Francisco Linhares, Roberto Solano, Ana C. Martín, Joaquín Iglesias, Antonio Leyva, and Javier Paz-Ares. 2001. 'A Conserved MYB Transcription Factor Involved in Phosphate Starvation

Signaling Both in Vascular Plants and in Unicellular Algae'. *Genes and Development* 15(16): 2122–33. doi:10.1101/gad.204401.

Salvioli di Fossalunga, Alessandra, Justine Lipuma, Francesco Venice, Laurence Dupont, and Paola Bonfante. 2017. 'The Endobacterium of an Arbuscular Mycorrhizal Fungus Modulates the Expression of Its Toxin-Antitoxin Systems during the Life Cycle of Its Host'. *The ISME Journal* 11(10): 2394–98. doi:10.1038/ismej.2017.84.

Samad, Abdul, Friederike Trognitz, Stéphane Compant, Livio Antonielli, and Angela Sessitsch. 2017. 'Shared and Host-Specific Microbiome Diversity and Functioning of Grapevine and Accompanying Weed Plants'. *Environmental Microbiology* 19(4): 1407–24. doi:10.1111/1462-2920.13618.

Sato, Takumi, Tatsuhiro Ezawa, Weiguo Cheng, and Keitaro Tawaraya. 2015. 'Release of Acid Phosphatase from Extraradical Hyphae of Arbuscular Mycorrhizal Fungus Rhizophagus clarus'. *Soil Science and Plant Nutrition* 61(2). : 269–74. doi:10.1080/00380768.2014.993298.

Sawers, Ruairidh J. H., Simon F. Svane, Clement Quan, Mette Grønlund, Barbara Wozniak, Mesfin-Nigussie Gebreselassie, Eliécer González-Muñoz, et al. 2017. 'Phosphorus Acquisition Efficiency in Arbuscular Mycorrhizal Maize Is Correlated with the Abundance of Root-External Hyphae and the Accumulation of Transcripts Encoding PHT1 Phosphate Transporters'. *New Phytologist* 214(2): 632–43. doi:10.1111/nph.14403.

Schachtman, Daniel P., Robert J. Reid, and Sarah M. Ayling. 1998. 'Phosphorus Uptake by Plants: From Soil to Cell'. *Plant Physiology* 116(2): 447–53. doi:10.1104/pp.116.2.447.

Schloss, Patrick D. 2018. 'Identifying and Overcoming Threats to Reproducibility, Replicability, Robustness, and Generalizability in Microbiome Research'. Edited by Jacques Ravel. *MBio* 9(3): e00525-18. doi:10.1128/mBio.00525-18.

Shears, Stephen B. 2015. 'Inositol Pyrophosphates: Why so Many Phosphates?' *Advances in Biological Regulation* 57: 203–16. doi:10.1016/j.jbior.2014.09.015.

Shi, Jincai, Boyu Zhao, Shuang Zheng, Xiaowei Zhang, Xiaolin Wang, Wentao Dong, Qiujin Xie, et al. 2021. 'A Phosphate Starvation Response-Centered Network Regulates Mycorrhizal Symbiosis'. *Cell* 184(22): 5527-5540.e18. doi:10.1016/j.cell.2021.09.030.

Shibata, Reiko, and Katsuya Yano. 2003. 'Phosphorus Acquisition from Non-labile Sources in Peanut and Pigeonpea with Mycorrhizal Interaction'. *Applied Soil Ecology* 24(2): 133–41. doi:10.1016/S0929-1393(03)00093-3.

Singh, Ajay, Nagina Parmar, and Ramesh C. Kuhad, eds. 2011. *Bioaugmentation, Biostimulation and Biocontrol*. Vol. 108. Soil Biology. Berlin, Heidelberg: Springer. doi:10.1007/978-3-642-19769-7.

Singh, Bijender, Pragya Ines Boukhris, Vinod Kumar, Ajar Nath Yadav, Ameny Farhat-Khemakhem, Anil Kumar, et al. 2020. 'Contribution of Microbial Phytases to the Improvement of Plant Growth and Nutrition: A Review'. *Pedosphere* 30(3): 295–313. doi:10.1016/S1002-0160(20)60010-8.

Stringlis, Ioannis A., Yu Ke, Kirstin Feussner, Ronnie de Jonge, Sietske Van Bentum, Marcel C. Van Verk, Roeland L. Berendsen, Peter A. H. M. Bakker, Ivo Feussner, and Corné M. J. Pieterse. 2018. 'MYB72-Dependent Coumarin Exudation Shapes Root Microbiome Assembly to Promote Plant Health'. *Proceedings of the National Academy of Sciences* 115(22). doi:10.1073/pnas.1722335115.

Suleman, Muhammad, Sumera Yasmin, Maria Rasul, Mahreen Yahya, Babar Manzoor Atta, and Muhammad Sajjad Mirza. 2018. 'Phosphate Solubilizing Bacteria with Glucose Dehydrogenase Gene for Phosphorus Uptake and Beneficial Effects on Wheat'. *PLOS One* 13(9): e0204408. doi:10.1371/journal.pone.0204408.

Thalineau, Elise, Carine Fournier, Sylvain Jeandroz, and Hoai-Nam Truong. 2020. 'Phosphorus Control of Plant Interactions with Mutualistic and Pathogenic Microorganisms: A Mini-Review and a Case Study of the *Medicago truncatula* B 9 Mutant'. In *The Model Legume Medicago Truncatula*, 346–54. Wiley. doi:10.1002/9781119409144.ch41.

Timmis, Kenneth, and Juan Luis Ramos. 2021. 'The Soil Crisis: The Need to Treat as a Global Health Problem and the Pivotal Role of Microbes in Prophylaxis and Therapy'. *Microbial Biotechnology* 14(3): 769–97. doi:10.1111/1751-7915.13771.

Tomasi, Nicola, Tobias Kretzschmar, Luca Espen, Laure Weisskopf, Anja Thoe Fuglsang, Michael Gjedde Palmgren, Günter Neumann, et al. 2009. 'Plasma Membrane H+-ATPase-Dependent Citrate Exudation from Cluster Roots of Phosphate-Deficient White Lupin'. *Plant, Cell and Environment* 32(5): 465–75. doi:10.1111/j.1365-3040.2009.01938.x.

Trivedi, Pankaj, Jan E. Leach, Susannah G. Tringe, Tongmin Sa, and Brajesh K. Singh. 2020. 'Plant–Microbiome Interactions: From Community Assembly to Plant Health'. *Nature Reviews in Microbiology* 18(11): 607–21. doi:10.1038/s41579-020-0412-1.

Vance, Carroll P., Claudia Uhde-Stone, and Deborah L. Allan. 2003. 'Phosphorus Acquisition and Use: Critical Adaptations by Plants for Securing a Nonrenewable Resource'. *New Phytologist* 157(3): 423–47. doi:10.1046/j.1469-8137.2003.00695.x.

Vandenkoornhuyse, Philippe, Achim Quaiser, Marie Duhamel, Amandine Le Van, and Alexis Dufresne. 2015. 'The Importance of the Microbiome of the Plant Holobiont'. *New Phytologist* 206(4): 1196–206. doi:10.1111/nph.13312.

Volpe, Veronica, Marco Giovannetti, Xue-Guang Sun, Valentina Fiorilli, and Paola Bonfante. 2016. 'The Phosphate Transporters LjPT4 and MtPT4 Mediate Early Root Responses to Phosphate Status in Non Mycorrhizal Roots'. *Plant, Cell and Environment* 39(3): 660–71. doi:10.1111/pce.12659.

Vorholt, Julia A., Christine Vogel, Charlotte I. Carlström, and Daniel B. Müller. 2017. 'Establishing Causality: Opportunities of Synthetic Communities for Plant Microbiome Research'. *Cell Host and Microbe* 22(2): 142–55. doi:10.1016/j.chom.2017.07.004.

Walters, William A., Zhao Jin, Nicholas Youngblut, Jason G. Wallace, Jessica Sutter, Wei Zhang, Antonio González-Peña, et al. 2018. 'Large-Scale Replicated Field Study of Maize Rhizosphere Identifies Heritable Microbes'. *Proceedings of the National Academy of Sciences* 115(28): 7368–73. doi:10.1073/pnas.1800918115.

Wang, Guiwei, Zexing Jin, Xinxin Wang, Timothy S. George, Gu Feng, and Lin Zhang. 2022. 'Simulated Root Exudates Stimulate the Abundance of Saccharimonadales to Improve the Alkaline Phosphatase Activity in Maize Rhizosphere'. *Applied Soil Ecology* 170: 104274. doi:10.1016/j.apsoil.2021.104274.

Wang, Peng, Roxane Snijders, Wouter Kohlen, Jieyu Liu, Ton Bisseling, and Erik Limpens. 2021. 'Medicago SPX1 and SPX3 Regulate Phosphate Homeostasis, Mycorrhizal Colonization, and Arbuscule Degradation'. *The Plant Cell* 33(11): 3470–86. doi:10.1093/plcell/koab206.

Weihrauch, Christoph, and Christian Opp. 2018. 'Ecologically Relevant Phosphorus Pools in Soils and Their Dynamics: The Story so Far'. *Geoderma* 325: 183–94. doi:10.1016/j.geoderma.2018.02.047.

Wild, Rebekka, Ruta Gerasimaite, Ji-Yul Jung, Vincent Truffault, Igor Pavlovic, Andrea Schmidt, Adolfo Saiardi, et al. 2016. 'Control of Eukaryotic Phosphate Homeostasis by Inositol Polyphosphate Sensor Domains'. *Science* 352(6288): 986–90. doi:10.1126/science.aad9858.

Xie, Xianan, Hui Lin, Xiaowei Peng, Congrui Xu, Zhongfeng Sun, Kexin Jiang, Antian Huang, et al. 2016. 'Arbuscular Mycorrhizal Symbiosis Requires a Phosphate Transceptor in the Gigaspora Margarita Fungal Symbiont'. *Molecular Plant* 9(12): 1583–608. doi:10.1016/j.molp.2016.08.011.

Yan, Yan, Eiko E. Kuramae, Mattias de Hollander, Peter G. L. Klinkhamer, and Johannes A. van Veen. 2017. 'Functional Traits Dominate the Diversity-Related Selection of Bacterial Communities in the Rhizosphere'. *The ISME Journal* 11(1): 56–66. doi:10.1038/ismej.2016.108.

Yang, Shu-Yi, Mette Grønlund, Iver Jakobsen, Marianne Suter Grotemeyer, Doris Rentsch, Akio Miyao, Hirohiko Hirochika, et al. 2012. 'Nonredundant Regulation of Rice Arbuscular Mycorrhizal Symbiosis by Two Members of the Phosphate Transporter1 Gene Family'. *The Plant Cell* 24(10): 4236–51. doi:10.1105/tpc.112.104901.

Zhang, Lin, Gu Feng, and Stéphane Declerck. 2018. 'Signal beyond Nutrient, Fructose, Exuded by an Arbuscular Mycorrhizal Fungus Triggers Phytate Mineralization by a Phosphate Solubilizing Bacterium'. *The ISME Journal* 12(10): 2339–51. doi:10.1038/s41396-018-0171-4.

Zhang, Lin, Minggang Xu, Yu Liu, Fusuo Zhang, Angela Hodge, and Gu Feng. 2016. 'Carbon and Phosphorus Exchange May Enable Cooperation between an Arbuscular Mycorrhizal Fungus and a Phosphate-Solubilizing Bacterium'. *New Phytologist* 210(3): 1022–32. doi:10.1111/nph.13838.

Zhou, Yi, Yanli Wei, Zhongjuan Zhao, Jishun Li, Hongmei Li, Peizhi Yang, Shenzhong Tian, et al. 2022. 'Microbial Communities along the Soil-Root Continuum Are Determined by Root Anatomical Boundaries, Soil Properties, and Root Exudation'. *Soil Biology and Biochemistry* 171: 108721. doi:10.1016/j.soilbio.2022.108721.

Zhuang, Lubo, Yan Li, Zhenshuo Wang, Yu Yue, Nan Zhang, Chang Yang, Qingchao Zeng, and Qi Wang. 2021. 'Synthetic Community with Six *Pseudomonas* Strains Screened from Garlic Rhizosphere Microbiome Promotes Plant Growth'. *Microbial Biotechnology* 14(2): 488–502. doi:10.1111/1751-7915.13640.

Ziegler, Jörg, Stephan Schmidt, Ranju Chutia, Jens Müller, Christoph Böttcher, Nadine Strehmel, Dierk Scheel, and Steffen Abel. 2016. 'Non-targeted Profiling of Semi-polar Metabolites in Arabidopsis Root Exudates Uncovers a Role for Coumarin Secretion and Lignification during the Local Response to Phosphate Limitation'. *Journal of Experimental Botany* 67(5): 1421–32. doi:10.1093/jxb/erv539.

7 Phosphorus and Plant Immunity

Anurag Kashyap, Swagata Saikia, Shenaz Sultana Ahmed, and Munmi Sarma

7.1 INTRODUCTION

In nature, plants live in an environment full of microbes, both beneficial as well as pathogenic. However, plant disease is an exception, as they are protected by an efficient immune system. Pathogens are perceived and countered by two different recognition systems that initiate the so-called pattern-triggered immunity (PTI) and effector-triggered immunity (ETI), both of which are accompanied by a set of induced defences that prevent pathogen infection (Jones and Dangl, 2006). In addition, plants have developed sophisticated networks of essential element uptake and microbial symbiotic association to ensure their nutrition. Phosphorus is one of the most essential macronutrients required by plants, which makes up 0.2% of a plant's dry weight (Prasad et al., 2019). It is a key component of molecules such as nucleic acids, phospholipids, and adenosine triphosphate (ATP), and, consequently, plants cannot grow without a reliable supply of this nutrient (Bechtaoui et al., 2021). It is essential for plant growth and development, and its ability to adapt to various stress responses. Furthermore, it controls key enzyme reactions and metabolic pathways, and drives energy-dependent processes in living cells. Although the total amount of phosphorus in the soil may be abundant, it is often unavailable to plants (Johan et al., 2021). In the natural environment, phosphorus uptake is mediated by phosphorus-solubilising microorganisms (PSM), which are involved in the mineralisation and transformation of insoluble soil nutrients to soluble form (Alori et al., 2017). Many of the soil fungi and bacteria species are able to solubilise phosphorus and can mobilise phosphorus in plants. Arbuscular mycorrhizal fungi (AMF) are one of the most well-known to maintain a symbiotic relationship with plants in a phosphorus-dependent manner (Alori et al., 2017). Furthermore, phosphorus deprivation stimulates the production of defence hormones (salicylic acid and jasmonic acid) (Khan et al., 2016) and plant secondary metabolites (flavonoids and glucosinolate) (Pant et al., 2015; Sharma et al., 2019), which activate the plant immune system. In plants, complex phosphate-starvation responses (PSRs) regulate morphological and physiological adaptive changes, which are critical for plant survival when phosphorus is limited, by promoting its acquisition and utilisation of symbiotic interaction (Shi et al., 2021).

In this chapter, we will discuss existing knowledge on mechanisms through which phosphorus interaction with plants shapes the microbial communities and their relationship with hosts, as well as the influence of phosphorus on the plant defence system.

7.2 RHIZOSPHERIC MICROBES AND PHOSPHORUS AVAILABILITY, AND THEIR INFLUENCE ON PLANT FITNESS AND DEFENCE

Phosphorus content in soil is about 0.05% (w/w), out of which only 0.1% is available for plant uptake (Illmer and Schinner, 1995). Earlier soil phosphorus deficiency problems were addressed by applying phosphorus fertilisers that are inorganic in nature; however, the majority of applied phosphorus fertiliser is not available to plants, and excessive addition has led to certain issues like groundwater contamination, waterway eutrophication, and health hazards to human and animals. Hence, emphasis is given to improving phosphorus fertilisation efficiency, reducing environmental hazards, and increasing crop yields (Postma et al., 2010). In the natural environment, PSMs enhance plant nutrient uptake, as they are involved in a wide range of biological processes including the mineralisation and transformation of insoluble soil nutrients to soluble form. A variety of microbes such as bacteria, fungi, actinomycetes, and algae have phosphorus solubilisation and mineralisation abilities, making it easy for plants to uptake (David et al., 2014). However, it is reported that fungi are more efficient in traversing phosphorus, as they typically produce and secrete more acids, such as gluconic, citric, lactic, 2-ketogluconic, oxalic, tartaric and acetic acids, than bacteria (Sharma et al., 2013). There are certain factors that affect the ability of PSM to transform insoluble organic and inorganic phosphorus. The major factors involve the physicochemical properties of soil (soil pH, temperature, richness of organic matter) and the physiological and growth status of the organism (Seshachala et al., 2012). Other factors that influence microbial phosphate solubilisation involve interaction with other microbes in the soil, the vegetation, prevailing ecological conditions, climatic zone, soil types, plant types, agronomic practices, land use systems, etc. (Seshachala and Tallapragada, 2012).

Plants absorb phosphorus in only two soluble forms, the monobasic (HPO_4^-) and the dibasic ($H_2PO_4^-$) (Kumar et al., 2018)). Phosphate-solubilising bacteria (PSB) solubilises insoluble phosphorus compounds through the release of organic acids and enzymes such as phosphatase and phytase. During the conversion process, a part of phosphorus is assimilated by the microbes, but the amount made soluble and released remains excess to the requirement of the microorganisms. That excess amount is made available for uptake by plants (Walpola and Yoon, 2012). Generally, bacteria solubilise the mineral phosphate by direct oxidation of glucose to gluconic acid where pyrroloquinoline quinone (PQQ) acts as a redox cofactor in glucose dehydrogenases (GDH), resulting in phosphate breakdown (Rodríguez and Fraga, 1999) . Equally important are nitric acid and sulphuric acid, and as a result, these organic and inorganic acids convert compounds like calcium phosphate to dibasic or monobasic phosphates and are then easily made available to the plant. The principal mechanism of solubilisation by PSM is the production of mineral-dissolving compounds such as organic acids, protons, hydroxyl ions, siderophores, and CO_2 (Rodríguez and Fraga, 1999; Sharma et al., 2013) The excretion of these organic acids is accompanied by a drop in pH, resulting in the acidification of the microbial cells and the surroundings; hence, phosphorus ions are released by substitution of H^+ for Ca^{2+}. The PSMs exhibit the ability to restore the productivity of degraded slightly productive and unproductive agricultural soils and thereby enhance plant growth via improving the phosphorus acquisition efficiency of plants, hence converting the insoluble forms of phosphorus to an accessible orthophosphate (Pi) form for plants. They also help to absorb the phosphorus from a wider area by developing an extended network around the plant root system. They even promote plant growth indirectly by increasing the accessibility of other trace elements such as siderophores (iron-chelating agents) (Walpola and Yoon, 2012). In addition, the PSMs also promote the efficiency of nitrogen fixation, thereby facilitating plant growth. The PSMs also provide protection to plants against various phytopathogens, typically owing to the production of antibiotics, hydrogen cyanate (HCN), organic acids like citric acid, malic acid, succinic acid, fumaric acid, tartaric acid, gluconic acid, and antifungal metabolites such as alkaloids, terpenoids, steroids, peptides, benzopyranones, ignocel, and isocoumarins (Gouda et al., 2016). For example, the co-culture of the phytopathogenic *Nigrospora oryzae* and endophytic *Irpex lacteus* on the same host *Dendrobium officinale* resulted in synthesis of a new tremulane sesquiterpene 5-demethyl conocenol C, conocenol B, and a new squalene irpenigirin B, presumed to have a role in defence response (Wu et al., 2019). Similarly, plants produce signalling molecules like plant growth hormones such as salicylic acid, jasmonic acid, cytokinin, and indole acetic acid in response to PSM, helping in their signalling and stress response, which is discussed later part in detail (Afzal and Bano, 2008).

7.3 PHOSPHATE-STARVATION RESPONSE (PSRS) SYSTEM IN PLANT DEFENCE

To adapt to Pi deficiency, land plants have evolved different strategies to deal with this deficiency. For example, plants activate the so-called phosphate-starvation response (PSR) system, which is activated by the transcription factor *PHOSPHATE STARVATION RESPONSE 1* (PHR1), to adjust plant growth and metabolic activity (Finkel et al., 2019; Chiou and Lin, 2011). Plants activate a battery of high-affinity H/Pi *PHOSPHATE TRANSPORTER 1* (*PHT1*) family co-transporters and *PHOSPHATE1* (PHO1) family Pi exporters to effectively take up solubilised Pi from the soil (Zhang et al., 2014). The E2 ubiquitin conjugase *PHOSPHATE2* (PHO2) and the E3 ubiquitin ligase *NITROGEN LIMITATION ADAPTATION 1* (*NLA1*) post-translationally regulate the activity of *PHT1* and *PHO1* (Lin et al., 2013). The expression of several protein-coding genes and microRNAs, generally known as Pi starvation-induced genes, is regulated in plants (Hernandez et al., 2007). These efforts led to the identification of PHR1, a regulator of a cohort of genes collectively named the PSR system (Rubio et al., 2001). A sophisticated network of signalling pathways controls the genetic mechanisms underpinning plant responses to Pi shortage (Medici et al., 2019). *PHR1* acts as the coordinating and integrating site for the interaction between Pi, other crucial nutrients (such as nitrogen), and both abiotic and biotic stimuli (Medici et al., 2019). *PHR1* is regarded as a master regulator of the plant responses to Pi shortage because of its adaptability and aptitude to integrate numerous signalling pathways. Additionally, many land plants have developed strategies to use advantageous microbes to acquire soluble and fixed phosphorus complexes (Hassani et al., 2018). Recently it has been shown that alterations in the plant immune system and plant microbiome have been connected to the PSR system and *PHR1* expression (Ma et al., 2020). The plant-associated microbiota are the microbes that colonise the roots and the phyllosphere, and both have a role in the host plant's ability to withstand diverse biotic and abiotic stresses. To produce a favourable result, land plants fine-tune their innate immune systems to create and control their microbiome (Ma et al., 2020). It is interesting to note that the *PHR1*-mediated PSR system and the plant host Pi state both influence plant innate immunity, where *PHR1* plays a central role in integrating the PSR and immune system outputs (Castrillo et al., 2017). The host plant Pi status affects symbiosis with Pi-uptake-improving fungi via *PHR1* and the PSR system, as well as shaping plant-associated microbiota through the regulation of the plant immune response (Table 7.1). For instance, the bacterial volatile chemical diacetyl mediates the beneficial-to-deleterious change in the *Bacillus amyloliquefaciens–Arabidopsis* interaction under Pi deprivation (Morcillo et al., 2020). Diacetyl decreases the production of reactive oxygen species and encourages *B. amyloliquefaciens* to colonise roots under ideal Pi conditions (Morcillo et al., 2020). However, in Pi-deficient *Arabidopsis* plants, diacetyl activates the PSR system and boosts phytohormone-mediated immunity (Morcillo et al., 2020). Likewise, legumes decrease the number of root nodules per root system when Pi levels are low (Hernandez et al., 2009). This reaction is believed to lower the high rhizobial requirement, because of the high carbon cost obtained from plants and ATP, to set up a dynamic equilibrium between the gains and the costs of root nodule symbiosis (Ferguson et al., 2019). Additionally, genome-wide transcriptome analysis shows that Pi-deficient *Proteus vulgaris* plants infected with rhizobia exhibit increased expression of phytohormone-mediated defence-related genes (Isidra Arellano et al., 2020). Based on these findings, it is evident that host plants can activate their immune systems to prevent symbiosis with soil bacteria that risk their Pi status when Pi is limited.

PHR1 and *PHR1-LIKE 1* (PHL1) work together to regulate the transcriptional activation of PSR (Bustos et al., 2010). *PHR1* transcription factors bind to the *PHR1* binding site, a conserved cis-motif found in the promoters of phosphate-starvation-induced (PSI) genes. The *PHR1* and *PSI* genes are primarily engaged in adaptation strategies to low Pi circumstances. In addition to *PSI* genes, *PHR1* has been shown to directly control the expression of genes involved in immune responses in *Arabidopsis thaliana* (Castrillo et al., 2017). Although plant PSR and soil P play a part in plant–microbe interactions, the direct link between Pi signalling and defence responses is receiving increased attention. For instance, the jasmonic acid (JA) signalling pathway is stimulated during P deprivation, increasing resilience to the insect pest *Spodoptera littoralis* (Khan et al., 2016). According to Khan et

TABLE 7.1
Phosphorus-Dependent Plant Defence Pathways Activated against Pathogens

Plant	Pathogen/Microbe	Gene/Protein/Metabolite Involved	References
Regulation of plant-associated microbial communities and their metabolites			
Arabidopsis	*Pseudomonas syringae* DC3000	*PHR1* and *PHL1*	Finkel et al. (2019)
Rice	*Magnaporthe oryzae*	MiR399f	Campos-Soriano et al., (2020)
Arabidopsis	*Bacillus amyloliquefaciens*	Diacetyl	Morcillo et al., (2020)
Activation of phosphate-starvation responses (PSRs) and associated defence responses			
Arabidopsis	*Heterodera sacchari*	Overexpression of *NITROGEN LIMITATION ADAPTATION (NLA)* gene	Hewezi et al. (2016)
Arabidopsis	*Heterodera schachtii*	*PHT1* and *PHO1*	Hernandez et al., (2007)
Pi transporters and their role in defence responses			
Arabidopsis	*Pseudomonas syringae* pv maculicola	*PHT4;1* (mutation-suppressed)	Wang et al. (2011)
Arabidopsis	*Pseudomonas syringae* DC3000	*PHT4;6* (SA- and SAG-induced)	Hassler et al. (2012)
Cross-talk with plant defence hormones			
Arabidopsis	*Bacillus amyloliquefaciens* strain GB03	SA-induced	Morcillo et al. (2020)
Cotton	*Verticillium dahliae*	JA-induced	Luo et al. (2021)
Arabidopsis	*Pseudomonas syringae* DC3000	SA-induced	Castrilo et al. (2017)
Arabidopsis, tomato, and tobacco	*Spodoptera littoralis*	JA-induced	Khan et al. (2016)
Arabidopsis	*Botrytis cinerea* and *Alternaria alternate*	Strigolactone-induced	Torres-Vera et al. (2014)
Phosphorylation and MAPK cascades			
Arabidopsis	*Pseudomonas aeruginosa*	FLS2	Gomez-Gomez and Boller (2000)
Arabidopsis	*Escherichia coli*	ELF18	Zipfel et al. (2006)
Arabidopsis	*Pseudomonas syringae*	avrRpt2	Underwood et al. (2007)

al. (2016), this process is conserved across plant species and can be seen in *Arabidopsis*, tomato (*Solanum lycopersicum*), and tobacco (*Nicotiana benthamiana*). It is interesting to note that Pi shortage induces *JASMONATE-ZIM-DOMAIN PROTEIN 10 (JAZ10)* expression, which is restricted to an early developmental stage (such as a seedling of seven days old), reliant on *PHR1* (Khan et al., 2016). As a result, *PHR1* might contribute in part to the JA-mediated immunity during Pi shortage. In another study, it was found that PSR, JA biosynthesis, and the SA-responsive defence are among the pathways whose expression profiles are altered by the phytoplasma effector SAP11 expressed in transgenic *Arabidopsis* plants (Lu et al., 2014). Also, Pi starvation-induced genes were strongly triggered in transgenic *Arabidopsis* expressing phytoplasma effector SAP11, in a *PHR1*-dependent manner, but these Pi-starvation-induced genes led to anthocyanin deficiency in the leaves and also made the transgenic plants more prone to *P. syringae* DC3000 infection, which might be partially due to downregulation of *LIPOXYGENASE 2* (JA biosynthesis), *PATHOGENESIS-RELATED GENE 1*, and ELICITOR *ACTIVATED GENE 3* (SA-responsive genes) (Lu et al., 2014).

7.4 ROLE OF PI TRANSPORTERS IN DEFENCE RESPONSES

Pi is one of the main mobile forms of phosphorus found in plants. Pi transporters, which are expressed in numerous cells and localised at various subcellular compartments, are in charge of its highly dynamic mobilisation. *Phosphate Transporter 1* (PHT1) family members, induced by Pi deprivation, are essential for both the initial uptake of Pi from the soil and subsequent Pi translocation inside the organs (Nussaume et al., 2011). Family members of *PHT2*, *PHT3*, *PHT4*, and *PHT5* are found in a variety of organelles, including chloroplasts, mitochondria, the Golgi apparatus, and vacuoles (Wang et al., 2018). Root-to-shoot Pi translocation and Pi loading into seeds are carried out by the *Phosphate 1* (*PHO1*) family of Pi transporters (Che et al., 2020). Pi must be distributed correctly for physiological function (Yang et al., 2017) and also to control defensive reactions. Potassium phosphate was first identified as a key player in plant immunity in a study (Inoue et al., 1994). Mutation in *PHT4;1* gene, a Pi transporter, makes plants more susceptible to *Pseudomonas syringae* strains that are virulent, down-streaming the SA signalling and thus demonstrating the Pi transporter's role in plant immunity. The susceptible phenotype can be rescued by the exogenous application of benzothiadiazole, an SA agonist, placing *PHT4;1* upstream of SA-mediated defence responses (Wang et al., 2011). The same authors subsequently reported that *PHT4;1* regulates SA-mediated defence by controlling important defence-related genes such as *SALICYLIC ACID INDUCTION DEFICIENT 2*, *AUXIN RESPONSE FACTOR-GAP DOMAIN 2-LIKE DEFENCE RESPONSE PROTEIN 1*, *ENHANCED DISEASE SUSCEPTIBILITY 5*, and *PHYTOALEXIN DEFICIENT 4* (Wang et al., 2014). Additionally, mutation of the Pi transporter *PHT4;6*, which is located in the Golgi apparatus, results in a constitutive defence response and increased resistance to *P. syringae* DC3000 (Hassler et al., 2012). NMR analysis detected up-regulation in Pi concentration in the acidic compartment of a *PHT4;6* mutant (Hassler et al., 2012). Since the total Pi level was normal in *PHT4;6* mutants, Pi compartmentation could be the primary cause of the developmental phenotype and enhanced immunity through the accumulation of SA and SA conjugate (Hassler et al., 2012). Very recently, *PHT1;4*-mediated phosphate uptake was shown to be repressed upon activation of PTI. This inhibition depended on the receptor-like cytoplasmic kinases *BOTRYTIS-INDUCED KINASE 1* (*BIK1*) and *PBS1-LIKE KINASE 1* (*PBL1*), which both phosphorylate *PHT1;4* acting as a negative regulation of phosphate transport by immune signalling (Dindas et al., 2022). Likewise, in a study to delineate the SA biosynthesis pathway, a *BENZOIC ACID HYPERSENSITIVE 1-DOMINANT* (*BAH1-D*) mutant defective in nitrogen limitation adaptation (*NLA*) was identified (Yaeno and Iba, 2008). *BAH1-D* or *NLA* mutants over-accumulated SA and exhibited enhanced resistance to *P. syringae* DC3000, suggesting that *NLA* functions as a negative regulator of the production of SA and defence responses (Yaeno and Iba, 2008). However, the opposite role of *NLA* in defence responses has also been cited in the literature. Overexpression of miR827 leads to cleavage of *NLA* transcript, which increased *Arabidopsis* susceptibility to *Heterodera schachtii* and *Pseudomonas fluorescens* strain EtHAn (Hewezi et al., 2016). The loss of function of *NLA* in *Arabidopsis* (nla-3) also showed enhanced susceptibility to *H. schachtii*, while overexpression of rNLA (a miR827-resistant version of NLA) conferred resistance (Hewezi et al., 2016). It's intriguing to note that plasma membrane-localised *PHT1* Pi transporters are degraded by an NLA that encodes a RING-type E3 ubiquitin ligase (Yang et al., 2020). These findings highlight the cross-talk of Pi transporters with the immune responses of plants, although the mechanism is far from well understood.

7.5 PHYTOHORMONE CROSS-TALK WITH PHOSPHORUS IN DEFENCE RESPONSES

Cros-regulation occurs between Pi starvation and other plant signalling pathways, such as in sugars and phytohormones (Rouached et al., 2010). Complex cell-to-cell and organ-to-organ communication networks allow plants to integrate their nutritional and symbiotic status. Pi starvation causes certain sugar accumulation in plants such as glucose and sucrose, and high sugar levels in roots induce root system architecture (RSA) changes under Pi deprivation (Hammond and White, 2008). Sugar accumulation is connected with various hormone signalling as well as influencing different functions. For example, more sucrose in the roots in response to Pi starvation increases auxin and ethylene levels, which are involved in RSA developmental changes (Jain et al., 2007). Pi and abscisic acid (ABA) signalling pathways mediate developmental changes, including increases in the root:shoot ratio as well as in the root hair density (Ciereszko et al.,, 2005). The stimulation of plant growth and nutrient acquisition by beneficial rhizobacteria is related to the biosynthesis of plant growth regulators, including auxins, gibberellins, cytokinins (CK), and ABA. These phytohormones are involved in plant responses to drought and P deficiency (Egamberdieva et al., 2017). For example, ethylene and ABA play an important role in the drought-stress adaptation of plants, as they cause various morphological and chemical changes in plants, ensuring plant survival under water-limited conditions. In addition, ABA is reported to induce stomatal closure, reduce leaf surface, and increase shoot:root ratio; thus, a reduction in transpiration and an increase in soil volume for water uptake by plants occur. Furthermore, indole-3-acetic acid increases root exudation for direct and indirect phosphorus mobilisation in soil (Yahya et al., 2021).

S-induced resistance is highly associated with the activation of downstream transcription factor NPR1 (Nonexpresser of Pathogenesis-Related gene 1), along with NPR2, NPR3, and NPR4, which dictate the activation of defence genes and synthesis of antimicrobial proteins and secondary metabolites (Ding and Ding, 2020). A change in SA level and SA-related signalling is generally observed in nutrient deficiency studies, particularly P deficiency (Hammond and White, 2008). SA-dependent defence responses against biotrophic pathogens that obtain nutrients from living plants can be modulated by cytokinin (Argueso et al., 2012), whose level is responsive to P starvation. For example, under P starvation, both CK level (kinetin equivalent in sunflower) (Ei-D et al., 1979) and the transcript of its receptor *CYTOKININ RESPONSE 1/ARABIDOPSIS HISTIDINE KINASE 4* (*CRE1/AHK4*) are suppressed (Franco-Zorrilla et al., 2002; Chiou and Lin, 2011). CK is reported to drastically increase the Pi root:shoot distribution ratio, in contrast to a relatively minor effect by other hormone treatments (Silva-Navas et al., 2019). It is known that the CK signalling pathway is also critical for defence in response to plant pathogens and herbivores (Cortleven et al., 2019). For example, in *Arabidopsis*, exogenous trans-Zeatin application is reported to promote plant resistance against *Pseudomonas syringae* DC3000 (Choi et al., 2010) and virulent oomycete *Hyaloperonospora arabidopsidis* isolate Noco2 (Argueso et al., 2012). CK treatment promoted both SA production and SA-induced gene expression after flg22 elicitation (Choi et al., 2010). JA, another phytohormone, is a lipid-derived phytohormone important for defence against insect attack and necrotrophic pathogens, but to date, knowledge about the P-starvation-induced defence response mediated by JA is limited. Khan et al. (2016) demonstrated that P deficiency induced JA and JA-Isoleucine accumulation in *Arabidopsis* leaves. Similarly, the leaves of *PHO1* mutants defective in root-to-shoot Pi translocation accumulate higher JA and JA-Isoleucine. Caterpillars feeding on *PHO1* mutant plants or plants grown in a P-deficient medium showed reduced survival rate and larvae weight (Khan et al., 2016). Very recently, additional proof for JA-mediated immunity under P limitation was reported in cotton, where JA biosynthesis was activated under phosphate deficiency, inducing flavonoid and lignin accumulation in cotton and thereby conferring resistance to *Verticillium dahlia*, which causes cotton wilt disease (Luo et al., 2021). In addition, the phytohormone strigolactones (SLs) have been reported to play a role in the JA-dependent defence response. For example, two reports indicated that SL is required for the full resistance against *Botrytis cinerea* and *Alternaria alternata* in tomato (Torres-Vera et al., 2014) and against *Sclerotinia sclerotiorum* in *Physcomitrella patens* (Decker et al., 2017). It is reported that SL synthesis is promoted under P starvation (Lopez-Raez et al., 2008). SL is a relatively "new hormone" with very limited information regarding its role in plant immunity. However, the positive correlation between JA level and JA marker gene expression makes it a promising candidate for studying PSR, plant immunity, and their cross-talk (Khan et al., 2016).

7.6 PHOSPHORYLATION AND MAPK CASCADES IN DEFENCE SIGNAL TRANSDUCTION

Apart from the role played by phosphorus in plant defence systems by way of influencing the microbial communities associated with plants, activating the PSR and cross-talk with hormones, phosphorylation also plays a predominant role in signal transduction after pathogen perception. Plants are protected against pathogens by a well-coordinated immune system (Jones and Dangl, 2006). PTI is initiated upon recognition of conserved microbial features known as microbe-associated molecular patterns (MAMPs) at the plasma membrane by pattern-recognition receptors (PRRs) in the plant. ETI is activated by the recognition of pathogen-secreted effectors via intracellular nucleotide-binding leucine-rich repeat proteins (NLRs). ETI is commonly known as a sturdier immune response, while PTI offers long-lasting and broad-spectrum resistance (Tsudaand Katagiri, 2010). However, these two layers of immunity operate synergistically, merging on several downstream responses including Ca^{2+} burst, stomatal closure to limit the entry of bacteria, restriction of nutrient transfer from the cytosol to the apoplast to limit bacterial multiplication, production and secretion

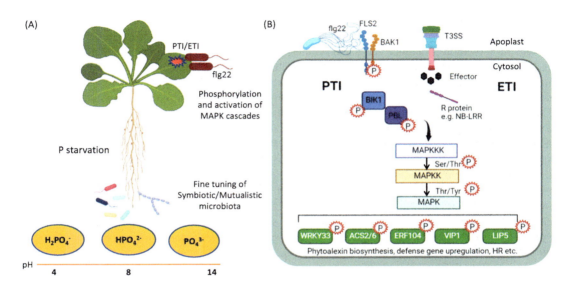

FIGURE 7.1 Phosphorus-dependent regulation of plant immune system. (A) Depending on phosphorus status in soil, plants have the ability to fine-tune the microbiota associated with them to derive optimum benefit. Plants have also evolved complex phosphate-starvation responses (PSRs), which regulate morphological and physiological adaptive changes when phosphorus availability is limited. In addition, a high degree of crosstalk of defence hormones and phosphorus availability is in place, which regulates the plant defence responses. (B) Further, phosphorylation also plays a predominant role in signal transduction after pathogen perception. Microbe-associated molecular pattern (MAMP) perception by pattern-recognition receptors (PRRs) and pathogen effector recognition by R gene products, such as the nucleotide-binding leucine-rich repeat proteins (NB-LRRs), lead to ligand-receptor complex formation resulting in different auto- and trans-phosphorylation reactions of the different players. *Arabidopsis FLAGELLIN-SENSITIVE 2* (*FLS2*) perceives the N terminus of flagellin and is represented by the 22-amino-acid-long epitope flg22, and *Arabidopsis EF-TU RECEPTOR* (*EFR*) perceives *ELONGATION FACTOR Tu* (EF-Tu) through an 18-amino-acid-long eliciting epitope known as elf18. *BRI1 ASSOCIATED RECEPTOR KINASE 1* (*BAK1*), which is a co-receptor, interacts with *FLS2* and can interact with different PRRs and binds flg22 in association with *FLS2*. In addition to *BAK1*, the receptor-like cytoplasmic kinase *BOTRYTIS-INDUCED KINASE 1* (*BIK1*) and related *PBL* (*PBS1-like*) *KINASES* associate with *FLS2* and *EFR* and become phosphorylated and unconstrained from the PRR complexes upon MAMP perception. Downstream of these, the activation of MAPKs takes place. Three tiers of these kinases are present in a cell, where activation of MAPKs is carried out by their upstream kinases, *MAPK KINASES* (*MAPKKs*), through the phosphorylation of a Thr and a Tyr residue. In turn, *MAPKKs* are activated by their upstream kinases, *MAPKK KINASES* (*MAPKKKs*), through the phosphorylation of two Ser/Thr residues in the *MAPKK* activation loop. These *MAPK* cascades phosphorylate target proteins, including transcription factors and enzymes, thus controlling the synthesis/signalling/up-regulation of defence hormones, the activation of defence genes, the synthesis of antimicrobial metabolites, the stomatal closure, and hypersensitive response (HR)-like cell death, along with an array of other defence responses. (Based on Meng and Zhang, 2013; Bigeard et al., 2015.)

of antimicrobial compounds including phytoalexins and defence-related proteins/peptides, generation of reactive oxygen species (ROS), which have toxic effects on pathogens, and a programmed cell death (PCD), referred to as the hypersensitive response – expression of defence-related genes and accumulation of physico-chemical barriers (Thomma et al., 2011; Ngou et al., 2021). One of the key drivers of these defences is the activation of receptor and co-receptor kinases upon pathogen perception, leading to phosphorylation of downstream targets and activation of mitogen-activated protein kinases (MAPKs). Protein phosphorylation, mediated by protein kinases and removed by protein phosphatases, is one the most abundant post-translational modifications found in eukaryotes and plays important functions in signal transduction of external stimuli into intracellular responses (Olsen and Mann, 2013). Phosphorylation of proteins can change several important properties, such as protein stability, enzyme activity, or subcellular localisation (Bigeard et al., 2015). Abundant phosphorylation occasions occur upon MAMP perception by PRRs (Rayapuram et al., 2018). PRRs are generally plasma membrane-located receptor-like kinases (RLKs) or receptor-like proteins with extracellular domains allowing MAMP perception (Bohm et al., 2014). A number of PRR/MAMP pairs have been reported, such as *Arabidopsis* FLAGELLIN-SENSITIVE 2 (*FLS2*), which perceives the N terminus of flagellin, represented by the 22-amino-acid-long epitope flg22 from *Pseudomonas aeruginosa* (Gomez-Gomez and Boller, 2000), and *Arabidopsis EF-TU RECEPTOR* (*EFR*), which perceives elongation factor Tu (*EF-Tu*) through an 18-amino-acid-long eliciting epitope known as elf18 from *Escherichia coli* (Zipfel et al., 2006). MAMP perception brings fast immune receptor complex formation at the plasma membrane and different auto- and trans-phosphorylation reactions of the different players (Macho and Zipfel, 2014). *BRI1 ASSOCIATED RECEPTOR KINASE 1* (*BAK1*) interacts with *FLS2*, and its absence reduces early flg22-dependent responses (Heese et al., 2007). *BAK1* can interact with different PRRs and binds flg22 in association with FLS2; hence, it is regarded as a co-receptor (Sun et al., 2013). In addition to *BAK1*, the receptor-like cytoplasmic kinase *BOTRYTIS-INDUCED KINASE 1* (*BIK1*) and related *PBL* (*PBS1-LIKE*) *KINASES* associate with *FLS2* and *EFR* and become phosphorylated and unconstrained from the PRR complexes upon MAMP perception (Zhang et al., 2014). Downstream of these, the activation of MAPKs takes place, which is one of the earliest signalling events after plant sensing of MAMPs and pathogen effectors. Three tiers of these kinases are present in a cell, where activation of MAPKs is carried out by their upstream kinases, MAPK kinases (MAPKKs), through the phosphorylation of a Thr and a Tyr residue in the Thr-X-Tyr activation motif of MAPKs. In turn, MAPKKs are activated by their upstream kinases, MAPKK kinases (MAPKKKs), through the phosphorylation of two Ser/Thr residues in the Ser/Thr-X3–5-Ser/Thr motif of the MAPKK activation loop. These three-kinase cascades, known as MAPK cascades, are important signalling modules that function downstream of receptor kinases (Figure 7.1). These MAPK cascades phosphorylate target proteins, including transcription factors and enzymes, thus controlling the synthesis/signalling/up-regulation of defence hormones, the activation of defence genes, the synthesis of antimicrobial metabolites, the stomatal closure, and hypersensitive response (HR)-like cell death, along with an array of other defence responses. To counter these immune responses, pathogens utilise effectors to inactivate plant MAPK cascades directly or suppress MAPK signalling by targeting upstream signalling components; thus a complex tug-of-war is at work between host and pathogens. Additionally, plant protein phosphatases act in dephosphorylating and inactivating MAPKs, regulating the magnitude, duration, and physiological result of MAPK activation, which might maintain a trade-off between plant development and defence response.

REFERENCES

Afzal, A. and Bano, A. (2008) Rhizobium and phosphate solubilizing bacteria improve the yield and phosphorus uptake in wheat (*Triticum aestivum* L.). *In:l J. Agric. Biol.* 10: 85–88.

Alori, E.T., Glick, B.R. and Babalola, O.O. (2017) Microbial phosphorus solubilization and its potential for use in sustainable agriculture. *Front. Microbiol.* 8(971): 1–8.

Argueso, C.T., Ferreira, F.J., Epple, P., To, J.P.C., Hutchison, C.E., Schaller, G.E. (2012) Two-component elements mediate interactions between cytokinin and salicylic acid in plant immunity. *PLOS Genet.* 8(1): e1002448.

Bechtaoui, N., Rabiu, M.K., Raklami, A., Oufdou, K., Hafidi, M. and Jemo, M. (2021) Phosphate-dependent regulation of growth and stresses management in plants. *Front. Plant Sci.* 12: 679916.

Bigeard, J., Colcombet, J. and Hirt, H. (2015) Signaling mechanisms in pattern-triggered immunity (PTI). *Mol. Plant* 8(4): 521–539.

Boehm, S.K. and Heike, F.B. (2013) Spotlight on age-diversity climate: The impact of age-inclusive HR practices on firm-level outcomes. *Pers. Psychol.* 67(2014): 667–704.

Bustos, R., Castrillo, G., Linhares, F., Puga, M.I., Rubio, V., Perez-Perez, J. (2010) A central regulatory system largely controls transcriptional activation and repression responses to phosphate starvation in *Arabidopsis*. *PLOS Genet.* 6(9): e1001102.

Campos-Soriano, L., Bundó, M., Bach-Pages, M., Chiang, S.F., Chiou, T.J. and San Segundo, B. (2020) Phosphate excess increases susceptibility to pathogen infection in rice. *Mol. Plant Pathol.* 21(4): 555–570.

Castrillo, G., Teixeira, P.J.P.L., Paredes, S.H., Law, T.F., de Lorenzo, L., Feltcher, M.E. et al. (2017) Root microbiota drive direct integration of phosphate stress and immunity. *Nature* 543(7646): 513–518.

Che, J., Yamaji, N., Miyaji, T., Mitani-Ueno, N., Kato, Y. and Shen, R.F. (2020) Node-localized transporters of phosphorus essential for seed development in rice. *Plant Cell Physiol.* 61(8): 1387–1398.

Chiou, T.J. and Lin, S.I. (2011) Signaling network in sensing phosphate availability in plants. *Annu. Rev. Plant Biol.* 62: 185–206.

Choi, J., Huh, S.U., Kojima, M., Sakakibara, H., Paek, K.H. and Hwang, I. (2010) The cytokinin-activated transcription factor ARR2 promotes plant immunity via TGA3/NPR1-dependent salicylic acid signaling in Arabidopsis. *Dev. Cell* 19(2): 284–295.

Ciereszko, I., Johansson, H. and Kleczkowski, L.A. (2005) Interactive effects of phosphate deficiency, sucrose and light/dark conditions on gene expression of UDP-glucose pyrophosphorylase in Arabidopsis. *J. Plant Physiol.* 162(3): 343–353.

Cortleven, A., Leuendorf, J.E., Frank, M., Pezzetta, D., Bolt, S. andSchmulling, T. (2019) Cytokinin action in response to abiotic and biotic stresses in plants. *Plant Cell Environ.* 42(3): 998–1018.

Daei-hassani, B., Nader, C., Leila, S. and Masumeh, A. (2016) Effects of phosphorus on antioxidant system in pepper cultivars under saline conditions. *Iran. J. Plant Physiol.* 7(1): 1935–1941.

David, P., Raj, R.S., Linda, R. and Rhema, S.B. (2014) Molecular characterization of phosphate solubilizing bacteria (PSB) and plant growth promoting rhizobacteria (PGPR) from pristine soils. *Int. J. Innov. Sci. Eng. Technol.* 1: 317–324.

Decker, E.L., Alder, A., Hunn, S., Ferguson, J., Lehtonen, M.T., Scheler, B., Kerres, K.L., Wiedemann, G., Safavi-Rizi, V., Nordzieke, S., Balakrishna, A., Baz, L., Avalos, J., Valkonen, J.P.T, Reski, R. and Al-Babili, S. (2017) Strigolactone biosynthesis is evolutionarily conserved, regulated by phosphate starvation and contributes to resistance against phytopathogenic fungi in a moss, Physcomitrella patens. *New Phytol.* 216(2): 455–468.

Dindas, J., Thomas, DeFalco, A., Yu, G., Zhang, L., David, P., Bjornson, T., M.C., Custódio, V., Castrillo, G., Nussaume, L., Macho, A.P. and Zipfel, C. (2022) Direct inhibition of phosphate transport by immune signaling in Arabidopsis. *Curr. Biol.* 32(2): 488–495.

Ding, P. and Ding, Y. (2020) Stories of salicylic acid: A plant defense hormone. *Trends Plant Sci.* 25(6): 549–565.

Egamberdieva, D., Wirth, S.J., Alqarawi, A.A. and Abd Allah, E. F., & Hashem, A. (2017). Phytohormones and Beneficial Microbes: Essential Components for Plants to Balance Stress and Fitness. *Front Microbial* 8: 2104.

Ei-D, A.M.S.A., Salama, A. and Wareing, P.F. (1979) Effects of mineral nutrition on endogenous cytokinins in plants of sunflower (*Helianthus annuus* L.). *J. Exp. Bot.* 30(5): 971–981.

Ferguson, B.J., Mens, C., Hastwell, A.H., Zhang, M., Su, H., Jones, C.H. (2019) Legume nodulation: The host controls the party. *Plant Cell Environ.* 42(1): 41–51.

Finkel, O.M., Salas-González, I., Castrillo, G., Spaepen, S., Law, T.F., Teixeira, P.J.P.L., et al. (2019) The effects of soil phosphorus content on plant microbiota are driven by the plant phosphate starvation response. *PLoS Biol.* 17(11): 1–34.

Franco-Zorrilla, J.M., Martin, A.C., Solano, R., Rubio, V., Leyva, A. and PazAres, J. (2002) Mutations at CRE1 impair cytokinin-induced repression of phosphate starvation responses in *Arabidopsis*. *Plant J.* 32(3): 353–360.

Gómez-Gómez, L. and Boller, T. (2000) FLS2: An LRR receptor-like kinase involved in the perception of the bacterial elicitor flagellin in Arabidopsis. *Mol. Cell.* 5(6): 1003–1011.

Gouda, S., Das, G., Sen, S.K., Shin, H.S. and Patra, J.K. (2016) Endophytes: A treasure house of bioactive compounds of medicinal importance. *Front. Microbiol.* 7: 1538.

Hammond, J.P. and White, P.J. (2008) Sucrose transport in the phloem: Integrating root responses to phosphorus starvation. *J. Exp. Bot.* 59(1): 93–109.

Hassani, M.A., Duran, P. and Hacquard, S. (2018) Microbial interactions within the plant holobiont. *Microbiome* 6(1): 58.

Hassler, S., Lemke, L., Jung, B., Mohlmann, T., Kruger, F. and Schumacher, K. (2012) Lack of the Golgi phosphate transporter PHT4;6 causes strong developmental defects, constitutively activated disease resistance mechanisms and altered intracellular phosphate compartmentation in *Arabidopsis*. *Plant J.* 72(5): 732–744.

Heese, A., Hann, D.R., Gimenez-Ibanez, S., Jones, A.M., He, K., Li, J., Schroeder, J.I., Peck, S.C. and Rathjen, J.P. (2007) The receptor-like kinase SERK3/BAK1 is a central regulator of innate immunity in plants. *Proc. Natl. Acad. Sci. U.S.A.* 104(29): 12217–12222.

Hernandez, G., Ramõrez, M., Valdes-Lopez, O., Tesfaye, M., Graham, M.A., Czechowski, T. (2007) Phosphorus stress in common bean: Root transcript and metabolic responses. *Plant Physiol.* 144(2): 752–767.

Hernandez, G., Valdes-Lopez, O., Ramõrez, M., Goffard, N., Weiller, G., Aparicio-Fabre, R. (2009) Global changes in the transcript and metabolic profiles during symbiotic nitrogen fixation in phosphorus-stressed common bean plants. *Plant Physiol.* 151(3): 1221–1238.

Hewezi, T., Piya, S., Qi, M., Balasubramaniam, M., Rice, J.H. and Baum, T.J. (2016) Arabidopsis miR827 mediates post-transcriptional gene silencing of its ubiquitin E3 ligase target gene in the syncytium of the cyst nematode Heterodera schachtii to enhance susceptibility. *Plant J.* 88(2): 179–192.

Illmer, P.A. and Schinner, F. (1995) Solubilization of inorganic calcium phosphates solubilization mechanisms. *Soil Biol. Biochem.* 27(3): 257–263.

Inoue, S., Macko, V. and Aist, J.R. (1994) Identification of the active component in the papilla-regulating extract from barley leaves. *Physiol. Mol. Plant Pathol.* 44(6): 441–453.

Isidra-Arellano, M.C., Pozas-Rodrõguez, E.A., Reyero-Saavedra, M.D.R., Arroyo-Canales, J., Ferrer-Orgaz, S., Sanchez-Correa, M.D.S., et al. (2020) Inhibition of legume nodulation by Pi deficiency is dependent on the autoregulation of nodulation (AON) pathway. *Plant J.* 103(3): 1125–1139.

Jain, A., Poling, M.D., Karthikeyan, A.S., Blakeslee, J.J., Peer, W.A., Titapiwatanakun, B., Murphy, A.S. and Raghothama, K.G. (2007) Differential effects of sucrose and auxin on localized phosphate deficiency-induced modulation of different traits of root system architecture in Arabidopsis. *Plant Physiol.* 144(1): 232–247.

Johan, P.D., Ahmed, O.H., Omar, L. and Hasbullah, N.A. (2021) Phosphorus transformation in soils following co-application of charcoal and wood ash. *Agronomy* 11(10), 2010.

Jones, J. and Dangl, J. (2006) The plant immune system. *Nature* 444(7117): 323–329

Khan, G.A., Vogiatzaki, E., Glauser, G. and Poirier, Y. (2016) Phosphate deficiency induces the jasmonate pathway and enhances resistance to insect herbivory. *Plant Physiol.* 171(1): 632–644.

Khan, G.A., Vogiatzaki, E., Glauser, G. and Poirier, Y. (2016) Phosphate deficiency induces the jasmonate pathway and enhances resistance to insect herbivory. *Plant Physiol.* 171(1): 632–644.

Kumar, A. and Patel, H. (2018) Role of microbes in phosphorus availability and acquisition by plants. *Int. J. Curr. Microbiol. Appl. Sci.* 7(5): 1344–1347.

Lin, W.Y., Huang, T.K. and Chiou, T.J. (2013) Nitrogen limitation adaptation, a target of MicroRNA827, mediates degradation of plasma membrane-localized phosphate transporters to maintain phosphate homeostasis in *Arabidopsis*. *Plant Cell* 25(10): 4061–4074.

Lopez-Raez, J.A., Charnikhova, T., Gomez-Roldan, V., Matusova, R., Kohlen, W., De Vos, R. (2008) Tomato strigolactones are derived from carotenoids and their biosynthesis is promoted by phosphate starvation. *New Phytol.* 178(4): 863–874.

Lu, Y.T., Li, M.Y., Cheng, K.T., Tan, C.M., Su, L.W., Lin, W.Y., et al. (2014) Transgenic plants that express the Phytoplasma effector SAP11 show altered phosphate starvation and defense responses. *Plant Physiol.* 164(3): 1456–1469.

Luo, X., Li, Z., Xiao, S., Ye, Z., Nie, X., Zhang, X. (2021) Phosphate deficiency enhances cotton resistance to *Verticillium dahliae* through activating jasmonic acid biosynthesis and phenylpropanoid pathway. *Plant Sci.* 302: 110724.

Ma, K.W., Niu, Y., Jia, Y., Ordon, J., Copeland, C. and Emonet, A. (2020) Coordination of microbe-host homeostasis via a crosstalk with plant innate immunity. *Res. Square.* doi: 10.21203/rs.3.rs-69445/v1.

Macho, A.P. and Zipfel, C. (2014) Plant PRRs and the activation of innate immune signaling. *Mol. Cell.* 54(2): 263–272.

Medici, A., Szponarski, W., Dangeville, P., Safi, A., Dissanayake, I.M., Saenchai, C. (2019) Identification of molecular integrators shows that nitrogen actively controls the phosphate starvation response in plants. *Plant Cell* 31(5): 1171–1184.

Meng, X. and Zhang, S. (2013) MAPK cascades in plant disease resistance signaling. *Annu. Rev. Phytopathol.* 51: 245–266.

Morcillo, R.J.L., Singh, S.K., He, D., An, G., Võlchez, J.I., Tang, K. (2020) Rhizobacterium-derived diacetyl modulates plant immunity in phosphate-dependent manner. *EMBO J.* 39(2): e102602.

Ngou, B.P.M., Ahn, H.K., Ding, P. and Jones, J.D.G. (2021) Mutual potentiation of plant immunity by cell-surface and intracellular receptors. *Nature.* 592 (7852): 110–115.

Nussaume, L., Kanno, S., Javot, H., Marin, E., Nakanishi, T.M. and Thibaud, M.-C. (2011) Phosphate import in plants: Focus on the PHT1 transporters. *Front. Plant Sci.* 2 (83):1-12.

Olsen, J.V. and Mann, M. (2013) Status of large-scale analysis of post-translational modifications by mass spectrometry. *Mol. Cell Proteomics.* 12(12): 3444–3452.

Pant, B.-D., Pant, P., Erban, A., Huhman, D., Kopka, J. and Scheible, W.-R. (2015) Identification of primary and secondary metabolites with phosphorus status-dependent abundance in Arabidopsis, and of the transcription factor PHR1 as a major regulator of metabolic changes during phosphorus limitation. *Plant Cell Environ.* 38(1): 172–187.

Postma, J., Nijhuis, E.H. and Someus, E. (2010) Selection of phosphorus solubilizing bacteria with biocontrol potential for growth in phosphorus rich animal bone charcoal. *Appl. Soil Ecol.* 46(3): 464–469.

Prasad, R. and Chakraborty, D. (2019) Phosphorus basics: Understanding phosphorus forms and their cycling in the soil. *Alabama Coop. Ext Syst.* https://www. aces. edu/blog/topics/crop-production/understanding-phosphorus-forms-and-their-cycling-in-the-soil.

Rayapuram, N., Bigeard, J., Alhoraibi, H., Bonhomme, L., Hesse, A.M., Vinh, J., Hirt, H. and Pflieger, D. (2018) Quantitative phosphoproteomic analysis reveals shared and specific targets of Arabidopsis mitogen-Activated Protein Kinases (MAPKs) MPK3, MPK4, and MPK6. *Mol. Cell Proteomics.* 17(1): 61–80.

Rodríguez, H. and Fraga, R. (1999) Phosphate solubilizing bacteria and their role in plant growth promotion. *Biotechnol. Adv.* 17(4–5): 319–339.

Rouached, H., Arpat, A.B. and Poirier, Y. (2010) Regulation of phosphate starvation responses in plants: Signaling players and cross-talks. *Mol. Plant* 3(2): 288–299.

Rubio, V., Linhares, F., Solano, R., Martõn, A.C., Iglesias, J. and Leyva, A. (2001) A conserved MYB transcription factor involved in phosphate starvation signaling both in vascular plants and in unicellular algae. *Genes Dev.* 15(16): 2122–2133.

Seshachala, U. and Tallapragada, P. (2012) Phosphate solubilizers from the rhizosphere of *Piper nigrum* L. in Karnataka, India. *Chil. J. Agric. Res.* 72(3): 397–403.

Sharma, S.B., Sayyed, R.Z., Trivedi, M.H. and Gobi, T.A. (2013) Phosphate solubilizing microbes: Sustainable approach for managing phosphorus deficiency in agricultural soils. *Springerplus* 2: 587–600.

Sharma, A., Shahzad, B., Rehman, A., Bhardwaj, R., Landi, M. and Zheng, B. (2019) Response of phenylpropanoid pathway and the role of polyphenols in plants under abiotic stress. *Molecules (Basel, Switzerland)* 24(13): 2452.

Shi, J., Zhao, B., Zheng, S., Zhang, X., Xiaolin, W., Dong, W., Xie, Q., Wang, G., Xiao, Y., Chen, F., Yu, N. and Wang, E. (2021) A phosphate starvation response-centered network regulates mycorrhizal symbiosis. *Cell* 184(22): 5527–5540.

Silva-Navas, J., Conesa, C.M., Saez, A., Navarro-Neila, S., Garcia-Mina, J.M., Zamarreño, A.M., Baigorri, R., Swarup, R. and Del Pozo, J.C. (2019) Role of cis-zeatin in root responses to phosphate starvation. *New Phytol.* 224(1): 242–257.

Sun, Y., Li, L., Macho, A.P., Han, Z., Hu, Z., Zipfel, C., Zhou, J.M. and Chai, J. (2013) Structural basis for flg22-induced activation of the Arabidopsis FLS2-BAK1 immune complex. *Science.* 342(6158): 624–628.

Thomma, B.P., Nürnberger, T. and Joosten, M.H. (2011) Of PAMPs and effectors: The blurred PTI-ETI dichotomy. *The Plant Cell.* 23(1): 4–15.

Torres-Vera, R., García, J.M., Pozo, M.J. and López-Ráez, J.A. (2014) Do strigolactones contribute to plant defence?. *Mol. Plant Pathol.* 15(2): 211–216.

Tsuda, K. and Katagiri, F. (2010) Comparing signaling mechanisms engaged in pattern-triggered and effector-triggered immunity. *Curr. Opin. Plant Biol.* 13(4): 459–465.

Underwood, W., Zhang, S. and Y. He., S. (2007) The Pseudomonas syringae type III effector tyrosine phosphatase HopAO1 suppresses innate immunity in Arabidopsis thaliana. *Plant J.* 52(4): 658–672.

Walpola, C. and Yoon, M. (2012) Prospectus of phosphate solubilizing microorganisms and phosphorus availability in agricultural soils: A review. *Afr. J. Microbiol. Res.* 6: 6600–6605.

Wang, F., Deng, M., Xu, J., Zhu, X. and Mao, C. (2018) Molecular mechanisms of phosphate transport and signaling in higher plants. *Semin. Cell Dev. Biol.* 74: 114–122.

Wang, G.-Y., Shi, J.-L., Ng, G., Battle, S.L., Zhang, C. and Lu, H. (2011) Circadian clock-regulated phosphate transporter PHT4;1 plays an important role in *Arabidopsis* defense. *Mol. Plant* 4(3): 516–526.

Wang, H., Xu, Q., Kong, Y.H., Chen, Y., Duan, J.Y., Wu, W.H. and Chen, Y.F. (2014) Arabidopsis WRKY45 transcription factor activates PHOSPHATE TRANSPORTER1;1 expression in response to phosphate starvation. *Plant Physiol.* 164: 2020–2029.

Wu, Y.M., Zhou, Q.Y., Yang, X.Q., Luo, Y.J., Qian, J.J., Liu, S.X., et al. (2019) Induction of antiphytopathogenic metabolite and squalene production and phytotoxin elimination by adjustment of the mode of fermentation in cocultures of phytopathogenic nigrospora oryzae and Irpex lacteus. *J. Agric. Food Chem.* 67(43): 11877–11882.

Yaeno, T. and Iba, K. (2008) BAH1/NLA, a RING-type ubiquitin E3 ligase, regulates the accumulation of salicylic acid and immune responses to *Pseudomonas syringae* DC3000. *Plant Physiol.* 148(2): 1032–1041.

Yahya, M., Rasul, M., Farooq, I., Mahreen, N., Tawab, A., Irfan, M., ... & Yasmin, S. (2021) Differential root exudation and architecture for improved growth of wheat mediated by phosphate solubilizing bacteria. *Front. Microbiol.* 12: 744094

Yang, S.Y., Huang, T.K., Kuo, H.F. and Chiou, T.J. (2017) Role of vacuoles in phosphorus storage and remobilization. *J. Exp. Bot.* 68(12): 3045–3055.

Yang, S.Y., Lu, W.C., Ko, S.S., Sun, C.M., Hung, J.C. and Chiou, T.J. (2020) Upstream open reading frame and phosphate-regulated expression of rice OsNLA1 controls phosphate transport and reproduction. *Plant Physiol.* 182(1): 393–407.

Zhang, Z., Liao, H. and Lucas, W.J. (2014) Molecular mechanism underlying phosphate sensing, signaling, and adaptation in plants. *J. Integr. Plant Biol.* 56(3): 192–220.

Zipfel, C., Kunze, G., Chinchilla, D., Caniard, A., Jones, J.D., Boller, T. and Felix, G. (2006) Perception of the bacterial PAMP EF-Tu by the receptor EFR restricts Agrobacterium-mediated transformation. *Cell.* 125(4): 749–760.

8 Biotechnological Approaches for Improving Phosphate Uptake and Assimilation in Plants

Rumi Rumi, Kanika Maurya, Mandavi Pandey, Pawandeep S. Kohli, Poonam Panchal, Alok K. Sinha, and Jitender Giri

8.1 INTRODUCTION

Phosphorus (P) is the eleventh most abundant element on Earth's crust and the second most important macronutrient for plant growth. Chemically, P is a member of group 5A, with a high affinity to gain electrons (Tiessen, 2008). To complete its valency, P tends to combine with various metals and non-metals. Thus, P is mostly found associated with iron (Fe), aluminium (Al), calcium (Ca), or fluorine (F) in phosphate (Pi) rocks on Earth (Tiessen, 2008; Shen et al., 2011). Massive non-renewable P-rock reservoirs exist in nature, but their unequal geographical distributions (Morocco has the largest reserves, and China is the biggest exporter of P fertilisers) have raised several economic and political concerns (Cooper et al., 2011; Filippelli, 2018; Zowada et al., 2020). Additionally, the slow weathering rate of P rocks (i.e. apatites) in soil formation, along with poor solubility due to soil pH (which ranges from 5.0 to 7.0) attribute to the slower diffusion rate and highly stable character of P, making it even less available for plant uptake and utilisation (Pandey et al., 2013; Shen et al., 2011; Tiessen, 2008; Toledo et al., 2021). Despite having a concentration range of 200–800 mg/Kg soil, P becomes the critical limiting nutrient for plant health, owing to its lower water solubility i.e., < 0.5 mg P/L in soil solution (Tiessen, 2008). Consequently, P-deficient soil accelerates the demand for Pi fertilisers synthesised from the mining of P rocks for agricultural use. Heavy application of these synthetic chemical fertilisers has raised global concern about sustainable agro-economic development (Peterson, 2022; Shen et al., 2011; Zowada et al., 2020). Still, the demand for more agricultural output remains unachieved due to the heavy input cost of fertilisers, among other reasons (Peterson, 2022; Tiessen, 2008). Biologically, P is indispensable to all living creatures because it forms the main part of the phospholipid cell membrane, constitutes the energy transfer molecule adenosine triphosphate (ATP), and forms the backbone of nucleic acids DNA and RNA (Liu, 2021; Shen et al., 2011).

P exists in two forms in soil i.e., inorganic Pi as $H_2PO_4^-$ or HPO_4^{2-} soluble ions, and organic P (Po) as inositol phosphates, phosphonates, mono- and diesters of orthophosphates, and organic polyphosphates (Shen et al., 2011; Toledo et al., 2021). In nature, the soil environment has ~10 µM of Pi concentration which is extremely low when compared with the Pi requirement of cell cytoplasm of 1,000 to 10,000 µM (Pandey et al., 2013; Shen et al., 2011). Plants usually fulfil their Pi requirements by undergoing various morphological, physiological, microbial, and metabolic adaptations regulated by comprehensive transcriptional alterations (Mehra et al., 2015). Morphological modifications in root growth/architecture such as root length, number, root angle, and root hair proliferation for enhanced soil exploration are well documented. For example, the development of a shallow root system in common bean helps in better Pi uptake from topsoil, known as "topsoil foraging" (Lynch et al., 2001), whereas the development of a dense and determinant lateral proteoid root system (cluster root) offers great help in acquiring Pi in white lupin (Neumann et al., 1999; Vance et al., 2003; Vance, 2008). Thus, root growth angle (RGA) acts as a dominant regulator of Pi acquisition efficiency (PAE) and plant yield (Liu, 2021). The formation of longer lateral roots (LRs) is yet another adaptation which helps better Pi acquisition and shoot growth in maize under Pi-deficient conditions (Jia et al., 2018). Under Pi deficiency, physiological processes such as secretion of mucilage, signalling molecules, organic acids, and acid phosphatases in the form of root exudates are important to deal with low Pi stress (Bhadouria & Giri, 2021; Panchal et al., 2021; Shen et al., 2011). These exudates solubilise Pi bound to organophosphates and metals in soil to make it available for root uptake (Liu, 2021). Pi release from membrane phospholipids helps in maintaining the intrinsic Pi equilibrium under stress conditions (Verma et al., 2021a, b). Similarly, various evidence of metabolic adaptations have been reported, where plants choose alternate glycolytic and cellular respiration pathways to survive in prolonged P-deficient conditions (Theodorou et al., 1993). Pi-solubilising microbes, (PSMs) such as Pi-solubilising fungi (PSF), form an extensive hyphal network with roots and assist in better Pi uptake, while Pi-solubilising bacteria (PSBs) help in efficient P acidification and mobilisation in soil (Krüger et al., 2015; Liu, 2021; Shen et al., 2011; Toledo et al., 2021).

Extensive studies to delineate the mechanisms behind Pi deficiency tolerance pathways have revealed a group of transcriptional regulators (PHR1, WRKY) involved in the regulation of Pi deficiency response *in planta*. For example, PHR1, an MYB transcription factor, is a master regulator of Pi deficiency response. *Arabidopsis* PHR1 and its homologs in other plants are mostly responsible for inducing transcriptional alterations for various morpho-physiological and metabolic adaptations in plants (Bustos et al., 2010; Rubio et al., 2001; Thibaud et al., 2010). Similarly, Jiang et al. (2017) reviewed the structural and functional role of WRKYs under different types of stresses, including low P stress (Jiang et al., 2017).

Despite facing protests from activists and bans in many countries, GM technology has been proven extremely useful in the case of Bt brinjal, Bt cotton, and a few other examples (Turnbull et al., 2021). Rapid development in the availability of gene mutants and the breakthrough discovery of CRISPR/Cas9 technology should increase the development of crops with new and improved traits, such as tolerance to P deficiency (Azadi et al., 2016; Bennett et al., 2004; Munive et al., 2018; Muthayya et al., 2012; Raman, 2017; Shelton et al., 2020). Slowly, gene-edited crops are being accepted in countries such as the United States, India, Japan, and Bangladesh, which is an encouraging development for the commercialisation of gene-edited crops with improved P-use-efficiency (PUE). Here, we discuss the available information on combating P-deficiency-stress by biotechnologically manipulating genes regulating Pi uptake and P recycling through the action of acid phosphatases, organic acids, and membrane phospholipids remodelling. Further, we introduce a few novel strategies involved in improving P acquisitions and propose how they can be targeted for crop improvement (Figure 8.1). Emerging genetic engineering tools, such as gene editing, would be of great help in improving crop plants for efficient P uptake and assimilation.

8.2 TARGETING P UPTAKE AND DISTRIBUTION IN PLANTS

8.2.1 PHOSPHATE TRANSPORTERS

A variety of membrane transporters are required to load and unload different mineral nutrients across the plant body to maintain mineral homeostasis for proper development. Tuning the expression of these Pi transporters at different growth stages and in different tissues is one of the important mechanisms to increase PAE and PUE by the plants. Primarily, the uptake of Pi from the soil in roots and its movement within the plant system are mediated by two transporter families: Pi transporters

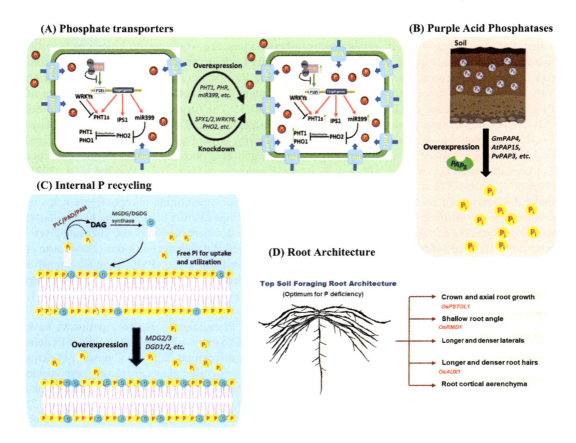

FIGURE 8.1 Different strategies for improving Pi acquisition and utilisation in plants. (A) Increasing Pi transport activity. Overexpression of plasma membrane-localised *PHT1* transporters lead to their abundance and more Pi import in cells. The binding of PHR to *P1BS* element in the promoter of target genes (*IPS1, miR399, PHT1s*) is inhibited via SPX proteins, knockout of which could lead to enhanced activity of targets and more Pi uptake. Overexpression of *PHR, miR399* (down-regulator of PHO2) enhances Pi transport. Few WRKYs (WRKY6) and PHO2 are known to downregulate *PHTs* and *PHO1*, the knockout of which can upregulate more Pi transport. Green arrows depict the increase in activity. (B) Enhancing secreted acid phosphatase activity. Overexpression of *PAPs* increases the hydrolysis of organic P (Po) in soil and releases more soluble Pi, which is available for root uptake. (C) Improving internal utilisation efficiency. Overexpression of lipid remodelling genes (*MGDs, DGDs*) that replace phospholipid (P in yellow circles) with galactolipids (G in blue circles) in cellular membranes using DAG as a substrate, releasing more free Pi and making it available for uptake and utilisation. (D) Optimising root system architecture. List of different traits to be considered to enhance topsoil foraging for improving Pi acquisition under Pi deficiency.

(PHTs) and Phosphate1 (PHO1) (Młodzińska & Zboińska, 2016). PHTs are broadly grouped into five classes – PHT1, PHT2, PHT3, PHT4, and PHT5 – based on differences in structure, localisation, and their associated functions (Srivastava et al., 2018; Wang et al., 2018). Most PHT1 class members are plasma membrane-localised high-affinity Pi transporters, which are expressed in the root epidermal cells. However, the other four (PHT2–PHT5) form a group of low-affinity transporters and are localised to the mitochondria, vacuole, chloroplast, and Golgi apparatus to maintain intracellular Pi homeostasis (Młodzińska & Zboińska, 2016; Nussaume et al., 2011). From roots, Pi is then supplied to shoot with the help of PHO1 transporters (Poirier et al., 1991). Discussion about further classification and the function of these transporters is beyond the focus of this chapter and has been wisely explained in various articles (Dissanayaka et al., 2021; Prathap et al., 2022; Lambers, 2022; Wang et al., 2021b, c).

Several approaches have been used in various studies, wherein overexpression or silencing of different Pi transport-related genes has been explored to increase the Pi uptake by the plants. Such studies under different Pi regimes helped in the better elucidation of the Pi transporter functions and their regulatory mechanisms in Pi uptake and acquisition in the plant system. This has allowed the understanding and development of different ways to target such transporters directly or indirectly to increase their activity, leading to enhanced Pi uptake and utilisation. As most soils are poor in P, many studies have focused on the overexpression of different Pi transporters, which could lead to enhanced Pi accumulation with better growth under different Pi conditions and less dependency on fertilisers. This is achieved through overexpressing genes under a variety of strong constitutive promoters such as *CaMV35S* and *Ubiquitin*. Most overexpression studies in rice and *Arabidopsis* have targeted *PHT1* gene family members, which are mainly expressed in roots. The overexpression of *PHT1* family members showed tolerance under Pi starvation and/or sufficient conditions by transgenic rice and *Arabidopsis*. This is accompanied by more Pi uptake, P accumulation, and growth parameters in terms of leaf number, shoot biomass, grain weight, and yield (Jia et al., 2011; Nagarajan et al., 2011; Sun et al., 2012b; Zhang et al., 2014a, 2014b; Zhang et al., 2015). Transgenic tobacco plants overexpressing soybean *GmPT1* showed a significant increase in seed weight, total weight, and PUE under Pi starvation conditions (Song et al., 2014). Overexpression of *GmPT7* also helped in promoting nodulation, Pi uptake, and overall soybean yield (Chen et al., 2019). Similarly, overexpression of maize *PHT1;7* also showed a higher growth rate, increased shoot biomass, larger young leaves, and elevated Pi levels compared to wild-type or knockout plants (Wang et al., 2020a).

Ma et al. (2021) recently illustrated that overexpression of *OsPHO1;2* helped in enhanced export of Pi from seeds, causing less Pi in seeds and husks with more Pi in flag leaf. This helps in recycling Pi under low Pi conditions, which improves PUE and plant overall stature and yield (Ma et al., 2021). Several studies also demonstrate changes in root architecture caused by overexpression of a particular *PHT* gene. In *Chrysanthemum*, an ornamental plant, overexpression of *PHT1;2* led to increased root length, root area, number of root tips, and plant length under Pi starvation, along with improved Pi uptake and accumulation (Liu et al., 2018).

Overexpression of the regulatory unit that targets Pi transporters at the transcriptional or translational level can also help in improving Pi uptake in plants, including PHR1 and PHL1, which function redundantly and act as transcriptional activators (Rubio et al., 2001; Sun et al., 2016). Under P starvation, these transcriptional factors become activated and bind to the palindromic region *P1BS* element present in phosphate starvation response (PSR) genes to initiate different P starvation responses. Such *P1BS* elements are also found in the promoter of different Pi transporter genes, and, thus, overexpressing these *PHRs* might affect the PHT activity. Studies found that overexpression of *AtPHR1* and its rice ortholog *OsPHR2* led to excessive Pi levels in the shoot. This might be due to the enhanced PHTs activity for more Pi uptake and accumulation (Guo et al., 2015; Nilsson et al., 2007; Zhou et al., 2008). Hence, overexpression of the master regulator may not always be a desirable strategy since it may affect other plant processes as well thus negating the overall benefits.

Similarly, WRKY domain transcription factors bind to the W-box region present in the promoter of genes associated with different stress conditions (Dong et al., 2003; Jiang et al., 2017). Studies found that these WRKY transcription factors bind to the promoter of *PHTs* and affect their expression in both a positive and negative manner. For example, AtWRKY42 acts as a transcriptional activator and suppressor for *PHT1;1* and *PHO1*, respectively (Su et al., 2015). Overexpression of *AtWRKY42* led to enhanced Pi uptake and accumulation within the plant. Similarly, AtWRKY42 and AtWRKY45 positively regulate the *AtPHT1* activity and increase Pi accumulation in transgenics (Su et al., 2015; Wang et al., 2014a). Studies in rice also connected WRKY transcription factors with Pi responses. For instance, overexpression of *OsWRKY74* influenced the expression of Pi transporters such as *PHT1;3*, *PHT1;9*, and *PHT1;10* and led to the accumulation of Pi in plants under normal and low P conditions (Dai et al., 2016). This was accompanied by more shoot and root biomass, and elongated and denser primary and adventitious roots under Pi deficiency. Moreover, such overexpression also led to more iron uptake. Recently, Zhang et al. (2021) found that OsWRKY21 and OsWRKY108 function redundantly in regulating Pi uptake, and their overexpression transgenics accumulate higher Pi in plants by enhancing the activity of PHT1;1 (Zhang et al., 2021).

Suppressing a few genes regulating the Pi transporter can also change the plant Pi accumulation. SPX-containing proteins are key regulators involved in maintaining Pi homeostasis in plants. Rice and *Arabidopsis* contain four and six SPX family members, respectively, some of which (*AtSPX1*) are upregulated under Pi starvation. Under Pi-sufficient conditions, Inositol pyrophosphate8 (IP$_8$)-mediated interaction of SPX with AtPHR1/OsPHR2 prevents the localisation of PHR to the nucleus and its activation of different PSI genes (Puga et al., 2014). In *Arabidopsis* and rice, knocking down or mutating SPX1 resulted in increased PHT1 transport activity (Puga et al., 2014; Wang et al., 2014b). Such transgenics accumulate high Pi in shoots under Pi-sufficient conditions. However, such plants showed impaired growth and reduced biomass, primarily because of Pi toxicity. Because of the observed functional redundancy among the PHT transporters, overexpressing or generating mutants can cause a Pi imbalance in different tissues.

Moreover, in many of the studies, plants overexpressing PHTs show Pi toxicity when grown in P-sufficient conditions, eventually leading to overall compromised plant biomass. In addition, a few transporters may function as multiple ion transporters, and overexpressing them might cause an imbalance in the different mineral ratios within the plants. To prevent the resultant Pi toxicity due to the constitutive overexpression of Pi transporters, Pi starvation-inducible promoters could be utilised to drive the expression of PHTs in transgenics. Similarly, as the activity of a particular transporter may depend upon a specific domain or amino acid present, mutating the particular amino acid in protein using the biotechnological tool is a better option. For instance, a study by Chen et al. (2015) focused on changing the Pi transporter's localisation to the plasma membrane (PM) to enhance its activity. In this study, replacing S517 with alanine at the OsPT8 C-terminal end (PT8-CT) prevented the phosphorylation of PT8 by casein kinase 2 (CK2), which allows the increased abundance of OsPT8 to PM by stepping up the localisation from endoplasmic reticulum (ER) to PM and thereby enhancing Pi uptake in rice (Chen et al., 2015). Maize *ZmPT7* and *Arabidopsis* PHT1;1 (major transporters for Pi uptake) are the closest homologs to *OsPT8* and undergo phosphorylation to prevent exit from ER to PM. It has been seen that modulating phosphorylating sites in these PTs to their non-phosphorylated form (ZmPT7S521E and AtPHT1;1S520E) significantly increased Pi uptake and showed a higher growth rate compared to wild type (Bayle et al., 2011; Wang et al., 2020a). As high homology between different Pi transporters and the phosphorylated sites in the C-terminal of PHTs are highly conserved among different plants, generating the mutants of PTs in crop plants carrying non-phosphorylation sites for CK2 will help in enhancing the PAE under different Pi regimes. The use of CRISPR Cas9 technology is an important solution to combat the problems associated with plant breeding or GM crops. Generating knockout, use of base editing, or transcriptionally activating particular genes can be used in crop improvement without leaving foreign trace elements (Tomlinson et al., 2019; Zsögön et al., 2018). For instance, base editing to cause C (Cytidine) to T (Thymidine) replacement at Thr327Met in NRT1 has improved N use efficiency (Lu & Zhu, 2017).

Similarly, manipulating Pi translocation to seed using node-based transporters has been demonstrated as an effective way forward for improving PUE (Yamaji et al., 2017). Earlier, a node-based transporter, SULTR-like P Distribution Transporter (SPDT), was identified to be responsible for the allocation of P in rice grains (Yamaji et al., 2017). Knocking out of the *Spdt* gene controls phytate accumulation in grains and less Pi accumulation in seeds without compromising the yield. Rather, newly absorbed P was distributed to the flag leaf and other parts of the plant. Many mutants associated with different transporters families such as ABC transporters, sulphate transporters, and PHTs have been identified for their functioning in seed's low phytic acid content (Cominelli et al., 2020). Recently, Kumar et al. (2021) identified a natural mutant that encodes for SPDT and shows low Pi transport to grain. It carries the single amino acid mutation (V330A), leading to lower total P accumulation in seeds and more mineral availability. Increased P content in leaves and straw can thus contribute to P resupply (manure) in the field and reduce dependency on external fertilisers (Kumar et al., 2021).

8.2.2 ROOT SYSTEM ARCHITECTURE

Roots are vital for nutrient acquisition and foraging, thus significantly controlling the uptake efficiency of the plant (Miguel et al., 2013; Sun et al., 2018a). In recent years, root research has seen major advancements (Ephrath et al., 2020); however, modern cultivars generally lack beneficial root traits due to shoot or yield-based artificial selection in fields with high resource input. Therefore, it is necessary to incorporate beneficial root traits into cultivars to improve nutrient acquisition efficiency.

As for P, topsoil foraging is one such trait to improve P acquisition, as most of the soil P is present in the upper soil layers (Lynch & Brown, 2001). Topsoil foraging can be enhanced by improving crown roots and axial root growth with a shallower angle and a high density of longer laterals and root hairs (Lynch, 2019). Molecular and genetic dissection of root traits for topsoil foraging led to the identification of important genes and mechanisms controlling these traits. In rice, a QTL PUP1 for P uptake is fine-mapped to a gene, *PSTOL1*, controlling crown root growth, depicting the association of P uptake with key root traits (Gamuyao et al., 2012; Heuer et al., 2009; Wissuwa et al., 2002). In addition, in rice, a low P-responsive RMD protein controls crown root angle by interacting with actin filaments and statoliths. Mutants of *rmd1* are unresponsive to low P in producing shallower crown roots (Huang et al., 2018). Further, the rice auxin transporter OsAUX1 facilitates the root hair response in P deficiency in rice, improving the acquisition and foraging of P (Giri et al., 2018). However, only a few of these findings are transferred to crop improvement for Pi acquisition. For example, overexpression of *PSTOL1* and introgression of PUP1 led to enhanced P uptake in rice varieties previously susceptible to P deficiency and lacking PUP1 (Chin et al., 2010, 2011; Gamuyao et al., 2012; Swamy et al., 2020). The approach to targeting single genes for enhancing root traits such as topsoil foraging has not been successful because root traits are quantitative and controlled with complex genetics involving environmental interactions and multiple component phenotypes (Lynch, 2019). Thus, the way forward is to improve each component phenotype and incorporate them into a single variety using gene pyramiding. For this, both marker-assisted selection and gene editing can be utilised.

8.2.3 ROLE OF ORGANIC ACIDS IN P UPTAKE AND ASSIMILATION

Complex P sources can also be utilised by plants through the release of organic acids at the root–soil interface (Wan et al., 2017). Organic acids are excellent chelating agents that form complexes with positively charged mineral nutrients through their carboxyl groups, concomitantly releasing the Pi for their uptake by plants (Panchal et al., 2021). The most common organic acids that are involved in mineral nutrients uptake and homeostasis are low molecular weight organic acids such as citrate, malate, oxalate, and gluconate (Casarin et al., 2004; Li et al., 2018; Ma et al., 1997; Shahbaz et al., 2006). Thus, targeting higher secretion of these low molecular weight organic acids at the root–soil interface is a promising approach to increase P uptake in plants. The genes related to organic acid biosynthesis and transport have been overexpressed in various plant species to solubilise complex P sources such as Al-P, Fe-P, and Ca-P in soil. Almost two decades ago, it was reported that overexpression of a gene responsible for citrate biosynthesis resulted in higher P uptake by utilising Ca-P, ectopically added to the soil (López-Bucio et al., 2000). After that, the other genes functional in organic acid biosynthesis were also analysed for their roles in P uptake by plants. Mitochondrial citrate synthase (*CS*) gene isolated from carrot (*Daucus carota*), when overexpressed in *Arabidopsis thaliana*, showed higher P uptake and improved growth parameters when grown in soil carrying high levels of Al-P (Koyama et al., 2000). Apart from citrate, malate exudation was also helpful for increasing P uptake in plants. Enhanced malate exudation due to the overexpression of the malate dehydrogenase (*MDH*) gene in tobacco helped the transgenics to take up P by solubilising it from Al-P, Fe-P, and Ca-P (Lü et al., 2012).

Previously, it was anticipated that the overexpression of genes involved in organic acid biosynthesis might not always result in enhanced citrate accumulation and secretion in plants, due to some of the inconsistencies observed by various scientific groups (Barone et al., 2008; Delhaize et al., 2001; Han et al., 2009). Thus, the organic acid transporter genes were targeted for their possible roles in P uptake. A malate efflux transporter gene, *Triticum aestivum Aluminium activated malate transporters 1 (TaALMT1)*, showed improved P uptake from acidic soils when overexpressed in barley (Delhaize et al., 2009). Earlier, the same gene provided tolerance to Al toxicity, a common abiotic stress that co-occurs with low P stress in acidic soils (Delhaize et al., 2004). A close homologue of TaALMT1 in soybean (GmALMT1) was responsible for Al toxicity tolerance through malate exudation from roots of a Pi-efficient cultivar, depending upon the concentration of P provided in the media (Liang et al., 2013). Transporters of both Al-activated malate transporters (ALMT) and the multidrug and toxic compound extrusion (MATE) family are known to alleviate Al toxicity through the secretion of either malate or citrate at the root–soil interface in crops. For instance, *transporters such as HvAACT1* in barley (Zhou et al., 2013), *ScFRDL2* in rye (Yokosho et al., 2010), *BnALMT1* and *BnALMT2* in rape (Ligaba et al., 2006), and *OsFRDL2* and *OsFRDL4* in rice (Yokosho et al., 2016a, 2016b) are important for secretion of organic acids that can bind to toxic Al^{3+} ions and in turn prevent the binding of these ions to the growing root tips. Further characterisation of these genes under low P stress using overexpression and knockdown/knockout lines might help understand their utility for providing tolerance to such stress in crop plants.

The secretion of organic acids by plants is often amplified by the presence of beneficial micro-organisms in the rhizosphere. For instance, the presence of *Burkholderia multivorans* WS-FJ9 in the rhizosphere of a perennial plant species, poplar, led to an increase in gluconic acid secretion by its root system (Li et al., 2018). In addition, a range of bacterial species, known as Pi-solubilising micro-organisms, secrete organic acids such as oxalic, citric, succinic, fumaric, and α-ketogutaric acid by themselves to solubilise P that the plants can utilise for their growth and development (Khan et al., 2006). In turn, organic acids released from plant roots attract micro-organisms by acting as chemo-attractants and further enhance their growth in the form of carbon and energy sources (Macias-Benitez et al., 2020; Zhang et al., 2014c). In this way, a mutual synergism might occur between plants and micro-organisms for better P uptake at the root–soil interface. A recent meta-analysis with perennial and annual plant species showed that the perennial legume species, but not the annual legume species, benefited from the P-solubilising advantages of micro-organisms (Primieri et al., 2022). The authors suggested that the reason for the conflicting results between the perennial and annual species could be the obvious differences between the life cycle patterns followed by the perennial and the annual species. The perennial species have more time to establish an association with their microbial partners, which could become more beneficial with time (Primieri et al., 2022). Therefore, manipulating genes involved in organic acid biosynthesis and transport in perennial crops might provide new avenues of P-deficiency-stress tolerance through enhanced organic acid secretion.

8.2.4 Secreted Acid Phosphatases

Acid phosphatases belong to a class of hydrolases (EC 3.1.3.2) that hydrolyse the P monoester bond present in organic substrates into alcohol and Pi (Bhadouria & Giri, 2021; Hurley et al., 2010; Veljanovski et al., 2006). Out of all isozymes present, purple acid phosphatases (PAPs) seem to play an active role in P metabolism under P-deficiency stress. On hydrolysis, the released product Pi helps in maintaining the P balance in plant cells (Wang et al., 2009). Thus, the enhanced activity of PAPs can be beneficial to crops growing in P-deficient soils. PAPs are binuclear metalloenzymes because their active site contains metal ions such as Fe coupled with manganese or zinc (Antonyuk et al., 2014). PAPs display a wide diversity in substrate specificity and can hydrolyse a wide variety of organic P substrates (Feder et al., 2020; Mehra et al., 2017; Olczak et al., 2003). PAPs exist in two forms, i.e. secreted PAPs and cytosolic PAPs. Cytosolic PAPs function to remobilise P from intracellular compartments of cells such as the vacuoles and lysosomes, while secreted ones are mostly confined to the root cell wall or secretory globules in the rhizosphere (Suen et al., 2015; Tian et al., 2015; Wang et al., 2009). Secreted PAPs thus help in Pi hydrolysis from soil organic P (Po) sources making it available for root uptake. Secreted PAPs are discussed here, and cytosolic will be discussed later in the chapter.

Plant PAPs have been widely studied in *Arabidopsis thaliana*, *Phaseolus vulgaris*, *Glycine max*, *Cicer arietinum*, *Solanum tuberosum*, *Oryza sativa*, *Lupinus albus*, *Populus* sp., and many more (Bhadouria et al., 2017; Gao et al., 2017; Kavka et al., 2021; Liang et al., 2010; Mehra et al., 2017; Miller et al., 2001; Robinson et al., 2012; Tian et al., 2012; Veljanovski et al., 2006; Wang et al., 2009, 2011; Zimmerman et al., 2004; Zhu et al., 2020). All secreted PAPs are functional in the extracellular spaces and possess a putative N-terminal signal sequence to become localised on the plasma membrane of root surface cells or become part of the root exudates in soil (Bhadouria & Giri, 2021; Zimmermann et al., 2004). Several studies from dicots to monocots have explored secreted PAPs to raise crops able to mitigate low Pi stress. For example, in dicots, heterologous overexpression of *GmPAP4* in *Arabidopsis* showed increased biomass accumulation through improved phytate utilisation in culture (Kong et al., 2014). Subsequently, in common bean (*Phaseolus vulgaris*), *PvPAP3* was overexpressed in the transgenic hairy root, which resulted in enhanced fresh weight and P content through better utilisation of exogenously applied ATP as the sole source of P (Liang et al., 2010). Also, when *GmPAP7a/7b* was overexpressed in soybean hairy root, a higher root-associated secreted APase activity for better extracellular ATP utilisation was reported (Zhu et al., 2020). Additionally, *GmPAP14* was analysed by overexpressing in *Arabidopsis*, which resulted in higher P accumulation and increased shoot weight suggesting better external phytate utilisation (Kong et al., 2018). More insights were gained through the successful exogenous application of *CaPAP7* purified protein on *Arabidopsis* seedlings, which enhanced phytate utilisation by the seedlings (Bhadouria et al., 2017). Studies suggest that *OsPAP10a* is a root-associated secreted APase, and its overexpression in rice improved the extracellular ATP hydrolysis and utilisation, therefore helping the plants to grow better (Tian et al., 2012). In continuation to this, overexpressing transgenics of *OsPAP10c* displayed higher tiller number and shorter plant height through increased APase activity of external Po in rice (Lu et al., 2016). Another study involving secreted PAP came from the overexpression of *OsPAP21b* in rice. The transgenics showed increased biomass and P content both in hydroponics supplemented with ATP/ADP and soil supplemented with manure as an organic P source (Mehra et al., 2017).

Constitutive overexpression of *AtPAP15*, containing a carrot (*Daucus carota*) extracellular targeting peptide, leads to an increase in APase activity in soybean hairy root. Further, it was noticed that improved biomass accumulation in overexpressed soybean plants when phytate was applied as the sole P source in the sand culture system (Wang et al., 2009). One of the adapted plants to P deficiency, white lupin (*Lupinus albus*) uses PAPs secreted through the proteoid roots for efficient P mobilisations for plant nutrition (Miller et al., 2001). In *Stylosanthes guianensis* sps., *SgPAP23* was reported to be the superior utiliser of extracellular phytate P, and its overexpression in both bean hairy root and *Arabidopsis* plants suggested increased extracellular phytate P utilisation in *Stylosanthes* (Liu et al., 2018). Recently, two *Stylosanthes* cultivars (RY2 and RY5) were compared for their P acquisition efficiency, and RY2 was shown to be better in acquiring P. A transcriptomic study on RY2 root tissues revealed that 21 PAPs were upregulated, including *purple acid phosphatase 2-, 22- like*, under low Pi conditions (Chen et al., 2021). Recently, in *Populus* sp., 33 putative PAPs were reported, and a secreted PAP, i.e., PtPAP1, was highly upregulated in the root apoplastic region under low Pi conditions (Kavka et al., 2021). Interestingly, no negative effects on plant growth were observed on PAP overexpression in most cases, thus making secreted PAPs a promising target for crop improvement.

8.3 TARGETING REALLOCATION AND RE-UTILISATION OF P IN PLANTS

For the last few decades, most studies have centred on PAE where Pi uptake from the soil is improved. This links to morphological changes in roots for topsoil foraging via enhancing lateral root density, number of root hairs, shallower root growth angles, etc. (Jia et al., 2018; Lynch, 2019). This is accompanied by the secretion of root exudates, enzymes, and organic acids to solubilise soil organic P and upregulate root Pi transporters to enhance P uptake (Panchal et al., 2021; Pang et al., 2018). Moreover, plants also undergo improvement in PUE where they utilise the acquired P via recycling and redistribution from various cellular and tissue pools. Several transporters (PHT1;4, PHT1;3, SPDT, etc.) are shown to reallocate the P from the source to sink organs within the plants under P deficiency (Chang et al., 2019; Yamaji et al., 2017). This includes the Pi redistribution from older to younger leaves and from flag leaves to panicles at the vegetative and reproductive stages, respectively. Overexpression of *GmPT1* in tobacco also enhanced plant dry biomass, PUE, and yield (Song et al., 2014). At the cellular level, vacuoles act as a storage and remobilisation system for Pi. To meet the P demand under Pi starvation, stored vascular Pi is consumed to maintain cytosolic and cellular Pi homeostasis. In rice, OsVPE1/2 serves as an effluxer for vacuolar Pi, where its expression increases under Pi starvation and its mutants showed early onset of PSR because of a delay in Pi release to the cytosol (Xu et al., 2019).

Improvement of PUE can also be achieved via utilising an organic P pool that is majorly present as P lipids, Pi esters, and nucleic acids. In a process known as membrane lipid remodelling, P-free lipids such as galactolipids and sulpholipids replace plasma membrane phospholipids and release Pi in a cytosolic pool to be utilised by plants (Siebers et al., 2015; Verma et al., 2021a, b). Similarly, PAPs target the ester P (Adenosine diphosphate, sugar phosphate) and release Pi during Pi starvation and improve PUE (Plaxton & Tran, 2011). These processes are discussed below.

8.3.1 Membrane Lipid Remodelling

Apart from improving Pi uptake, plant cells can recycle P from P-lipid cellular membranes. Cellular membrane lipids (phospholipids) represent one-third of the total cellular organic P (Pant et al., 2015; Poirier et al., 1991) and the largest pool of P that can be internally channelled (Nakamura, 2013; Veneklaas et al., 2012). These phospholipids are exchanged with galactolipids such as monogalactosyldiacylglycerol (MGDG), digalactosyldiacylglycerol (DGDG), and sulfolipid sulfoquinovosyldiacylglycerol (SQDG), a process that has been referred to as membrane lipid remodelling (Härtel et al., 2000; Nakamura, 2013; Zhang et al., 2014d).

The membrane phospholipid remodelling comprises two steps: (i) hydrolysis of phospholipids and (ii) their replacement by galactolipids and sulfolipids (Nakamura, 2013), which are synthesised using diacylglycerol (DAG) as a substrate. The cascades of genes involved in lipid remodelling were found to be correlated with P-deficiency response. Phospholipases and phosphatidic acid phosphatases, the genes involved in the biosynthesis of galactolipids MGDG, DGDG, and SQDG, were all found to be upregulated during Pi deficiency in various plants (Cruz-Ramirez et al., 2006; Gaude et al., 2008; Lan et al., 2012; Nakamura et al., 2005, 2009). Some phospholipases are specific for P starvation, e.g., in *Arabidopsis*, non-specific lipases NPC4 and NPC5 (Gaude et al., 2008; Nakamura et al., 2005). Similarly, phospholipases D zeta 1 and 2 (PLDζ1 and PLDζ2) hydrolyse major phospholipids, and then there are phosphatidate phosphohydrolases 1 and 2 (PAH1 and PAH2) that hydrolyse phosphatidic acid to produce DAG, which acts as a substrate for galactolipid biosynthesis. MGDG synthase MGD1-3 acts on DAG to synthesise MGDGs. DGDG synthases, DGD1 and DGD2, account for the increase in DGDG. DGD2 is activated during Pi starvation (Kelly et al., 2003). SQD2 is the major final enzyme involved in sulpholipid biosynthesis but is also employed in glucuronosyl diacylglycerol synthesis (Okazaki et al., 2013). It has now known that SQD2 is only the major enzyme that has a role under Pi deficiency stress, while SQD1 has some other roles (Okazaki et al., 2013; Yu et al., 2002).

Efficient replacement of P lipids with non-P lipids can enhance PUE (Dissanayaka et al., 2018; Verma et al., 2021a). PUE and related measures are important parameters that depict the level of P availability and its utilisation and recycling in plants (Heuer et al., 2017; Han et al., 2022). Many studies have been done on crop species such as oat, soybean, maize, and especially rice to understand lipid remodelling in relation to PUE (Andersson et al., 2003; Mehra & Giri, 2016, Mehra et al., 2019; Tawaraya et al., 2018; Verma et al., 2022). Rice overexpressing glycerophosphodiester phosphodiesterases *OsGDPD-2* had increased biomass and P content under P-deficient conditions owing to a higher level of galactolipids (Mehra et al., 2019). Yuan et al. (2019) generated *DGK1* knockout and overexpression lines to show their role in regulating root architecture and thus nutrient uptake efficiency in plants (Yuan et al., 2019). Diacylglycerol kinase (DGK) enzyme phosphorylates DAG to form phosphatidic acid. The glycerophosphodiester phosphodiesterase (GPX-PDE) enzyme involved in phospholipid catabolism was identified in maize by Wang et al. (2021a) and generated mutant lines using CRISPR/Cas9 (Wang et al., 2021a). It was found to be involved in improving Pi recycling from senescent leaves to young leaves. Recently, Verma et al. (2022) showed that *OsMGD3* affects Pi utilisation and acquisition in rice via lipid remodelling. Thus, modulating lipid-related genes may help improve P utilisation and assimilation but may also have repercussions, as targeting lipids may hinder further signalling. The authors discussed the connection between phosphatidic acid (PA) and P utilisation efficiency. PA is found to regulate root architecture and thus nutrient uptake and utilisation (Verma et al., 2022). However, extensive reports are still lacking with respect to the signalling roles of lipid molecules under P deficiency in plants. In future, further research in this field may better establish the role of membrane lipid remodelling in improving PUE.

8.3.2 Cytosolic PAPs in P Recycling

Cytosolic PAPs are believed to be the primary responders in plants facing low Pi stress or senescence, when plants utilise strategies like P recycling or Pi scavenging mechanisms to cope with Pi deprivation (Bhadouria & Giri, 2021; Tian et al., 2015). For example, vacuolar PAPs are well known to recycle Pi from expendable sources of cytosolic phosphomonoesters and anhydrides in Pi-deprived plant cells (Tran et al., 2010). However, prolonged Pi-deprived conditions cause significant reductions in cytoplasmic P metabolite content and impairs many P-requiring biochemical processes (Tian et al., 2015; Tran et al., 2010). Although cytosolic PAPs are involved in P recycling, molecular evidence discussing the underlying mechanism remains undetermined. Nevertheless, some biotechnological attempts in *Arabidopsis thaliana*, *Glycine max*, and *Oryza sativa* have been made, which shed some light on the functional role of cytosolic PAPs. A study conducted on *GmPAP21* revealed that its overexpression in the transgenic hairy root improved root growth under Pi-deficient conditions and effectively enhanced Pi utilisation in the root cells (Li et al., 2017). Not only this, the overexpression of *GmPAP12* within the nodules of transformed hairy roots of soybean showed the importance of Pi remobilisation and recycling activities through increased phytase and APase activity (Wang et al., 2020b). Further reports on internal Pi remobilisation for efficient Pi utilisation was discussed in the senescing leaves of *Arabidopsis*, which were experimentally analysed by the creation of T-DNA insertion *atpap26* mutants (Robinson et al., 2012). A similar study was conducted in rice, where the overexpression of *OsPAP26* was reported for efficient Pi remobilisation from senescing leaves to young leaves, which helped the plants to grow better (Gao et al., 2017).

Additionally, some PAPs are known as "dual PAPs" because they have both secreted and cytosolic modes of action. For example, in *Glycine max*, when *GmPAP7a/7b* was overexpressed, it showed a significant increase in the cytosolic root-associated APase activity (in response to Pi starvation conditions) as compared with the leaf (Zhu et al., 2020). Adding to this, when *AtPAP2* was overexpressed in *Arabidopsis*, it provided better seed yield and more biomass accumulation through increased cytosolic phytate utilisation (Sun et al., 2012a, 2018b). Another study in *Arabidopsis* involved the T-DNA insertion lines of *AtPAP15*, which delineated its role in pollen germination through an increased cytosolic P-reflux mechanism (Kuang et al., 2009). Moreover, studies conducted on *AtPAP26* discussed the cytosolic APase activity that could efficiently reflux vacuolar Pi under P-starved conditions in the *Arabidopsis* cell suspension system (Veljanovski et al., 2006). In continuation to this, a highly compromised phenotype was observed in the T-DNA insertion mutant of *atpap26* in *Arabidopsis* with decreased root and shoot cytosolic APase activity (Hurley et al., 2010). Cytosolic PAPs can help recycle plant P in both a development and Pi deficiency-dependent manner. However, PAP proteins are often degraded in the presence of sufficient P. The use of gene editing in mutating the responsible amino acids can help improve PAP stability and enhancement of internal P recycling, thus improving PUE.

8.4 FUTURE PERSPECTIVE: NOVEL P-ASSIMILATING STRATEGIES

In addition to targeting the major pathways and processes for improving P acquisition and utilisation efficiency, novel strategies can be employed. One such strategy is using a non-metabolisable form of P, phosphite (PO_3^{3-}), along with phosphite metabolising enzymes from microbes. Transgenic tobacco and *Arabidopsis* with *ptxD* (phosphite oxidoreductase) can grow normally after phosphite application, as foreign phosphite oxidoreductase can now convert phosphite to orthophosphate (López-Arredondo & Herrera-Estrella, 2012). Interestingly, phosphite is also an excellent weed control system, as weeds cannot grow in the presence of phosphite, while transgenic plants can (López-Arredondo & Herrera-Estrella, 2012). Similarly, using genetic engineering, phosphite metabolism genes have been introduced in cotton and rice to incorporate phosphite metabolising properties (Pandeya et al., 2018; Ram et al., 2019). Apart from non-metabolising forms, a major portion of soil P is in an insoluble form that must be solubilised before root uptake. For this conversion, root-secreted APases and organic acids can help. However, the production and release of acid phosphatases/organic acids is an energy-consuming process and can increase the metabolic cost of plants in P-stressed environments. Therefore, introducing PSMs in the soil can be a better approach. PSMs act on the insoluble P forms by releasing acid, alkaline phosphatases, or phytase enzymes to convert organic or inorganic insoluble forms to plant-absorbable forms (Alori et al., 2017). Various genes responsible for encoding these enzymes are being characterised, which could further be used to bioengineer microbes or plants using a transgenic approach (Sharma et al., 2013).

Under P stress, reducing the metabolic cost of soil exploration is a necessary adaptation for continual survival. For this, plants develop root cortical aerenchyma (RCA), which significantly reduces the metabolic cost of soil exploration (Lynch, 2019). Maize genotypes with higher RCA and lower living cells displayed increased yield and growth in P-deficient soil (Galindo-Castañeda et al., 2018). However, trade-offs, such as reduced radial nutrient transport and mycorrhizal association, are also associated with more RCA (Galindo-Castañeda et al., 2018; Hu et al., 2014). Thus, crop-specific and locally adapted approaches would be more beneficial. For example, crops such as *Lupinus albus*, *Macadamia integrifolia*, and *Cucurbita pepo* form short-lived cluster roots that can increase the surface for P acquisition and are involved in the exudation of carboxylates and phosphatases for P solubilisation (Shane & Lambers, 2005). Improving the longevity and growth of cluster roots in these crops would be a great crop-specific strategy for low P conditions. All of these novel approaches are beneficial in improving the uptake and availability of P in a deficient environment. For improving PUE, lipid remodelling pathways are the most promising targets. However, P recycling can also happen from plant cell walls through the depolymerisation of cell wall pectates. For depolymerisation, pectin methyl esterases are induced in P-deficient conditions (Tao et al., 2022). Enhancing P recycling via overexpressing pectin catabolic genes could result in higher PUE.

8.5 ACKNOWLEDGEMENTS

RR and KM acknowledge the CSIR-Senior Research Fellowship (SRF). PSK and MP acknowledge DBT. PP acknowledges UGC and NIPGR. JG acknowledges the Indo-Swiss Joint Research Programme (BT/IN/Swiss/46/JG/2018-2019).

REFERENCES

Alori, E. T., Glick, B. R., & Babalola, O. O. (2017). Microbial phosphorus solubilization and its potential for use in sustainable agriculture. *Frontiers in Microbiology*, 8, 971.

Andersson, M. X., Stridh, M. H., Larsson, K. E., Liljenberg, C., & Sandelius, A. S. (2003). Phosphate-deficient oat replaces a major portion of the plasma membrane phospholipids with the galactolipid digalactosyldiacylglycerol. *FEBS Letters*, 537(1–3), 128–132.

Antonyuk, S. V., Olczak, M., Olczak, T., Ciuraszkiewicz, J., & Strange, R. W. (2014). The structure of a purple acid phosphatase involved in plant growth and pathogen defence exhibits a novel immunoglobulin-like fold. *IUCrJ*, 1(2), 101–109.

Azadi, H., Samiee, A., Mahmoudi, H., Jouzi, Z., Rafiaani Khachak, P., De Maeyer, P., & Witlox, F. (2016). Genetically modified crops and small-scale farmers: Main opportunities and challenges. *Critical Reviews in Biotechnology*, 36(3), 434–446.

Barone, P., Rosellini, D., LaFayette, P., Bouton, J., Veronesi, F., & Parrott, W. (2008). Bacterial citrate synthase expression and soil aluminum tolerance in transgenic alfalfa. *Plant Cell Reports*, 27(5), 893–901.

Bayle, V., Arrighi, J.-F., Creff, A., Nespoulous, C., Vialaret, J., Rossignol, M., Gonzalez, E., Paz-Ares, J., & Nussaume, L. (2011). *Arabidopsis thaliana* high-affinity phosphate transporters exhibit multiple levels of posttranslational regulation. *The Plant Cell*, 23(4), 1523–1535.

Bennett, R. M., Ismael, Y., Morse, S., & Kambhampati, U. S. (2004). Economic impact of genetically modified cotton in India. *AgBioForum*, 7(3): 96–100.

Bhadouria, J., & Giri, J. (2021). Purple acid phosphatases: Roles in phosphate utilization and new emerging functions. *Plant Cell Reports*, 1–19.

Bhadouria, J., Singh, A. P., Mehra, P., Verma, L., Srivastawa, R., Parida, S. K., & Giri, J. (2017). Identification of purple acid phosphatases in chickpea and potential roles of CaPAP7 in seed phytate accumulation. *Scientific Reports*, 7(1), 1–12.

Bustos, R., Castrillo, G., Linhares, F., Puga, M. I., Rubio, V., Perez-Perez, J., ... Paz-Ares, J. (2010). A central regulatory system largely controls transcriptional activation and repression responses to phosphate starvation in Arabidopsis. *PLOS Genetics*, 6(9), e1001102.

Casarin, V., Plassard, C., Hinsinger, P., & Arvieu, J. C. (2004). Quantification of ectomycorrhizal fungal effects on the bioavailability and mobilization of soil P in the rhizosphere of *Pinus pinaster*. *New Phytologist*, 163(1), 177–185.

Chang, M. X., Gu, M., Xia, Y. W., Dai, X. L., Dai, C. R., Zhang, J., Wang, S. C., Qu, H. Y., Yamaji, N., Feng Ma, J., & Xu, G. H. (2019). OsPHT1;3 mediates uptake, translocation, and remobilization of phosphate under extremely low phosphate regimes. *Plant Physiology*, 179(2), 656–670.

Chen, J., Wang, Y., Wang, F., Yang, J., Gao, M., Li, C., ... & Wu, P. (2015). The rice CK2 kinase regulates trafficking of phosphate transporters in response to phosphate levels. *The Plant Cell*, 27(3), 711–723.

Chen, L., Qin, L., Zhou, L., Li, X., Chen, Z., Sun, L., Wang, W., Lin, Z., Zhao, J., Yamaji, N., Ma, J. F., Gu, M., Xu, G., & Liao, H. (2019). A nodule-localized phosphate transporter GmPT7 plays an important role in enhancing symbiotic N2 fixation and yield in soybean. *New Phytologist*, 221(4), 2013–2025.

Chen, Z., Song, J., Li, X., Arango, J., Cardoso, J. A., Rao, I., ... Liu, G. (2021). Physiological responses and transcriptomic changes reveal the mechanisms underlying adaptation of *Stylosanthes guianensis* to phosphorus deficiency. *BMC Plant Biology*, 21(1), 1–15.

Chin, J. H., Gamuyao, R., Dalid, C., Bustamam, M., Prasetiyono, J., Moeljopawiro, S., Wissuwa, M., et al. (2011). Developing rice with high yield under phosphorus deficiency: Pup1 sequence to application. *Plant Physiology*, 156(3), 1202–1216.

Chin, J. H., Lu, X., Haefele, S. M., Gamuyao, R., Ismail, A., Wissuwa, M., & Heuer, S. (2010). Development and application of gene-based markers for the major rice QTL phosphorus uptake 1. *Theoretical and Applied Genetics*, 120(6), 1073–1086.

Cominelli, E., Pilu, R., & Sparvoli, F. (2020). Phytic acid and Transporters: What Can We Learn from low phytic acid Mutants. *Plants (Basel, Switzerland)*, 9(1), 69.

Cooper, J., Lombardi, R., Boardman, D., & Carliell-Marquet, C. (2011). The future distribution and production of global phosphate rock reserves. *Resources, Conservation and Recycling*, 57, 78–86.

Cruz-Ramírez, A., Oropeza-Aburto, A., Razo-Hernández, F., Ramírez-Chávez, E., & Herrera-Estrella, L. (2006). Phospholipase DZ2 plays an important role in extraplastidic galactolipid biosynthesis and phosphate recycling in Arabidopsis roots. *Proceedings of the National Academy of Sciences*, 103(17), 6765–6770.

Dai, X., Wang, Y., & Zhang, W.-H. (2016). OsWRKY74, a WRKY transcription factor, modulates tolerance to phosphate starvation in rice. *Journal of Experimental Botany*, 67(3), 947–960.

Delhaize, E., Hebb, D. M., & Ryan, P. R. (2001). Expression of a *Pseudomonas aeruginosa* citrate synthase gene in tobacco is not associated with either enhanced citrate accumulation or efflux. *Plant Physiology*, 125(4), 2059–2067.

Delhaize, E., Ryan, P. R., Hebb, D. M., Yamamoto, Y., Sasaki, T., & Matsumoto, H. (2004). Engineering high-level aluminum tolerance in barley with the ALMT1 gene. *Proceedings of the National Academy of Sciences*, 101(42), 15249–15254.

Delhaize, E., Taylor, P., Hocking, P. J., Simpson, R. J., Ryan, P. R., & Richardson, A. E. (2009). Transgenic barley (*Hordeum vulgare* L.) expressing the wheat aluminium resistance gene (*TaALMT1*) shows enhanced phosphorus nutrition and grain production when grown on an acid soil. *Plant Biotechnology Journal*, 7(5), 391–400.

Dissanayaka, D. M. S. B., Ghahremani, M., Siebers, M., Wasaki, J., & Plaxton, W. C. (2021). Recent insights into the metabolic adaptations of phosphorus-deprived plants. *Journal of Experimental Botany*, 72(2), 199–223.

Dissanayaka, D. M. S. B., Plaxton, W. C., Lambers, H., Siebers, M., Marambe, B., & Wasaki, J. (2018). Molecular mechanisms underpinning phosphorus-use efficiency in rice. *Plant, Cell and Environment*, 41(7), 1483–1496.

Dong, J., Chen, C., & Chen, Z. (2003). Expression profiles of the Arabidopsis WRKY gene superfamily during plant defense response. *Plant Molecular Biology*, 51(1), 21–37.

Ephrath, J. E., Klein, T., Sharp, R. E., & Lazarovitch, N. (2020). Exposing the hidden half: Root research at the forefront of science. *Plant and Soil*, 447(1–2), 1–5.

Feder, D., McGeary, R. P., Mitić, N., Lonhienne, T., Furtado, A., Schulz, B. L., ... & Schenk, G. (2020). Structural elements that modulate the substrate specificity of plant purple acid phosphatases: Avenues for improved phosphorus acquisition in crops. *Plant Science*, 294, 110445.

Filippelli, G. M. (2018). Balancing the global distribution of phosphorus with a view toward sustainability and equity. *Global Biogeochemical Cycles*, 32(6), 904–908.

Galindo-Castañeda, T., Brown, K. M., & Lynch, J. P. (2018). Reduced root cortical burden improves growth and grain yield under low phosphorus availability in maize. *Plant, Cell and Environment*, 41(7), 1579–1592.

Gamuyao, R., Chin, J. H., Pariasca-Tanaka, J., Pesaresi, P., Catausan, S., Dalid, C., ... Heuer, S. (2012). The protein kinase Pstol1 from traditional rice confers tolerance of phosphorus deficiency. *Nature*, 488(7412), 535–539.

Gao, W., Lu, L., Qiu, W., Wang, C., & Shou, H. (2017). *OsPAP26* encodes a major purple acid phosphatase and regulates phosphate remobilization in rice. *Plant and Cell Physiology*, 58(5), 885–892.

Gaude, N., Nakamura, Y., Scheible, W. R., Ohta, H., & Dörmann, P. (2008). Phospholipase C5 (NPC5) is involved in galactolipid accumulation during phosphate limitation in leaves of Arabidopsis. *The Plant Journal*, 56(1), 28–39.

Giri, J., Bhosale, R., Huang, G., Pandey, B. K., Parker, H., Zappala, S., ... Bennett, M. J. (2018). Rice auxin influx carrier OsAUX1 facilitates root hair elongation in response to low external phosphate. *Nature Communications*, 9(1), 1–7.

Guo, M., Ruan, W., Li, C., Huang, F., Zeng, M., Liu, Y., Yu, Y., Ding, X., Wu, Y., Wu, Z., Mao, C., Yi, K., Wu, P., & Mo, X. (2015). Integrative comparison of the role of the Phosphate Response1 subfamily in phosphate signaling and homeostasis in rice. *Plant Physiology*, 168(4), 1762–1776.

Han, Y., White, P. J., & Cheng, L. (2022). Mechanisms for improving phosphorus utilization efficiency in plants. *Annals of Botany*, 129(3), 247–258.

Han, Y., Zhang, W., Zhang, B., Zhang, S., Wang, W., & Ming, F. (2009). One novel mitochondrial citrate synthase from *Oryza sativa* L. can enhance aluminum tolerance in transgenic tobacco. *Molecular Biotechnology*, 42(3), 299–305.

Härtel, H., Dörmann, P., & Benning, C. (2000). DGD1-independent biosynthesis of extraplastidic galactolipids after phosphate deprivation in Arabidopsis. *Proceedings of the National Academy of Sciences*, 97(19), 10649–10654.

Heuer, S., Gaxiola, R., Schilling, R., Herrera-Estrella, L., López-Arredondo, D., Wissuwa, M., ... Rouached, H. (2017). Improving phosphorus use efficiency: A complex trait with emerging opportunities. *The Plant Journal*, 90(5), 868–885.

Heuer, S., Lu, X., Chin, J. H., Tanaka, J. P., Kanamori, H., Matsumoto, T., De Leon, T., et al. (2009). Comparative sequence analyses of the major quantitative trait locus phosphorus uptake 1 (Pup1) reveal a complex genetic structure. *Plant Biotechnology Journal*, 7(5), 456–471.

Hu, B., Henry, A., Brown, K. M., & Lynch, J. P. (2014). Root cortical aerenchyma inhibits radial nutrient transport in maize (*Zea mays*). *Annals of Botany*, 113(1), 181–189.

Huang, G., Liang, W., Sturrock, C. J., Pandey, B. K., Giri, J., Mairhofer, S., ... Zhang, D. (2018). Rice actin binding protein RMD controls crown root angle in response to external phosphate. *Nature Communications*, 9(1), 1–9.

Hurley, B. A., Tran, H. T., Marty, N. J., Park, J., Snedden, W. A., Mullen, R. T., & Plaxton, W. C. (2010). The dual-targeted purple acid phosphatase isozyme AtPAP26 is essential for efficient acclimation of Arabidopsis to nutritional phosphate deprivation. *Plant Physiology*, 153(3), 1112–1122.

Jia, H., Ren, H., Gu, M., Zhao, J., Sun, S., Zhang, X., ... Xu, G. (2011). The phosphate transporter gene OsPht1;8 is involved in phosphate homeostasis in rice. *Plant Physiology*, 156(3), 1164–1175.

Jia, X., Liu, P., & Lynch, J. P. (2018). Greater lateral root branching density in maize improves phosphorus acquisition from low phosphorus soil. *Journal of Experimental Botany*, 69(20), 4961–4970.

Jiang, J., Ma, S., Ye, N., Jiang, M., Cao, J., & Zhang, J. (2017). WRKY transcription factors in plant responses to stresses. *Journal of Integrative Plant Biology*, 59(2), 86–101.

Kavka, M., Majcherczyk, A., Kües, U., & Polle, A. (2021). Phylogeny, tissue-specific expression, and activities of root-secreted purple acid phosphatases for P uptake from ATP in P starved poplar. *Plant Science*, 307, 110906.

Kelly, A. A., Froehlich, J. E., & Dörmann, P. (2003). Disruption of the two digalactosyldiacylglycerol synthase genes DGD1 and DGD2 in Arabidopsis reveals the existence of an additional enzyme of galactolipid synthesis. *The Plant Cell*, 15(11), 2694–2706.

Khan, M. S., Zaidi, A., & Wani, P. A. (2006). Review article Methods for studying root colonization by introduced. *Agronomie*, 23, 407–418.

Kong, Y., Li, X., Ma, J., Li, W., Yan, G., & Zhang, C. (2014). GmPAP4, a novel purple acid phosphatase gene isolated from soybean (*Glycine max*), enhanced extracellular phytate utilization in *Arabidopsis thaliana*. *Plant Cell Reports*, 33(4), 655–667.

Kong, Y., Li, X., Wang, B., Li, W., Du, H., & Zhang, C. (2018). The soybean purple acid phosphatase GmPAP14 predominantly enhances external phytate utilization in plants. *Frontiers in Plant Science*, 9, 292.

Koyama, H., Kawamura, A., Kihara, T., Hara, T., Takita, E., & Shibata, D. (2000). Overexpression of mitochondrial citrate synthase in *Arabidopsis thaliana* improved growth on a phosphorus-limited soil. *Plant and Cell Physiology*, 41(9), 1030–1037.

Krüger, M., Teste, F. P., Laliberté, E., Lambers, H., Coghlan, M., Zemunik, G., & Bunce, M. (2015). The rise and fall of arbuscular mycorrhizal fungal diversity during ecosystem retrogression. *Molecular Ecology*, 24(19), 4912–4930.

Kuang, R., Chan, K. H., Yeung, E., & Lim, B. L. (2009). Molecular and biochemical characterization of AtPAP15, a purple acid phosphatase with phytase activity, in Arabidopsis. *Plant Physiology*, 151(1), 199–209.

Kumar, A., Nayak, S., Ngangkham, U., Sah, R. P., Lal, M. K., Tp, A., ... Sharma, S. (2021). A single nucleotide substitution in the SPDT transporter gene reduced phytic acid and increased mineral bioavailability from Rice grain (Oryza sativa L.). *Journal of Food Biochemistry*, 45(7), e13822.

Lambers, H. (2022). Phosphorus acquisition and utilization in plants. *Annual Review of Plant Biology*, 73, 11–126.

Lan, P., Li, W., & Schmidt, W. (2012). Complementary proteome and transcriptome profiling in phosphate-deficient Arabidopsis roots reveals multiple levels of gene regulation. *Molecular and Cellular Proteomics*, 11(11), 1156–1166.

Li, C., Li, C., Zhang, H., Liao, H., & Wang, X. (2017). The purple acid phosphatase GmPAP21 enhances internal phosphorus utilization and possibly plays a role in symbiosis with rhizobia in soybean. *Physiologia Plantarum*, 159(2), 215–227.

Li, G. X., Wu, X. Q., Ye, J. R., & Yang, H. C. (2018). Characteristics of organic acid secretion associated with the interaction between *Burkholderia multivorans* WS-FJ9 and poplar root system. *BioMed Research International*, 2018.

Liang, C., Piñeros, M. A., Tian, J., Yao, Z., Sun, L., Liu, J., ... Liao, H. (2013). Low pH, aluminum, and phosphorus co-ordinately regulate malate exudation through *GmALMT1* to improve soybean adaptation to acid soils. *Plant Physiology*, 161(3), 1347–1361.

Liang, C., Tian, J., Lam, H. M., Lim, B. L., Yan, X., & Liao, H. (2010). Biochemical and molecular characterization of PvPAP3, a novel purple acid phosphatase isolated from common bean enhancing extracellular ATP utilization. *Plant Physiology*, 152(2), 854–865.

Ligaba, A., Katsuhara, M., Ryan, P. R., Shibasaka, M., & Matsumoto, H. (2006). The *BnALMT1* and *BnALMT2* genes from rape encode aluminum-activated malate transporters that enhance the aluminum resistance of plant cells. *Plant Physiology*, 142(3), 1294–1303.

Liu, C., Su, J., Stephen, G. U. K., Wang, H., Song, A., Chen, F., ... Jiang, J. (2018). Overexpression of phosphate transporter gene *CmPht1*; 2 facilitated Pi uptake and alternated the metabolic profiles of Chrysanthemum under phosphate deficiency. *Frontiers in Plant Science, 9,* 686.

Liu, D. (2021). Root developmental responses to phosphorus nutrition. *Journal of Integrative Plant Biology, 63*(6), 1065–1090.

Liu, P., Cai, Z., Chen, Z., Mo, X., Ding, X., Liang, C., ... Tian, J. (2018). A root-associated purple acid phosphatase, SgPAP23, mediates extracellular phytate-P utilization in *Stylosanthes guianensis*. *Plant, Cell and Environment, 41*(12), 2821–2834.

Lizcano-Toledo, R., Reyes-Martín, M. P., Celi, L., & Fernández-Ondoño, E. (2021). Phosphorus dynamics in the Soil–Plant–Environment relationship in cropping systems: A review. *Applied Sciences, 11*(23), 11133.

López-Arredondo, D. L., & Herrera-Estrella, L. (2012). Engineering phosphorus metabolism in plants to produce a dual fertilization and weed control system. *Nature Biotechnology, 30*(9), 889–893.

Lopez-Bucio, J., De la Vega, O. M., Guevara-Garcia, A., & Herrera-Estrella, L. (2000). Enhanced phosphorus uptake in transgenic tobacco plants that overproduce citrate. *Nature Biotechnology, 18*(4), 450–453.

Lü, J., Gao, X., Dong, Z., Yi, J., & An, L. (2012). Improved phosphorus acquisition by tobacco through transgenic expression of mitochondrial malate dehydrogenase from *Penicillium oxalicum*. *Plant Cell Reports, 31*(1), 49–56.

Lu, L., Qiu, W., Gao, W., Tyerman, S. D., Shou, H., & Wang, C. (2016). OsPAP10c, a novel secreted acid phosphatase in rice, plays an important role in the utilization of external organic phosphorus. *Plant, Cell and Environment, 39*(10), 2247–2259.

Lu, Y., & Zhu, J. K. (2017). Precise editing of a target base in the rice genome using a modified CRISPR/Cas9 system. *Molecular Plant, 10*(3), 523–525.

Lynch, J. P. (2019). Root phenotypes for improved nutrient capture: An underexploited opportunity for global agriculture. *New Phytologist, 223*(2), 548–564.

Lynch, J. P., & Brown, K. M. (2001). Topsoil foraging - An architectural adaptation of plants to low phosphorus availability. *Plant and Soil, 237*(2), 225–237.

Ma, B., Zhang, L., Gao, Q., Wang, J., Li, X., Wang, H., ... & He, Z. (2021). A plasma membrane transporter coordinates phosphate reallocation and grain filling in cereals. *Nature Genetics, 53*(6), 906–915.

Ma, J. F., Zheng, S. J., Matsumoto, H., & Hiradate, S. (1997). Detoxifying aluminium with buckwheat. *Nature, 390*(6660), 569–570.

Macias-Benitez, S., Garcia-Martinez, A. M., Caballero Jimenez, P., Gonzalez, J. M., Tejada Moral, M., & Parrado Rubio, J. (2020). Rhizospheric organic acids as biostimulants: Monitoring feedbacks on soil microorganisms and biochemical properties. *Frontiers in Plant Science, 11,* 633.

Mehra, P., & Giri, J. (2016). Rice and chickpea *GDPDs* are preferentially influenced by low phosphate and CaGDPD1 encodes an active glycerophosphodiester phosphodiesterase enzyme. *Plant Cell Reports, 35*(8), 1699–1717.

Mehra, P., Pandey, B. K., & Giri, J. (2015). Genome-wide DNA polymorphisms in low phosphate tolerant and sensitive rice genotypes. *Scientific Reports, 5*(1), 1–14.

Mehra, P., Pandey, B. K., & Giri, J. (2017). Improvement in phosphate acquisition and utilization by a secretory purple acid phosphatase (OsPAP21b) in rice. *Plant Biotechnology Journal, 15*(8), 1054–1067.

Mehra, P., Pandey, B. K., Verma, L., & Giri, J. (2019). A novel glycerophosphodiester phosphodiesterase improves phosphate deficiency tolerance in rice. *Plant, Cell and Environment, 42*(4), 1167–1179.

Miguel, M. A., Widrig, A., Vieira, R. F., Brown, K. M., & Lynch, J. P. (2013). Basal root whorl number: A modulator of phosphorus acquisition in common bean (*Phaseolus vulgaris*). *Annals of Botany, 112*(6), 973–982.

Miller, S. S., Liu, J., Allan, D. L., Menzhuber, C. J., Fedorova, M., & Vance, C. P. (2001). Molecular control of acid phosphatase secretion into the rhizosphere of proteoid roots from phosphorus-stressed white lupin. *Plant Physiology, 127*(2), 594–606.

Młodzińska, E., & Zboińska, M. (2016). Phosphate uptake and allocation–a closer look at *Arabidopsis thaliana* L. and *Oryza sativa* L. *Frontiers in Plant Science, 7,* 1198.

Muthayya, S., Hall, J., Bagriansky, J., Sugimoto, J., Gundry, D., Matthias, D., ... Maberly, G. (2012). Rice fortification: An emerging opportunity to contribute to the elimination of vitamin and mineral deficiency worldwide. *Food and Nutrition Bulletin, 33*(4), 296–307.

Nagarajan, V. K., Jain, A., Poling, M. D., Lewis, A. J., Raghothama, K. G., & Smith, A. P. (2011). Arabidopsis Pht1; 5 mobilizes phosphate between source and sink organs and influences the interaction between phosphate homeostasis and ethylene signaling. *Plant Physiology, 156*(3), 1149–1163.

Nakamura, Y. (2013). Phosphate starvation and membrane lipid remodeling in seed plants. *Progress in Lipid Research, 52*(1), 43–50.

Nakamura, Y., Awai, K., Masuda, T., Yoshioka, Y., Takamiya, K. I., & Ohta, H. (2005). A novel phosphatidylcholine-hydrolyzing phospholipase C induced by phosphate starvation in Arabidopsis. *Journal of Biological Chemistry, 280*(9), 7469–7476.

Nakamura, Y., Koizumi, R., Shui, G., Shimojima, M., Wenk, M. R., Ito, T., & Ohta, H. (2009). Arabidopsis lipins mediate eukaryotic pathway of lipid metabolism and cope critically with phosphate starvation. *Proceedings of the National Academy of Sciences, 106*(49), 20978–20983.

Neumann, G., Massonneau, A., Martinoia, E., & Römheld, V. (1999). Physiological adaptations to phosphorus deficiency during proteoid root development in white lupin. *Planta, 208*(3), 373–382.

Nilsson, L., Müller, R., & Nielsen, T. H. (2007). Increased expression of the MYB-related transcription factor, PHR1, leads to enhanced phosphate uptake in *Arabidopsis thaliana*. *Plant, Cell and Environment, 30*(12), 1499–1512.

Nussaume, L., Kanno, S., Javot, H., Marin, E., Pochon, N., Ayadi, A., ... Thibaud, M. C. (2011). Phosphate import in plants: Focus on the PHT1 transporters. *Frontiers in Plant Science, 2*, 83.

Okazaki, Y., Otsuki, H., Narisawa, T., Kobayashi, M., Sawai, S., Kamide, Y., ... Saito, K. (2013). A new class of plant lipid is essential for protection against phosphorus depletion. *Nature Communications, 4*(1), 1–10.

Olczak, M., Morawiecka, B., & Watorek, W. (2003). Plant purple acid phosphatases-genes, structures and biological function. *Acta Biochimica Polonica, 50*(4), 1245–1256.

Panchal, P., Miller, A. J., & Giri, J. (2021). Organic acids: Versatile stress-response roles in plants. *Journal of Experimental Botany, 72*(11), 4038–4052.

Pandey, B. K., Mehra, P., & Giri, J. (2013). Phosphorus starvation response in plants and opportunities for crop improvement. *Climate Change and Plant Abiotic Stress Tolerance*, 991–1012.

Pandeya, D., López-Arredondo, D. L., Janga, M. R., Campbell, L. M., Estrella-Hernández, P., Bagavathiannan, M. V., ... Rathore, K. S. (2018). Selective fertilization with phosphite allows unhindered growth of cotton plants expressing the *ptxD* gene while suppressing weeds. *Proceedings of the National Academy of Sciences, 115*(29), E6946–E6955.

Pang, J., Bansal, R., Zhao, H., Bohuon, E., Lambers, H., Ryan, M. H., Ranathunge, K., & Siddique, K. H. M. (2018). The carboxylate-releasing phosphorus-mobilizing strategy can be proxied by foliar manganese concentration in a large set of chickpea germplasm under low phosphorus supply. *New Phytologist, 219*(2), 518–529.

Pant, B. D., Burgos, A., Pant, P., Cuadros-Inostroza, A., Willmitzer, L., & Scheible, W. R. (2015). The transcription factor PHR1 regulates lipid remodeling and triacylglycerol accumulation in *Arabidopsis thaliana* during phosphorus starvation. *Journal of Experimental Botany, 66*(7), 1907–1918.

Peterson, E. (2022). The coming global food crisis. Cornhusker Economics.

Plaxton, W. C., & Tran, H. T. (2011). Metabolic adaptations of phosphate-starved plants. *Plant Physiology, 156*(3), 1006–1015.

Poirier, Y., Thoma, S., Somerville, C., & Schiefelbein, J. (1991). Mutant of Arabidopsis deficient in xylem loading of phosphate. *Plant Physiology, 97*(3), 1087–1093.

Prathap, V., Kumar, A., Maheshwari, C., & Tyagi, A. (2022). Phosphorus homeostasis: Acquisition, sensing, and long-distance signaling in plants. *Molecular Biology Reports*, 1–16.

Primieri, S., Magnoli, S. M., Koffel, T., Stürmer, S. L., & Bever, J. D. (2022). Perennial, but not annual legumes synergistically benefit from infection with arbuscular mycorrhizal fungi and rhizobia: A meta-analysis. *New Phytologist, 233*(1), 505–514.

Puga, M. I., Mateos, I., Charukesi, R., Wang, Z., Franco-Zorrilla, J. M., de Lorenzo, L., ... Paz-Ares, J. (2014). SPX1 is a phosphate-dependent inhibitor of Phosphate Starvation Response 1 in Arabidopsis. *Proceedings of the National Academy of Sciences of the United States of America, 111*(41), 14947–14952.

Ram, B., Fartyal, D., Sheri, V., Varakumar, P., Borphukan, B., James, D., ... Reddy, M. K. (2019). Characterization of *phoA*, a bacterial alkaline phosphatase for Phi use efficiency in rice plant. *Frontiers in Plant Science, 10*, 37.

Raman, R. (2017). The impact of Genetically Modified (GM) crops in modern agriculture: A review. *GM Crops and Food, 8*(4), 195–208.

Robinson, W. D., Carson, I., Ying, S., Ellis, K., & Plaxton, W. C. (2012). Eliminating the purple acid phosphatase AtPAP 26 in *Arabidopsis thaliana* delays leaf senescence and impairs phosphorus remobilization. *New Phytologist, 196*(4), 1024–1029.

Rocha-Munive, M. G., Soberón, M., Castañeda, S., Niaves, E., Scheinvar, E., Eguiarte, L. E., Souza, V. (2018). Evaluation of the impact of genetically modified cotton after 20 years of cultivation in Mexico. *Frontiers in Bioengineering and Biotechnology, 6*, 82.

Rubio, V., Linhares, F., Solano, R., Martín, A. C., Iglesias, J., Leyva, A., & Paz-Ares, J. (2001). A conserved MYB transcription factor involved in phosphate starvation signaling both in vascular plants and in unicellular algae. *Genes and Development, 15*(16), 2122–2133.

Shahbaz, A. M., Oki, Y., Adachi, T., Murata, Y., & Khan, M. H. R. (2006). Phosphorus starvation induced root-mediated pH changes in solublization and acquisition of sparingly soluble P sources and organic acids exudation by Brassica cultivars. *Soil Science and Plant Nutrition*, 52(5), 623–633.

Shane, M. W., & Lambers, H. (2005). Cluster roots: A curiosity in context. *Plant and Soil*, 274(1–2), 101–125.

Sharma, S. B., Sayyed, R. Z., Trivedi, M. H., & Gobi, T. A. (2013). Phosphate solubilizing microbes: Sustainable approach for managing phosphorus deficiency in agricultural soils. *SpringerPlus*, 2(1), 1–14.

Shelton, A. M., Sarwer, S. H., Hossain, M. J., Brookes, G., & Paranjape, V. (2020). Impact of Bt brinjal cultivation in the market value chain in five districts of Bangladesh. *Frontiers in Bioengineering and Biotechnology*, 8, 498.

Shen, J., Yuan, L., Zhang, J., Li, H., Bai, Z., Chen, X., ... & Zhang, F. (2011). Phosphorus dynamics: from soil to plant. *Plant Physiology*, 156(3), 997–1005.

Siebers, M., Dörmann, P., & Hölzl, G. (2015). Membrane remodelling in phosphorus-deficient plants. *Annual Plant Reviews*, 48, 237–263.

Song, H., Yin, Z., Chao, M., Ning, L., Zhang, D., & Yu, D. (2014). Functional properties and expression quantitative trait loci for phosphate transporter GmPT1 in soybean. *Plant, Cell and Environment*, 37(2), 462–472.

Srivastava, S., Upadhyay, M. K., Srivastava, A. K., Abdelrahman, M., Suprasanna, P., & Tran, L. S. P. (2018). Cellular and subcellular phosphate transport machinery in plants. *International Journal of Molecular Sciences*, 19(7), 1914.

Su, T., Xu, Q., Zhang, F. C., Chen, Y., Li, L. Q., Wu, W. H., & Chen, Y. F. (2015). WRKY42 modulates phosphate homeostasis through regulating phosphate translocation and acquisition in Arabidopsis. *Plant Physiology*, 167(4), 1579–1591.

Suen, P. K., Zhang, S., & Sun, S. S. M. (2015). Molecular characterization of a tomato purple acid phosphatase during seed germination and seedling growth under phosphate stress. *Plant Cell Reports*, 34(6), 981–992.

Sun, B., Gao, Y., & Lynch, J. P. (2018a). Large crown root number improves topsoil foraging and phosphorus acquisition. *Plant Physiology*, 177(1), 90–104.

Sun, F., Suen, P. K., Zhang, Y., Liang, C., Carrie, C., Whelan, J., ... Lim, B. L. (2012a). A dual-targeted purple acid phosphatase in *Arabidopsis thaliana* moderates carbon metabolism and its overexpression leads to faster plant growth and higher seed yield. *New Phytologist*, 194(1), 206–219.

Sun, L., Song, L., Zhang, Y., Zheng, Z., & Liu, D. (2016). Arabidopsis PHL2 and PHR1 act redundantly as the key components of the central regulatory system controlling transcriptional responses to phosphate starvation. *Plant Physiology*, 170(1), 499–514.

Sun, Q., Li, J., Cheng, W., Guo, H., Liu, X., & Gao, H. (2018b). AtPAP2, a unique member of the PAP family, functions in the plasma membrane. *Genes*, 9(5), 257.

Sun, S., Gu, M., Cao, Y., Huang, X., Zhang, X., Ai, P., ... & Xu, G. (2012b). A constitutive expressed phosphate transporter, OsPht1; 1, modulates phosphate uptake and translocation in phosphate-replete rice. *Plant Physiology*, 159(4), 1571–1581.

Swamy, H. K., Anila, M., Kale, R. R., Rekha, G., Bhadana, V. P., Anantha, M. S., ... & Sundaram, R. M. (2020). Marker assisted improvement of low soil phosphorus tolerance in the bacterial blight resistant, fine-grain type rice variety, Improved Samba Mahsuri. *Scientific Reports*, 10(1), 1–14.

Tao, Y., Huang, J., Jing, H. K., Shen, R. F., & Zhu, X. F. (2022). Jasmonic acid is involved in root cell wall phosphorus remobilization through the nitric oxide dependent pathway in rice (G. Xu, Ed.). *Journal of Experimental Botany*, 73(8), 2618–2630.

Tawaraya, K., Honda, S., Cheng, W., Chuba, M., Okazaki, Y., Saito, K., ... & Wagatsuma, T. (2018). Ancient rice cultivar extensively replaces phospholipids with non-phosphorus glycolipid under phosphorus deficiency. *Physiologia Plantarum*, 163(3), 297–305.

Theodorou, M. E., & Plaxton, W. C. (1993). Metabolic adaptations of plant respiration to nutritional phosphate deprivation. *Plant Physiology*, 101(2), 339–344.

Thibaud, M. C., Arrighi, J. F., Bayle, V., Chiarenza, S., Creff, A., Bustos, R., ... & Nussaume, L. (2010). Dissection of local and systemic transcriptional responses to phosphate starvation in Arabidopsis. *The Plant Journal*, 64(5), 775–789.

Tian, J., & Liao, H. (2015). The role of intracellular and secreted purple acid phosphatases in plant phosphorus scavenging and recycling. *Annual Plant Reviews*, 48, 265–287.

Tian, J., Wang, C., Zhang, Q., He, X., Whelan, J., & Shou, H. (2012). Overexpression of *OsPAP10a*, a root-associated acid phosphatase, increased extracellular organic phosphorus utilization in rice. *Journal of Integrative Plant Biology*, 54(9), 631–639.

Tiessen, H. (2008). Phosphorus in the global environment. In: White, P.J., Hammond, J.P. (eds). The ecophysiology of plant-phosphorus interactions. In *Plant Ecophysiology* (pp. 1–7). Dordrecht: Springer.

Tomlinson, L., Yang, Y., Emenecker, R., Smoker, M., Taylor, J., Perkins, S., ... & Jones, J. D. (2019). Using CRISPR/Cas9 genome editing in tomato to create a gibberellin-responsive dominant dwarf della allele. *Plant Biotechnology Journal*, 17(1), 132–140.

Tran, H. T., Hurley, B. A., & Plaxton, W. C. (2010). Feeding hungry plants: The role of purple acid phosphatases in phosphate nutrition. *Plant Science*, 179(1–2), 14–27.

Turnbull, C., Lillemo, M., & Hvoslef-Eide, T. A. (2021). Global regulation of genetically modified crops amid the gene edited crop boom–a review. *Frontiers in Plant Science*, 12, 630396.

Vance, C. P. (2008). Plants without arbuscular mycorrhizae. In Philip J. White, John P. Hammond (Eds.) *The Ecophysiology of Plant-Phosphorus Interactions* (pp. 117–142). Dordrecht: Springer.

Vance, C. P., Uhde-Stone, C., & Allan, D. L. (2003). Phosphorus acquisition and use: Critical adaptations by plants for securing a non-renewable resource. *New Phytologist*, 157(3), 423–447.

Veljanovski, V., Vanderbeld, B., Knowles, V. L., Snedden, W. A., & Plaxton, W. C. (2006). Biochemical and molecular characterization of AtPAP26, a vacuolar purple acid phosphatase up-regulated in phosphate-deprived Arabidopsis suspension cells and seedlings. *Plant Physiology*, 142(3), 1282–1293.

Veneklaas, E. J., Lambers, H., Bragg, J., Finnegan, P. M., Lovelock, C. E., Plaxton, W. C., ... & Raven, J. A. (2012). Opportunities for improving phosphorus-use efficiency in crop plants. *New Phytologist*, 195(2), 306–320.

Verma, L., Bhadouria, J., Rupam, B. K., Singh, S., Panchal, P., Bhatia, C., ... & Giri, J. (2022). Monogalactosyl diacylglycerol synthase 3 (OsMGD3) affects phosphate utilization and acquisition in rice. *Journal of Experimental Botany*, erac192.

Verma, L., Kohli, P. S., Maurya, K., Abhijith, K. B., Thakur, J. K., & Giri, J. (2021a). Specific galactolipids species correlate with rice genotypic variability for phosphate utilization efficiency. *Plant Physiology and Biochemistry*, 168, 105–115.

Verma, L., Rumi, Sinha, A. K., & Giri, J. (2021b). Phosphate deficiency response and membrane lipid remodeling in plants. *Plant Physiology Reports*, 26(4), 614–625.

Wan, T. E. N. G., Xue, H. E., & Tong, Y. P. (2017). Transgenic approaches for improving use efficiency of nitrogen, phosphorus and potassium in crops. *Journal of Integrative Agriculture*, 16(12), 2657–2673.

Wang, F., Cui, P. J., Tian, Y., Huang, Y., Wang, H. F., Liu, F., & Chen, Y. F. (2020a). Maize ZmPT7 regulates Pi uptake and redistribution which is modulated by phosphorylation. *Plant Biotechnology Journal*, 18(12), 2406–2419.

Wang, F., Deng, M., Xu, J., Zhu, X., & Mao, C. (2018). Molecular mechanisms of phosphate transport and signaling in higher plant. In *Seminars in Cell and Developmental Biology* (Vol. 74, pp. 114–122). Academic Press, Cambridge.

Wang, H., Xu, Q., Kong, Y. H., Chen, Y., Duan, J. Y., Wu, W. H., & Chen, Y. F. (2014a). Arabidopsis WRKY45 transcription factor activates phosphate transporter1; 1 expression in response to phosphate starvation. *Plant Physiology*, 164(4), 2020–2029.

Wang, J., Pan, W., Nikiforov, A., King, W., Hong, W., Li, W., & Cheng, L. (2021a). Identification of two glycerophosphodiester phosphodiesterase genes in maize leaf phosphorus remobilization. *The Crop Journal*, 9(1), 95–108.

Wang, L., Li, Z., Qian, W., Guo, W., Gao, X., Huang, L., ... & Liu, D. (2011). The Arabidopsis purple acid phosphatase AtPAP10 is predominantly associated with the root surface and plays an important role in plant tolerance to phosphate limitation. *Plant Physiology*, 157(3), 1283–1299.

Wang, X., Wang, Y., Tian, J., Lim, B. L., Yan, X., & Liao, H. (2009). Overexpressing *AtPAP15* enhances phosphorus efficiency in soybean. *Plant Physiology*, 151(1), 233–240.

Wang, Y., Wang, F., Lu, H., Liu, Y., & Mao, C. (2021b). Phosphate uptake and transport in plants: An elaborate regulatory system. *Plant and Cell Physiology*, 62(4), 564–572.

Wang, Y., Yang, Z., Kong, Y., Li, X., Li, W., Du, H., & Zhang, C. (2020b). *GmPAP12* is required for nodule development and nitrogen fixation under phosphorus starvation in soybean. *Frontiers in Plant Science*, 11, 450.

Wang, Z., Kuo, H. F., & Chiou, T. J. (2021c). Intracellular phosphate sensing and regulation of phosphate transport systems in plants. *Plant Physiology*, 187(4), 2043–2055.

Wang, Z., Ruan, W., Shi, J., Zhang, L., Xiang, D., Yang, C., ... & Wu, P. (2014b). Rice SPX1 and SPX2 inhibit phosphate starvation responses through interacting with PHR2 in a phosphate-dependent manner. *Proceedings of the National Academy of Sciences*, 111(41), 14953–14958.

Wissuwa, M., Wegner, J., Ae, N., & Yano, M. (2002). Substitution mapping of *Pup1*: A major QTL increasing phosphorus uptake of rice from a phosphorus-deficient soil. *Theoretical and Applied Genetics*, 105(6–7), 890–897.

Xu, L., Zhao, H., Wan, R., Liu, Y., Xu, Z., Tian, W., Ruan, W., Wang, F., Deng, M., Wang, J., Dolan, L., Luan, S., Xue, S., & Yi, K. (2019). Identification of vacuolar phosphate efflux transporters in land plants. *Nature Plants*, 5(1), 84–94.

Yamaji, N., Takemoto, Y., Miyaji, T., Mitani-Ueno, N., Yoshida, K. T., & Ma, J. F. (2017). Reducing phosphorus accumulation in rice grains with an impaired transporter in the node. *Nature*, 541(7635), 92–95.

Yokosho, K., Yamaji, N., & Ma, J. F. (2010). Isolation and characterisation of two *MATE* genes in rye. *Functional Plant Biology*, 37(4), 296–303.

Yokosho, K., Yamaji, N., Fujii-Kashino, M., & Ma, J. F. (2016a). Functional analysis of a MATE gene *OsFRDL2* revealed its involvement in Al-induced secretion of citrate, but a lower contribution to Al tolerance in rice. *Plant and Cell Physiology*, 57(5), 976–985.

Yokosho, K., Yamaji, N., Fujii-Kashino, M., & Ma, J. F. (2016b). Retrotransposon-mediated aluminum tolerance through enhanced expression of the citrate transporter *OsFRDL4*. *Plant Physiology*, 172(4), 2327–2336.

Yu, B., Xu, C., & Benning, C. (2002). Arabidopsis disrupted in *SQD2* encoding sulfolipid synthase is impaired in phosphate-limited growth. *Proceedings of the National Academy of Sciences*, 99(8), 5732–5737.

Yuan, S., Kim, S. C., Deng, X., Hong, Y., & Wang, X. (2019). Diacylglycerol kinase and associated lipid mediators modulate rice root architecture. *New Phytologist*, 223(1), 261–276.

Zhang, F., Sun, Y., Pei, W., Jain, A., Sun, R., Cao, Y., ... & Sun, S. (2015). Involvement of *Os Pht1; 4* in phosphate acquisition and mobilization facilitates embryo development in rice. *The Plant Journal*, 82(4), 556–569.

Zhang, F., Wu, X. N., Zhou, H. M., Wang, D. F., Jiang, T. T., Sun, Y. F., ... Xu, G. H. (2014a). Overexpression of rice phosphate transporter gene *OsPT6* enhances phosphate uptake and accumulation in transgenic rice plants. *Plant and Soil*, 384(1), 259–270.

Zhang, J., Gu, M., Liang, R., Shi, X., Chen, L., Hu, X., ... Xu, G. (2021). OsWRKY21 and OsWRKY108 function redundantly to promote phosphate accumulation through maintaining the constitutive expression of *OsPHT1; 1* under phosphate-replete conditions. *New Phytologist*, 229(3), 1598–1614.

Zhang, L., Hu, B., Li, W., Che, R., Deng, K., Li, H., ... & Chu, C. (2014b). Os PT 2, a phosphate transporter, is involved in the active uptake of selenite in rice. *New Phytologist*, 201(4), 1183–1191.

Zhang, N., Wang, D., Liu, Y., Li, S., Shen, Q., & Zhang, R. (2014c). Effects of different plant root exudates and their organic acid components on chemotaxis, biofilm formation and colonization by beneficial rhizosphere-associated bacterial strains. *Plant and Soil*, 374(1), 689–700.

Zhang, Z., Liao, H., & Lucas, W. J. (2014d). Molecular mechanisms underlying phosphate sensing, signaling, and adaptation in plants. *Journal of Integrative Plant Biology*, 56(3), 192–220.

Zhou, G., Delhaize, E., Zhou, M., & Ryan, P. R. (2013). The barley *MATE* gene, *HvAACT1*, increases citrate efflux and Al^{3+} tolerance when expressed in wheat and barley. *Annals of Botany*, 112(3), 603–612.

Zhou, J., Jiao, F., Wu, Z., Li, Y., Wang, X., He, X., ... Wu, P. (2008). *OsPHR2* is involved in phosphate-starvation signaling and excessive phosphate accumulation in shoots of plants. *Plant Physiology*, 146(4), 1673–1686.

Zhu, S., Chen, M., Liang, C., Xue, Y., Lin, S., & Tian, J. (2020). Characterization of purple acid phosphatase family and functional analysis of *GmPAP7a/7b* involved in extracellular ATP utilization in soybean. *Frontiers in Plant Science*, 11, 661.

Zimmermann, P., Regierer, B., Kossmann, J., Frossard, E., Amrhein, N., & Bucher, M. (2004). Differential expression of three purple acid phosphatases from potato. *Plant Biology*, 6(05), 519–528.

Zowada, C., Gulacar, O., Siol, A., & Eilks, I. (2019). Phosphorus–a "political" element for transdisciplinary chemistry education. *Chemistry Teacher International*, 2(1), 20180020.

Zsögön, A., Čermák, T., Naves, E. R., Notini, M. M., Edel, K. H., Weinl, S., ... Peres, L. E. P. (2018). De novo domestication of wild tomato using genome editing. *Nature Biotechnology*, 36(12), 1211–1216.

9 Analysis and Comparison of Alphafold-Structure Predictions between Pi-Uptake Transporters Recovering Phosphate in Natural Environments

Nussaume Laurent, Desnos Thierry, Jinsheng Zhu, David Pascale, Kumiko Miwa, and Kanno Satomi

9.1 PHOSPHATE ORIGIN AND DISTRIBUTION

9.1.1 Phosphate through the Ages

Geological surveys indicate that 80 to 90% of earth's 4.5 billion years of history was characterised by a limited presence of phosphorus (P) in the form of orthophosphate (Pi) burial in nearshore sediments due to iron-rich waters (Bjerrum and Canfield, 2002).

According to this analysis, the concentration of Pi during the Archean and Proterozoic eras (3.2 to 1.9 billion years ago) reached only 10 to 25% compared to the current period. In such conditions, the archaea and bacteria were the sole living organism forms present on the earth. The low P availability limited photosynthesis development and reduced the long-term oxygen production during early earth periods.

This view is nevertheless controversial due to abundant dissolved silica assumed to be present in the Archean oceans. If present, such an element would have been a strong competitor of phosphate for binding with iron ferrihydrite particles, therefore, limiting the possibility that phosphate would have been limiting for phytoplankton productivity (Konhauser et al., 2007)

The detection of cyanobacteria presence started only 2.7 billion years ago (Brocks et al., 1999, 2017). Surprisingly, this is before the increase of oxygen in the atmosphere, estimated 2.4 billion years ago. In addition, even in this period, the phenomenon remains modest (below 1–2%). A striking shift occurred only 540–1000 million years ago in the Neoproterozoic era and is correlated with the increase of phosphate present in the sediments. This indicated that anoxygenic photosynthesis started first to modulate oxygen during the Proterozoic area (Johnston et al., 2009). The cyanobacteria activity then raised the oxygen level in the atmosphere (Falkowski et al., 2004; Johnston et al., 2009), thus promoting the selection of mechanisms for P acquisition and nitrogen-fixation as developed by early animals and complex ecosystems (Reinhard et al., 2017). This provides an appealing hypothesis to explain why P is mostly recovered by photosynthetic organisms as Pi.

9.1.2 PRESENT DISTRIBUTION OF PHOSPHATE ON EARTH

The phosphate concentration observed at the sea surface (Levitus et al., 1993) is extremely low (0,04–0,2 µM in the Pacific), as this element is massively recovered by phytoplankton to sustain their growth (Perry and Eppley, 1981). Interestingly, some of these organisms have capacities to adapt to very low P, as the organisms can fulfil their P needs with as low as 10–20 nM (Perry and Eppley, 1981; Lomas et al., 2014) using Pi or hydrolysed dissolved organic phosphorus (DOP). DOP, reported to be present in a range of 0.06–0.6 µM (Karl, 2014), can represent a source of Pi for the microbial community after enzymatic digestion.

In the oceans, the Pi is more abundant around the poles. On the Antarctic continent, "hot" spots with concentrations higher than 2 µM have been observed, whereas the subarctic Pacific has concentrations starting at 0,4 µM and increasing to 1.4 µM. Some sea or ocean waters are nitrate depleted, but this is not the case for Pi, which is present on average in concentrations varying from 0.1–0.4 µM below the sea surface. These concentrations rise very quickly according to depth (ranging from 1.2 to 3.2 µM at 1,000 metres depth) due to sediment presence. Such values are also influenced by the age of the ocean; for example, in young oceans, such as the Atlantic, Pi levels are typically ~1.5 µm, whereas in the older Pacific Ocean, ~2.5 µM of Pi is found on average (Levitus et al., 1993). This is also due to the sedimentation process promoting Pi accumulation at the bottom of the oceans. In the Atlantic Ocean, the North Atlantic deep-water stream promotes the presence of a phosphate minimum (with a 50% reduction of Pi concentration present compared to the rest of the Atlantic).

On land, soil total phosphorus concentration presents significant differences. In the first 30 cm (topsoil layer), the average P level found varies from ~160 (South America, Africa, Oceania) to 657 (North America, Europe) mg kg^{-1} (He et al., 2021). The next soil layer (30–100 cm) exhibits a reduction to 30 mg kg^{-1} on average of these values. But this data is far from reflecting the huge discrepancies that exist. An analysis of 5,275 soils distributed worldwide identified a range of 1.4–9630 mg kg^{-1} (He et al., 2021). In addition, many soils are used for agriculture and receive fertilisers. As a consequence, the soils of natural ecosystems (excluding Antarctica) contain much less phosphorus (26.8–62.2 mg kg^{-1} on average).

The majority of P present in soils is associated with organic matter or when present in inorganic forms is chelated by cations (Fe, Al, Ca, Cd, etc.) or clays. Only free orthophosphate ion present in the soil solution (liquid surrounding the roots) can supply P for plants (Nussaume et al., 2011). This represents less than 0.1% of the total soil P (Fardeau, 1995).

Therefore, for the soils analysed above, Pi availability for plants indicates important variations ranging from 0.032–310 µM. However, on average, the Pi concentration in the soil solution is only a few µM. For a natural ecosystem, this represents on average an amount of 0.8–2 µM Pi to sustain plant development. This is close to the value found in the agricultural soils of the United States before the green revolution and the use of fertilisers. Indeed, in 1927, an extensive US soil analysis taking into account a broad variety of textures and organic matter revealed a 3 µM Pi average concentration in the soil solution (Pierre and Parker, 1927). Therefore, plant Pi uptake creates a depleted area around the roots (Furihata et al., 1992; Schachtman et al., 1998), which can be visualised by radio imaging experiments (Kanno et al., 2012) and measured by different techniques (Kanno et al., 2016a). This explains why Pi is often considered a rare macronutrient despite the abundant presence of phosphorus (the 11th most common element on earth).

We will now focus on the different Pi-uptake systems developed by living organisms to recover available phosphate in their environment. We will take advantage of the recent advances in bioinformatics combined with deep learning resulting in novel software, AlphaFold 2, to predict protein structures with a very high level of confidence (Jumper et al., 2021; Skolnick et al., 2021). This provides the opportunity to compare the structure of proteins, which can be highly conserved despite the divergence of their sequences analysed with MultAlin software (Corpet, 1988; Corpet et al., 1999) to try to classify the different Pi transporters.

9.2 UPTAKE SYSTEMS FOR PHOSPHATE PRESENT IN DIFFERENT ORGANISMS

9.2.1 PROKARYOTES

9.2.1.1 Bacteria

In bacteria, phosphate uptake has been extensively studied in *Escherichia coli* and *Bacillus subtilis* (Willsky and Malamy, 1980; Rao and Torriani, 1990; Novak et al., 1999; Zheng et al., 2016). It relies mostly on two distinct Pi-uptake systems.

9.2.1.1.1 High-Affinity Pi Transporters: The Pst System

The first system is high affinity and low velocity transporter, known as the phosphate-specific transporter (Pst) system. It is an ATP-binding cassette-type (ABC) Pi transporter, characterised by a Km in the range of 0.25–0.43 µM Pi (Willsky and Malamy, 1980; Rao and Torriani, 1990). It is made up of at least four components: PstA, PstB, PstC, and PstS/PiBP. PstA and PstC form the transmembrane channel with six and five helix domains, respectively. These two proteins belong to homologue genes and present a very conserved structure, despite limited similarities (35.5% and 22.6% identity) of their sequences (Figure 9.1). PstB is located on the cytoplasmic side. It carries an ATPase site providing the energy required by the transporter to free Pi. PiBP/PstS is an extracellular Pi-binding protein located in the periplasm. Activated in the situations of Pi deficiency, it is repressed by Pi. The bacterial Pst transporters are part of the PHO regulon. Their vast majority is regulated by a two-component regulatory system (TCS) at the transcriptional level. Under Pi deprivation, the transcription factor PhoB is phosphorylated by the histidine kinase PhoR and binds to the promoters exhibiting consensus PHO box. Such motif, present in the genes of the Pst complex, promotes their upregulation. In *Escherichia coli*, this increases by 100-fold (Willsky and Malamy, 1980). Furthermore, in many bacteria, an additional inhibitory regulatory protein, known as PhoU, contributes to the regulation of the Pst complex to limit excess Pi toxicity (Haldimann et al., 1998; Novak et al., 1999). Interestingly, PhoU homologues identified in *Thermotoga maritima* and *Aquifex*

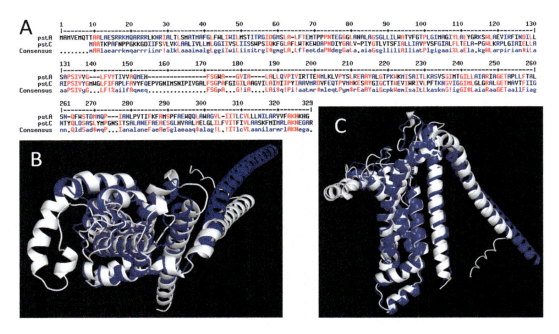

FIGURE 9.1 Comparison of Pst A (bleu) and Pst C (white). A: Amino acid sequences aligned with MultAlin software (Corpet, 1988, 1999); B–C: Structure predicted by AlphaFold (Jumper et al., 2021) and superimposed using PyMOL (version 2.1.1, Schrodinger).

aeolicus exhibited homologies with eukaryotic chaperone Hsp70, suggesting a chaperone role in bacteria (Liu et al., 2005; Oganesyan et al., 2005) and also acting to dephosphorylate PhoB protein-linking transport and transcription regulation (Hsieh and Wanner, 2010).

9.2.1.1.2 Low-Affinity Pi Transporters: The Pit System

The second transport system identified in bacteria is known as phosphate transport (Pit). It is a constitutive transport system exhibiting low-affinity and high-velocity features with a Km of ~10–40 μM Pi (Medveczky and Rosenberg, 1971;. Willsky and Malamy, 1980). In *E. coli*, two proteins, PitA and PitB, have been identified. They are very closely related, with 90% similarity (81% of identity) and not surprisingly have a very homologue structure (Figure 9.2). They are symporters of divalent cations, such as Mg^{2+}, Zn^{2+}, and Ca^{2+}. Identification of PitA activation by Zn^{2+}, where Pi limitation has no effect, suggests that this transporter acts primarily to transport divalent metal cations rather than Pi (Jackson et al., 2008; Graham et al., 2009). PitB, not expressed in normal conditions, probably plays no role, or a minor role, in Pi transport.

The flux of metal/HPO_4^{2-} complex is co-transported with one proton (1:1 stoichiometry) as revealed by the importance of extracellular pH dependence for the Pi-metal uptake (Rosenberg et al., 1982; Vanveen et al., 1994). It is interesting to note that most homologues of the Pit system in other organisms (yeast, animals) use sodium ions instead of protons.

We will not detail here the *nptA* transporter exhibiting even lower affinity (Km ~300μM) for Pi, identified in pathogen bacteria such as *Vibrio cholerae*, *Streptococcus pneumomiae*, and *Staphyloccocus aureus* (Lebens et al., 2002). It belongs to the Sodium Phosphate 2 type of eukaryotic

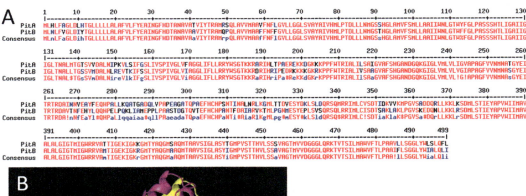

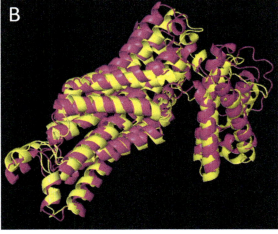

FIGURE 9.2 Comparison of Pit A (yellow) and Pit B (magenta) transporters from *E. coli*. A: Amino acid alignment; B: Structural superimposition.

transporter, extensively studied in higher organisms (Werner and Kinne, 2001). Its origin (ancestral gene or acquired from a eukaryotic source) remains a source of debate.

An outer membrane porin PhoE (Km 1 µM) can also recover organic Pi in external media, which is then degraded in the periplasm by phosphatase to provide a source of Pi (Rao and Torriani, 1990).

Compared to bacteria and eukaryotes, very little is known about the archaeal PHO regulon, but it does not seem to differ greatly from bacteria. For example, the genome of the halophilic euryarchaeon *Halobacterium salinarum* encodes two copies of the Pi high-affinity *Pst* systems with respective Km of 0.1–0.7 µM and Vmax of 30–85 pmol Pi min^{-1} (Furtwangler et al., 2010). This range of values is comparatively similar to those identified in *E. coli* (see above), *B. subtilis* (Km: 0.5 µM Pi (Qi et al., 1997), and *Sinorhizobium meliloti* (Km: 0.2 µM Pi (Yuan et al., 2006). Nevertheless, a lower value has been published for *Methanobacterium thermoautotrophicum*, with a 10-fold reduced Km value (25 nM Pi), but the transporting system in this organism was not identified (Krueger et al., 1986).

9.2.1.2 Cyanobacteria

Phosphate import has been studied in various cyanobacteria due to the importance of Pi for their development and the resulting impact on lake and river eutrophication. The analysis of marine cyanobacteria Synechococcus sp WH7803, Synechococcus sp PCC7942, and freshwater Synechocystic sp PCC6803 (Donald et al., 1997; Ritchie et al., 1997; Burut-Archanai et al., 2011) identified different kinetic parameters for Pi uptake corresponding to Pit and Pts systems. The parameters vary according to the conditions of Pi supply and illumination. For example, in conditions of Pi starvation, the Km measured is in a range of 2–3 µM with illumination, and it increases four to five times in the dark. In Pi-replete conditions, when low-affinity transporters are active, these values rise six to ten times in light and 20–30 times in the dark. As most of the measurements were generally made on the whole organism, we could not access the specific Km of the various transporters.

Few Synechocystis, like many other cyanobacteria, lack the low-affinity Pit-type transporter identified in most bacteria. For the Synechocystis sp. PCC6803 strain, harbouring two Pts systems, genetic experiments have been performed (Burut-Archanai et al., 2011). The experiments revealed that the first gene, known as Pst1, is Pi inducible and responsible for 90% of Pi influx but presents a lower affinity for Pi (Km 5,16 µM) than the constitutive Pst2 transporter (Km 0,13 µM). Both transporters appear favoured by sodium ions. Interestingly, while acidic pH strongly inhibited Pi uptake, no significant difference was observed between pH 7 and 10. This suggests that the transport system can recognise both $H_2PO_4^-$ and HPO_4^{2-}, the predominant forms at pH 7 and 10, respectively (Ritchie et al., 1997; Burut-Archanai et al., 2011).

It should be noted that there exist examples, such as in *Raphidiopsis raciborskii*, where low-affinity Pi transporter Pit can also be regulated by the level of Pi supplied (Willis et al., 2019).

The strong impact of light on the high-affinity system of cyanobacteria suggests an ATP-dependent mechanism similar to the bacteria system described above.

All these organisms can also use organic P compounds and cleave the Pi groups to use them.

9.2.2 FUNGI

9.2.2.1 Yeast

Extensive data on P acquisition are the result of extensive work on the yeast *Saccharomyces cerevisiae* (for review, see Samyn and Persson, 2016). Similar to bacteria, dual active transporters exhibiting high- and low-affinity properties control the transport across the membrane of phosphate in this organism.

9.2.2.1.1 High-Affinity Pi Transporters: Pho84, Pho89

In Pi-limiting conditions, the main Pi transporter is Pho84 (Bun-Ya et al., 1991). It presents a maximum transport activity at acidic pH (4.5), where monobasic ion $H_2PO_4^-$ is predominant with a Km ranging from 8 to 60 µM (Wykoff and O'Shea, 2001). Transport of one Pi ion resulted from symport with two to three protons (Roomans and Borst-Pauwels, 1979). Pho84 is downregulated in the presence of Pi after phosphorylation by Protein Kinase A (PKA) followed by a ubiquitination process leading to the degradation of the protein in the vacuole (Lundh et al., 2009). The structure of the protein has been revealed (Pedersen et al., 2013) in *Piriformospora indica* fungi, showing three distinct portions involved, respectively, in a lid sealing the transportation pathway located on the extracellular side of the transporter, a second part shuttling the protons, and a third part binding Pi. Interestingly, the growth arrest observed during Pi deprivation is due to the downregulation of the PKA pathway controlled by Pho84 acting as a transceptor (Lundh et al., 2009; Popova et al., 2010). Pi addition resulted in the activation of a PKA pathway by Pho84, but in the absence of this protein, such a sensor role is taken up by Pho87 (Samyn and Persson, 2016; Michigami et al., 2018).

The transporter Pho89 is also a high-affinity Pi transporter exhibiting 12 transmembrane domains. Genetic experiments indicated that it contributes 100-fold less to Pi transport than Pho84 (Pattison-Granberg and Persson, 2000). Like Pho84, it is also downregulated by Pi presence but with distinct kinetics, suggesting a fine-tuning of these transporters (Pattison-Granberg and Persson, 2000). Pho89 exhibited high homologies with the type III Na$^+$:Pi transporters identified in mammals and Pit bacterial transporters. It promotes preferential symport of $H_2PO_4^-$ with Na$^+$ according to a 2:1 Na/Pi stoichiometry (and, to a lesser extent, K$^+$ or Li$^+$) with a Km of 0.5–38µM reported for Pi according to different studies and an optimum pH above 7.5 (Roomans and Borst-Pauwels, 1979; Zvyagilskaya et al., 2008; Sengottaiyan et al., 2013). This offers an opportunity for the yeast, which prefers acidic or neutral pH, to extend its capacity to sustain development in an alkaline environment exhibiting a Pi depletion feature. Amino acid analysis revealed homologies with the Pit family, indeed, conserved amino acids crucial for Pi transport in Pit proteins such as Glu55 or Glu491 from Pho89 are conserved. Unlike Pho84 and Pho87, Pho 89 is not involved in the activation of the PKA pathway.

9.2.2.1.2 *Low-Affinity Pi Transporters: Pho87, Pho90*

These transporters belong to the same protein family of H$^+$/Pi symporters and exhibit a Km value ranging from 200 µM to 1 mM (Wykoff and O'Shea, 2001). Their optimum activity is observed at acidic pH as they use proton gradient force to drive Pi uptake. Both are very closely related with 66% identical amino acids and 77% similarities. Their structure is very well conserved (Figure 9.3) and contains an SPX domain (Figure 9.3 D). This controls transporter endocytosis and degradation through interactions with distinct protein partners for Pho87 and Pho90 (Ghillebert et al., 2011).

9.2.2.2 Mycorrhizal Fungi

These organisms have created mutualistic symbiotic interactions for 450 million years with plants. Mycorrhizal fungi concern more than 80% of the root system and play an essential role to extend root capacities to recover nutrients from soil (Ferrol et al., 2019). The first fungi transporter identified belongs to *Glomus versiforme* (Harrison and van Buuren, 1995). It has been found to be a high-affinity Pi/H$^+$ co-transporter with a Km of 18 µM and a homologue of the yeast Pho84 protein. It is expressed in the mycelium part located outside of the root, suggesting a role in Pi uptake. An orthologue of this transporter has been identified in multiple species and has been shown to be present in the arbuscule, suggesting its plays an important role in Pi transfer from fungus to plant. Recent progress in genomics has revealed the presence of Pho89 and Pho 91 in different arbuscular mycorrhizal fungi such as *Glomus intraradices* (Tisserant et al., 2012), but further study will be required to unravel their exact physiological roles. This pinpoints the strong conservation of Pi transporters between the different fungal species.

Analysis and Comparison of Alphafold-Structure Predictions 135

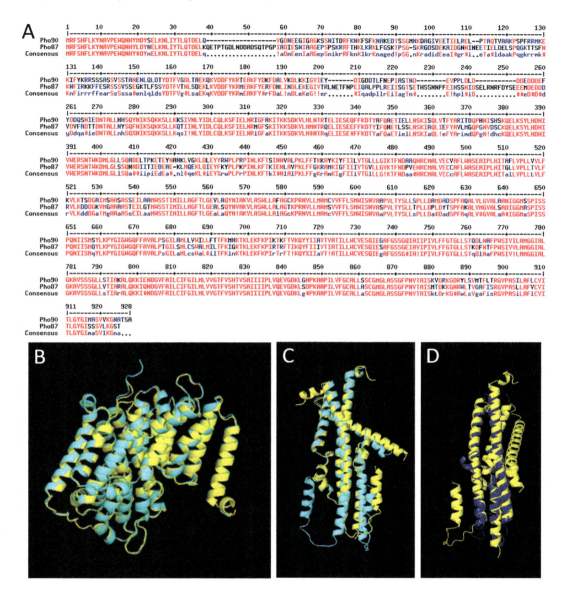

FIGURE 9.3 Comparison of Pho87 (yellow) and Pho90 (cyan) transporters from yeast. A: Amino acid alignment; B: Structural superimposition of transmembrane domains; C: Structural superimposition of cytoplasmic moieties containing the SPX domain (the long unstructured part of the protein has been deleted to simplify the figure); D: Structural superimposition of the cytoplasmic part of Pho87 and rice SPX domain (blue) of SPX1 protein (PDB: 7E40).

9.2.3 GREEN ALGAE

Pi-uptake studies in algae remain very limited and have mostly focused on green algae. They are considered closely related to land plants and also belong to Streptophyta, one of the two green algae taxa with Chlorophyta.

Green algae possess two types of inorganic Pi transporters: (i) a H⁺/Pi symport system known as PHOSPHATE TRANSPORTER A (PTA) and (ii) a PHOSPHATE TRANSPORTER B (PTB) assumed to be an Na$^+$/Pi symporter. PTB has been identified in a broad number of species such as *Dunaliella salina* (Weiss et al., 2001), *Chara corallina* (Reid et al., 2000; Mimura et al., 2002), *Chlamydomonas reinhardtii* (Kobayashi et al., 2003; Wang et al., 2020), *Tetraselmis chui* (Chung et al., 2003), *Dunaliella viridis* (Li et al., 2006), *Coccomyxa subellipsoidea*, *Ostreoccocus taurii* (Djouani-Tahri et al., 2011), *Klebsormidium flaccidum*, and *Coleochaete nitellarum* (Bonnot et al., 2017). Different studies have revealed the induction of most PTA and PTB by Pi starvation (Chung et al., 2003; Kobayashi et al., 2003; Li et al., 2006; Moseley et al., 2006; Djouani-Tahri et al., 2011; Bonnot et al., 2017). It is, therefore, more likely that both H$^+$/Pi and Na$^+$/Pi symports are functional in green algae (Bonnot et al., 2017). Like many organisms, these genes are often present in multicopies. For example, in *Chlamydomonas reinhardi*, 4 sequences encoding PTA genes and 11 for PTB have been identified (Wang et al., 2020). Transcriptome analysis revealed that CrPTA1 and CrPTA3 are downregulated by Pi starvation and may be candidates for the low-affinity Pi-uptake system. Nevertheless, this assertion should be taken with caution, as some transporters like PHT1;1 (Shin et al., 2004) in plants are very poorly regulated by Pi level and remain, nevertheless, high-affinity transporters. It may be noticed also that for marine algae, due to the low Pi concentration present in the sea, the role of a low-Pi-affinity transporter would be difficult to envisage.

The expressions of CrPTB2, CrPTB3, CrPTB4, CrPTB5, CrPTB7, CrPTB8, and CrPTB12 were induced by Pi starvation and could correspond to a high-affinity Pi transport system (Wang et al., 2020). The conservation of homologous genes to the Na$^+$/Pi symport system in metazoan Pit and fungal Pit (51.6% similarities for the case of *T. chui* (Chung et al., 2003), suggested similar functionality for these PTB transporters. A view reinforced by physiological experiments in *C. corallina* indicated that sodium uptake was requested for Pi influx during P starvation (Reid et al., 2000; Mimura et al., 2002).

The transcriptional control required, as in higher plants, a homologue of plant transcription factor PHR1 gene, known as PSR, in algae providing evidence of the strong conservation of Pi regulation (cf the plant part below). In the Chlamydomonas study, the sequence analysis predicted a similar structure for PTA proteins with 11 transmembrane domains, while PTB proteins exhibited more variations ranging from 8 to 15 (Wang et al., 2020). Due to the lack of data on these individual transporters, we do not have access to the characterisation of their specific biochemical parameters. The only values reported are limited in describing Pi uptake taking place at the whole organism level, for example in Chlorophyta *Ulva intestinalis* with a Km of ~2 µM (Douglas et al., 2014). Further analysis is therefore required to detail this family of proteins.

A study of red algae (*Melanthalia abscissa* and *Pterocladia lucida*) and brown algae (*Cystophora torulosa*, *Zonaria turneriana*, and *Xiphophora chondrophylla*) identified a Km of 0.6/1.6 µM and ~5/10 µM, respectively (Douglas et al., 2014). This revealed the presence of a high-affinity transporter in Rodophyta and Phaeophyceae algae. This view was confirmed recently by the identification of Pho89 homologues in those organisms (Jiang et al., 2019).

9.2.4 BRYOPHYTES

Bryophytes are early diverging land plants, which inherited both H$^+$/Pi and Na$^+$/Pi symporters from their ancestors (Bonnot et al., 2017). The presence of H+/Pi homologues to plant PHT1 transporters has been reported in different species, such as *Marchantia polymorpha* and *Physcomitrella. patens* (Wang et al., 2008; Saint-Marcoux et al., 2015). For PTB proteins (Na$^+$/Pi symporters), Bonnot et al. demonstrated that the heterologous expression of *Marchantia polymorpha* core PTB proteins restores Pi uptake in a *Saccharomyces cerevisiae* mutant defective for high-affinity Pi transporters, demonstrating their functionalities. In addition to *Marchantia,* PTB genes were also identified (based on sequence homologies) in *P. patens*, *Encalypta streptocarpa*, and *Phaeoceros carolinianus* (Bonnot et al., 2017).

The vast majority of these genes have been shown to be induced by Pi starvation, suggesting they may play a role in high-affinity Pi transport. Nevertheless, the Na$^+$ concentration is very different on land compared to sea, and usage of the different Pi transporters requires additional study to clearly decipher their precise physiological functions.

9.2.5 Higher Plants

Pi is the exclusive source of phosphorus for plants with dihydrogen phosphate (H_2PO4^-). The form present at pH 5–6 is favoured, as the maximum Pi uptake is observed within this pH range (Furihata et al., 1992; Schachtman et al., 1998). If we exclude very rare cases of marine angiosperms such as *Zostera marina* (where electrophysiological experiments revealed the presence of a sodium-dependent high-affinity phosphate transporter(s) presenting a Km estimated at 1,5 µM (Rubio et al., 2005)), the Pi transport in plant relies on a single multigenic family of plasma membrane proteins known as PHT1 or Phosphate transporter 1 (Nussaume et al., 2011; Lambers, 2022). Sequence analysis revealed strong conservation from fungi to plants. For example, the identities of 76% of amino acids across several plant species have been observed and 50% of *Arabidopsis* PHT1 and the yeast PHO84 proteins (Nussaume et al., 2011).

The analysis of their structure was undertaken in the endophytic fungus *Piriformospora indica*. The presence of 12 membrane-spanning domains was identified, with both C and N ends localised into the cytosol (Pedersen et al., 2013). Several experiments using western blot analysis revealed the presence of multiple bands, suggesting a multimerisation process (Chiou et al., 2001; Nussaume et al., 2011; Fontenot et al., 2015). A conclusion supported by genetic experiments identified the semi-dominant mutation of *pht1;3* in *Arabidopsis* (Catarecha et al., 2007). A trait associated with the presence of inactive subunits in a multimeric transporter results in an inactive structure. The Pi uptake requires energy and promotes alkalinisation of the medium, suggesting a co-transport between Pi and protons (Ullrich and Novacky, 1990). Measurements of proton fluxes and Pi absorption revealed a stoichiometry of two to four protons per Pi ion crossing the plasma membrane ((Ullrich-Eberius et al., 1981; Ullrich-Eberius et al., 1984; Sakano, 1990). Recent electrophysiological experiments confirmed this view in *Arabidopsis*, indicating that more than three protons per Pi ion were requested (Dindas et al., 2022).

Similar to many plant nutrients, uptake kinetics realised with radioactive ^{33}Pi or ^{32}Pi isotopes revealed the co-existence of low- and high-affinity systems (Cogliati and Clarkson, 1983; Drew et al., 1984). The Km values identified for the high-affinity system measured on whole plants were in the range of 2–13 µM, while the low-affinity system indicated values between 300 and 1000 µM (for a review, see Nussaume et al. (2011) and Ayadi et al., (2015)). It was originally believed that high or low affinity corresponds to specific members within this family. Genetic experiments resulting in the knock-out of several PHT1 members in *Arabidopsis* (reducing until 96% of Pi flux) modified this view, as both high- and low-Pi-affinity systems were found co-affected (Ayadi et al., 2015). This provides strong experimental evidence that, in plants, PHT1 members can be both high- or low-affinity transporters as a consequence of post-translational modifications (Ayadi et al., 2015). Nevertheless, individual characteristics (K_m, V_{max}) of the transporters remain difficult to measure *in planta* due to the presence of other transporters. The complementation of the yeast double mutant *pho84/pho89* (the two main high-affinity transporters of Pi present in yeast) provide values that most of the time exceed those identified in plants. This is probably due to the additional component requested for plant protein regulation, which is missing in these organisms. Indeed, several regulatory networks have been identified so far that affect PHT1 transporters. Very tight regulatory networks downregulate the level and activity of the different PHT1s present in the plant in the presence of Pi. Transcriptional control is one of the main regulatory networks (Misson et al., 2005; Morcuende et al., 2007; Bustos et al., 2010; Thibaud et al., 2010), as the vast majority of PHT1 transporters exhibited the P1BS binding site for Myb transcription factor PHR1 (Phosphate Starvation Response 1 (Rubio et al., 2001). This transcription factor is constitutively expressed and is inactivated by SPX1 family proteins (Lv et al., 2014; Puga et al., 2014; Wang et al., 2014) in the presence of Pi. This fast process takes place within minutes (three to five minutes in *Arabidopsis* root) following Pi addition (Hani et al., 2021). Pi derivative metabolites, known as inositol pyrophosphate (Wild et al., 2016; Zhu et al., 2019), promote the dimerisation of SPX1 proteins, which then bind PHR1 family members, thereby inhibiting their activity (Ried et al., 2021; Zhou et al., 2021).

It is interesting to note that the loss of the downregulation of PHT1 by Pi observed in several native Australian species (Shane et al., 2004; Lambers et al., 2013) is one of the plant strategies developed to cope with very poor soil (Lambers, 2022). As a consequence, such plants became hypersensitive to Pi and exhibited important Pi-toxicity symptoms when germinated on soils containing Pi (Lambers et al., 2013).

The spatial distribution of the different PHT1 members is also transcriptionally controlled in the plant. The vast majority of PHT1 family members exhibited specific expression patterns with high redundancy in the different plant tissues (Misson et al., 2005; Thibaud et al., 2010; Nussaume et al., 2011). Interestingly, most of them are expressed in the root, which is in agreement with their crucial role in Pi uptake: eight out of nine, for example, in *Arabidopsis* (Karthikeyan et al., 2002; Mudge et al., 2002; Misson et al., 2004; Nussaume et al., 2011). It is also noteworthy that the root apex and the root cap contain a very high number of Pi transporters (Arnaud et al., 2010; Kanno et al., 2016b). This may be due to the low mobility of Pi in soils, providing the opportunity for the plant apex to experience "first come, first served".

Interestingly, for plants exhibiting mycorrhizal symbiosis, the expression of PHT1 induced in the root to recover Pi provided by the fungi appears to be controlled at the transcriptional level. Indeed, in *Medicago*, the polar targeting of MtPT4 (the PHT1 transporter associated with the mycorrhization process) to the peri-arbuscular membrane resulted from specific temporal promoter activation during symbiosis. At this stage, a transient reorientation of the secretion and the alterations in the protein cargo entering the secretory system of the colonised root cell take place (Pumplin et al., 2012).

In addition to transcriptional control, multiple post-translational mechanisms affecting membrane targeting, accumulation, recycling, and degradation have been identified. For example, to cross the endoplasmic reticulum, PHT1 transporters required a conserved protein known as PHF1 (PHOSPHATE TRANSPORTER TRAFFIC FACILITATOR1). Identified genetically, its mutation strongly reduced Pi flux (Gonzalez et al., 2005; Bayle et al., 2011; Chen et al., 2011). The proper subcellular targeting also involves an ESCRT-III-Associated Protein known as ALIX (Cardona-Lopez et al., 2015, 2022).

Tight regulation of the recycling and degradation at plasma membranes is also observed. This implies a phosphorylation process of PHT1 proteins (Bayle et al., 2011) resulting from CK2 kinase activity (Chen et al., 2015). The degradation in the ER also required the E2 ligase PHO2 (Aung et al., 2006; Bari et al., 2006), identified by the overaccumulation of Pi in the aerial part of *Arabidopsis* mutants (Delhaize and Randall, 1995). Another crucial element of this cascade is the E3 ligase NLA, which controls the level of PHT1 accumulation taking place at the plasma membrane level (Lin et al., 2013; Park et al., 2014).

Recently, additional regulation was identified, revealing the integration of immune responses to alleviate phosphate starvation stress during the association with beneficial microbes. This occurs through a phosphorylation process, repressing PHT1 transporters and involving two receptor-like cytoplasmic kinases, BOTRYTIS-INDUCED KINASE 1 (BIK1) and PBS1-LIKE KINASE 1 (PBL1) (Dindas et al., 2022).

9.3 CONCLUSION

Based on sequence and structure analysis, we distinguished four different groups of Pi transporters exhibiting high-level homologies.

The first one contained the high-affinity Pi transporter homologues from the high-affinity crystalised *Piriformospora indica* Pi transporter (Pedersen et al., 2013), yeast PHO84, and plant PHT1

Analysis and Comparison of Alphafold-Structure Predictions 139

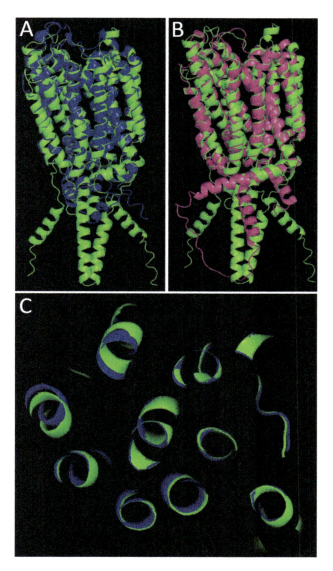

FIGURE 9.4 A–B: Structural superimpositions of Pho84 (green) and PHT1;1 from *Arabidopsis* (blue); C: with *Piriformospora indica* Pi transporter (pink). C: cross-section from (A).

family. They all present very closely related structures (Figure 9.4). As expected, this family of transporters can also be easily identified by BLAST analysis in several algae, such as *Parachlorella kessieri*, *Chlorella sorokiniana*, *Coccomyxa sp. obi*, and *Chlamydomonas incerta* (Figure 9.5). Structural and genetic experiments in yeast and *P indica* fungi (Pedersen et al., 2013; Samyn and Persson, 2016) pinpointed several crucial amino acids for Pi-binding, proton-tunnelling, and opening of the Pi exit pathways (Figure 9.5). These amino acids and the transporter structure appear well conserved between all fungi and photosynthetic organisms investigated here (Figures 9.4 and 9.5).

It is surprising that all these transporters are only poorly related to the membrane subunit of their putative bacterial ancestor Pst (Figure 9.6 A, B), which will therefore be classified here in a distinct group. A search for a bacterial structural homologue of PHT1;1 using Foldseek search (van Kempen et al., 2011) identified distinct putative candidates such as the *E. Coli* sialic acid transporter *nanT*, which appears much more related (Figure 9.6 C, D). A search using SMARTBLAST

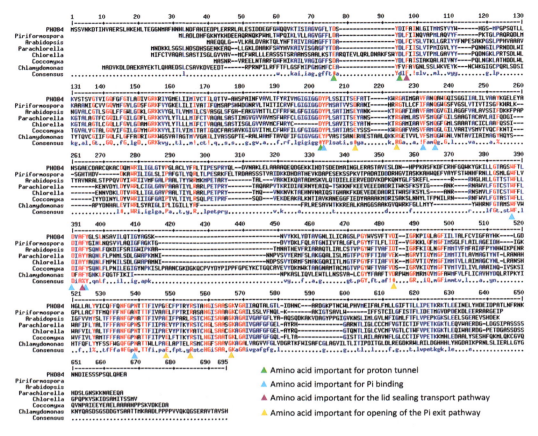

FIGURE 9.5 Conservation in plants and algae of important structural amino acids for high-affinity phosphate/proton symport identified in fungi (Pedersen et al., 2013; Samyn and Peerson, 2016). *Arabidopsis* thaliana (PHT1;1, AT5G43350), PHO84 from *Saccharomyces cerevisiae* (NCBI NP_013583), Pirisformosa indica (A8N031), *Parachlorella kessleri* (BAU71131.1), *Chlorella sorokiniana* (PRW20180.1), *Coccomyxa sp.* Obi (BDA45410.1), and *Chlamydomonas incerta* (KAG2435777.1). Alignment realised with MultAlin software.

also identified a transporter of the major facilitator superfamily (NCBI: WP_011277753.1) from the thermoacidophilic archaeon *Sulfolobus acidocaldarius* (Figure 9.6 E). This raises serious doubts about the evolutionary relationship between Pst and PHT1;1. It will be therefore interesting to search for a PHT1;1 ancestor more closely. Another line of study is the investigation of putative horizontal transfer between fungi and photosynthetic bacteria. This line of research is supported by the close homologies between transporters present in these organisms (Figure 9.5), which appear more closely related than the bacterial homologues identified here (Figure 9.6 E).

The third group includes (i) the low-affinity chloroplast Pi transporter PHT2;1 (Daram et al., 1999; Versaw and Harrison, 2002), which is classified as a proton/phosphate transporter based on yeast complementation experiments (Daram et al., 1999; Versaw and Harrison, 2002); (ii) the yeast Pho89 (symport Pi/Na+); (iii) the *C. reinhardi* PTB genes; and (iv) the *Escherichia coli* PitA and PitB. Figure 9.7 (A–I) presents the conservation of the structures. Although very well conserved in some cases, such as PitB and PitA, variations can be still found. They affect mainly the number of transmembrane domains between proteins of this group (Figure 9.7 A–F). The importance of comparing the structure is crucial here, as the amino acid alignment does not provide much valuable information (Figure 9.7 K). It is interesting to note that this group includes Pi symporters that use very distinct cations according to species – proton (plant), sodium (algae/fungi), and metal

Analysis and Comparison of Alphafold-Structure Predictions 141

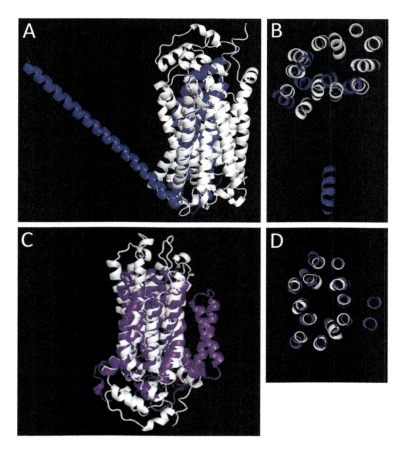

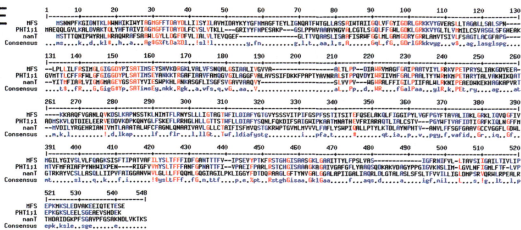

FIGURE 9.6 Structural superimpositions of PHT1;1 from *Arabidopsis* (white) with PstA (blue; A-B) or with sialic acid transporter nanT (UniPROT B5YSV1; purple; C-D) from *E coli*. (E): Sequence alignment between PHT1;1, nanT, and MSF transporter (NCBI: WP_011277753.1) from a thermoacidophilic archaeon (*Sulfolobus acidocaldarius*) cross-section of structures superimposed in A and C, respectively. Note that in A and B, the two proteins have a different number of transmembrane domains.

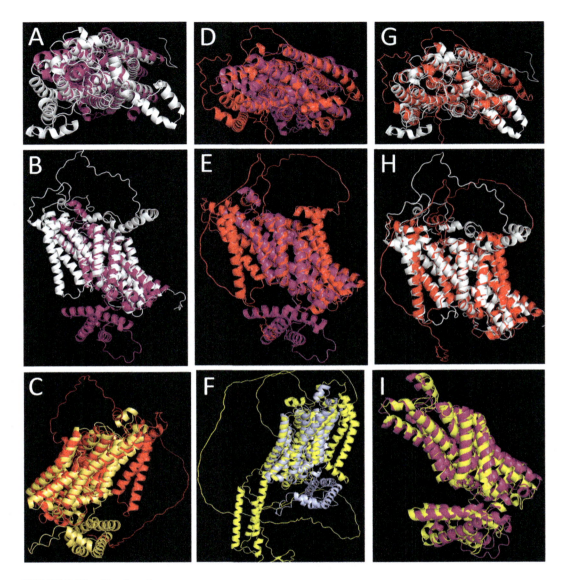

FIGURE 9.7A Continued

(bacteria). They also exhibit very different affinity, ranging from high (such as PHO89) to low (bacterial Pit transporters or plant PHT2).

The fourth group contains the significantly related low-affinity Pi/H+ transporters PHO87 and PHO90, which do not present any homologues with the three other groups of transporters (Figure 9.3).

Figure 9.8 summarises this view and presents the different systems identified in living organisms that recover inorganic sources of Pi from water or soil environments. They all present high- and low-affinity transporters for fine-tuning Pi uptake. But this function devoted to distinct proteins in bacteria and yeast has evolved: plant post-translation regulation modulates the activity of the PHT1 family from high to low affinity according to Pi supply (Ayadi et al., 2015). It is interesting to note

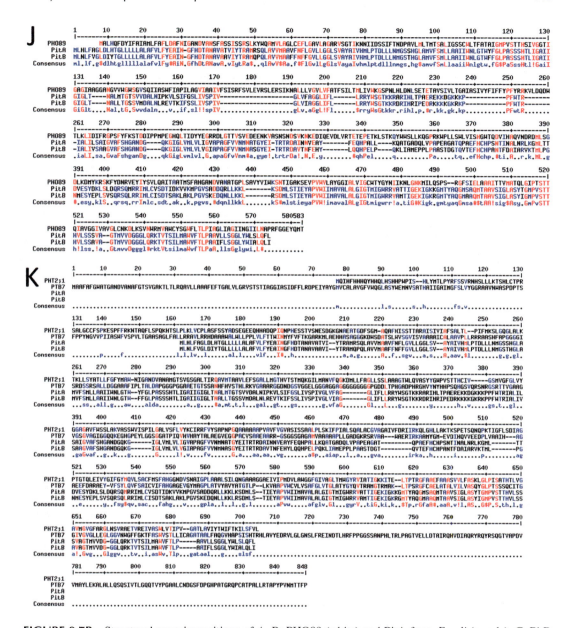

FIGURE 9.7B Structural superimpositions of A–B: PHO89 (white) and PitA from *E.coli* (purple); C: PitB *E.coli* (pale yellow) and PHT2-1 *Arabidopsis* (red); D–E: PitA (purple) and PHT2-1 (red); F: PitB (grey) and PTB7 from *C. reinhardtii* (yellow); G–H: PHO89 (white) and PHT2.1 (red); I: PitA (purple) and Pit B (yellow); J–K sequence alignment of the same proteins.

that in all organisms, the high-affinity Pi transporter presents similar Km values (of a few micromolars), but the more sensitive ones reported so far are in unicellular organisms (Km of 0.1–0.5 μM for yeast or bacteria and capacities for some phytoplankton organisms to survive with only 10–20 nM Pi). More extensive study is required of unicellular algae to ascertain whether more efficient Pi-uptake mechanisms are selected in those organisms.

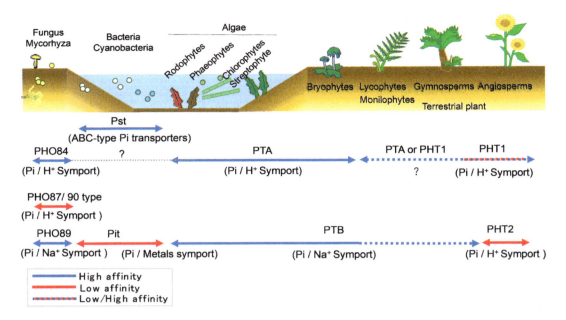

FIGURE 9.8 Different related Pi-uptake systems present in fungi, bacteria, algae, and terrestrial plants.

9.4 ACKNOWLEDGEMENT

T.D. thanks L. Blanchard for introducing him to PyMOL.

BIBLIOGRAPHY

Arnaud, C., Bonnot, C., Desnos, T., and Nussaume, L. (2010). The root cap at the forefront. *C R Biol* 333(4), 335–343.

Aung, K., Lin, S.I., Wu, C.C., Huang, Y.T., Su, C.L., and Chiou, T.J. (2006). pho2, a phosphate overaccumulator, is caused by a nonsense mutation in a microRNA399 target gene. *Plant Physiol* 141(3), 1000–1011.

Ayadi, A., David, P., Arrighi, J.F., Chiarenza, S., Thibaud, M.C., Nussaume, L., and Marin, E. (2015). Reducing the genetic redundancy of Arabidopsis phosphate transporter1 transporters to study phosphate uptake and signaling. *Plant Physiol* 167(4), 1511–1526.

Bari, R., Datt Pant, B., Stitt, M., and Scheible, W.R. (2006). PHO2, microRNA399, and PHR1 define a phosphate-signaling pathway in plants. *Plant Physiol* 141(3), 988–999.

Bayle, V., Arrighi, J.F., Creff, A., Nespoulous, C., Vialaret, J., Rossignol, M., Gonzalez, E., Paz-Ares, J., and Nussaume, L. (2011). Arabidopsis thaliana high-affinity phosphate transporters exhibit multiple levels of posttranslational regulation. *Plant Cell* 23(4), 1523–1535.

Bjerrum, C.J., and Canfield, D.E. (2002). Ocean productivity before about 1.9 Gyr ago limited by phosphorus adsorption onto iron oxides. *Nature* 417(6885), 159–162.

Bonnot, C., Proust, H., Pinson, B., Colbalchini, F.P., Lesly-Veillard, A., Breuninger, H., Champion, C., Hetherington, A.J., Kelly, S., and Dolan, L. (2017). Functional PTB phosphate transporters are present in streptophyte algae and early diverging land plants. *New Phytol* 214(3), 1158–1171.

Brocks, J.J., Logan, G.A., Buick, R., and Summons, R.E. (1999). Archean molecular fossils and the early rise of eukaryotes. *Science* 285(5430), 1033–1036.

Brocks, J.J., Jarrett, A.J.M., Sirantoine, E., Hallmann, C., Hoshino, Y., and Liyanage, T. (2017). The rise of algae in Cryogenian oceans and the emergence of animals. *Nature* 548(7669), 578.

Bun-Ya, M., Nishimura, M., Harashima, S., and Oshima, Y. (1991). The PHO84 gene of Saccharomyces cerevisiae encodes an inorganic phosphate transporter. *Mol Cell Biol* 11(6), 3229–3238.

Burut-Archanai, S., Eaton-Rye, J.J., and Incharoensakdi, A. (2011). Na+-stimulated phosphate uptake system in synechocystis sp. PCC 6803 with Pst1 as a main transporter. *BMC Microbiol* 11, 225.

Bustos, R., Castrillo, G., Linhares, F., Puga, M.I., Rubio, V., Perez-Perez, J., Solano, R., Leyva, A., and Paz-Ares, J. (2010). A central regulatory system largely controls transcriptional activation and repression responses to phosphate starvation in Arabidopsis. *PLOS Genet* 6(9), e1001102.

Cardona-Lopez, X., Cuyas, L., Marin, E., Rajulu, C., Irigoyen, M.L., Gil, E., Puga, M.I., Bligny, R., Nussaume, L., Geldner, N., Paz-Ares, J., and Rubio, V. (2015). ESCRT-III-associated protein ALIX mediates high-affinity phosphate transporter trafficking to maintain phosphate homeostasis in Arabidopsis. *Plant Cell* 27(9), 2560–2581.

Cardona-Lopez, X., Cuyas, L., Marin, E., Rajulu, C., Irigoyen, M.L., Gil, E., Puga, M.I., Bligny, R., Nussaume, L., Geldner, N., Paz-Ares, J., and Rubio, V. (2022). Correction to: ESCRT-III-associated protein ALIX mediates high-affinity phosphate transporter trafficking to maintain phosphate homeostasis in Arabidopsis. *Plant Cell* 34, 2809.

Catarecha, P., Segura, M.D., Franco-Zorrilla, J.M., Garcia-Ponce, B., Lanza, M., Solano, R., Paz-Ares, J., and Leyva, A. (2007). A mutant of the Arabidopsis phosphate transporter PHT1;1 displays enhanced arsenic accumulation. *Plant Cell* 19(3), 1123–1133.

Chen, J., Liu, Y., Ni, J., Wang, Y., Bai, Y., Shi, J., Gan, J., Wu, Z., and Wu, P. (2011). OsPHF1 regulates the plasma membrane localization of low- and high-affinity inorganic phosphate transporters and determines inorganic phosphate uptake and translocation in rice. *Plant Physiol* 157(1), 269–278.

Chen, J., Wang, Y., Wang, F., Yang, J., Gao, M., Li, C., Liu, Y., Liu, Y., Yamaji, N., Ma, J.F., Paz-Ares, J., Nussaume, L., Zhang, S., Yi, K., Wu, Z., and Wu, P. (2015). The rice CK2 kinase regulates trafficking of phosphate transporters in response to phosphate levels. *Plant Cell* 27(3), 711–723.

Chiou, T.J., Liu, H., and Harrison, M.J. (2001). The spatial expression patterns of a phosphate transporter (MtPT1) from Medicago truncatula indicate a role in phosphate transport at the root/soil interface. *Plant J* 25(3), 281–293.

Chung, C.C., Hwang, S.P., and Chang, J. (2003). Identification of a high-affinity phosphate transporter gene in a prasinophyte alga, Tetraselmis chui, and its expression under nutrient limitation. *Appl Environ Microbiol* 69(2), 754–759.

Cogliati, D.H., and Clarkson, D.T. (1983). Physiological changes in, and phosphate uptake by potato plants during development of and recovery from phosphate deficiency. *Physiol Plant* 58(3), 287–294.

Corpet, F. (1988). Multiple sequence alignment with hierarchical clustering. *Nucleic Acids Res* 16(22), 10881–10890.

Corpet, F., Gouzy, J., and Kahn, D. (1999). Browsing protein families via the 'Rich Family Description' format. *Bioinformatics* 15(12), 1020–1027.

Daram, P., Brunner, S., Rausch, C., Steiner, C., Amrhein, N., and Bucher, M. (1999). Pht2;1 encodes a low-affinity phosphate transporter from Arabidopsis. *Plant Cell* 11, 2153–2166.

Delhaize, E., and Randall, P.J. (1995). Characterization of a phosphate-accumulator mutant of Arabidopsis thaliana. *Plant Physiol* 107(1), 207–213.

Dindas, J., DeFalco, T.A., Yu, G., Zhang, L., David, P., Bjornson, M., Thibaud, M.C., Custodio, V., Castrillo, G., Nussaume, L., Macho, A.P., and Zipfel, C. (2022). Direct inhibition of phosphate transport by immune signaling in Arabidopsis. *Curr Biol* 32(2), 488–495 e485.

Djouani-Tahri, E.B., Sanchez, F., Lozano, J.C., and Bouget, F.Y. (2011). A phosphate-regulated promoter for fine-tuned and reversible overexpression in Ostreococcus: Application to circadian clock functional analysis. *PLOS One* 6(12), e28471.

Donald, K.M., Scanlan, D.J., Carr, N.G., Mann, N.H., and Joint, I. (1997). Comparative phosphorus nutrition of the marine cyanobacterium Synechococcus WH7803 and the marine diatom Thalassiosira weissflogii. *J Plankton Res* 19(12), 1793–1813.

Douglas, E.J., Haggitt, T.R., and Rees, T.A.V. (2014). Supply- and demand-driven phosphate uptake and tissue phosphorus in temperate seaweeds. *Aquat Biol* 23(1), 49–60.

Drew, M.C., Saker, L.R., Barber, S.A., and Jenkins, W. (1984). Changes in the kinetics of phosphate and potassium absorption in nutrient-deficient barley roots measured by a solution-depletion technique. *Planta* 160(6), 490–499.

Falkowski, P.G., Katz, M.E., Knoll, A.H., Quigg, A., Raven, J.A., Schofield, O., and Taylor, F.J.R. (2004). The evolution of modern eukaryotic phytoplankton. *Science* 305(5682), 354–360.

Fardeau, J.C. (1995). Dynamics of phosphate in soils. An isotopic outlook. *Fert Res* 45(2), 91–100.

Ferrol, N., Azcon-Aguilar, C., and Perez-Tienda, J. (2019). Review: Arbuscular mycorrhizas as key players in sustainable plant phosphorus acquisition: An overview on the mechanisms involved. *Plant Sci* 280, 441–447.

Fontenot, E.B., Ditusa, S.F., Kato, N., Olivier, D.M., Dale, R., Lin, W.Y., Chiou, T.J., Macnaughtan, M.A., and Smith, A.P. (2015). Increased phosphate transport of Arabidopsis thaliana Pht1;1 by site-directed mutagenesis of tyrosine 312 may be attributed to the disruption of homomeric interactions. *Plant Cell Environ* 38(10), 2012–2022.

Furihata, T., Suzuki, M., and Sakurai, H. (1992). Kinetic characterization of two phosphate uptake systems with different affinities in suspension-cultured Catharanthus roseus protoplants. *Plant Cell Physiol* 33, 1151–1157.

Furtwangler, K., Tarasov, V., Wende, A., Schwarz, C., and Oesterhelt, D. (2010). Regulation of phosphate uptake via Pst transporters in Halobacterium salinarum R1. *Mol Microbiol* 76(2), 378–392.

Ghillebert, R., Swinnen, E., De Snijder, P., Smets, B., and Winderickx, J. (2011). Differential roles for the low-affinity phosphate transporters Pho87 and Pho90 in Saccharomyces cerevisiae. *Biochem J* 434(2), 243–251.

Gonzalez, E., Solano, R., Rubio, V., Leyva, A., and Paz-Ares, J. (2005). Phosphate Transporter traffic facilitator1 is a plant-specific SEC12-related protein that enables the endoplasmic reticulum exit of a high-affinity phosphate transporter in Arabidopsis. *Plant Cell* 17(12), 3500–3512.

Graham, A.I., Hunt, S., Stokes, S.L., Bramall, N., Bunch, J., Cox, A.G., McLeod, C.W., and Poole, R.K. (2009). Severe zinc depletion of Escherichia coli roles for high affinity zinc binding by zint, zinc transport and zinc-independent proteins. *J Biol Chem* 284(27), 18377–18389.

Haldimann, A., Daniels, L.L., and Wanner, B.L. (1998). Use of new methods for construction of tightly regulated arabinose and rhamnose promoter fusions in studies of the Escherichia coli phosphate regulon. *J Bacteriol* 180(5), 1277–1286.

Hani, S., Cuyas, L., David, P., Secco, D., Whelan, J., Thibaud, M.C., Merret, R., Mueller, F., Pochon, N., Javot, H., Faklaris, O., Marechal, E., Bertrand, E., and Nussaume, L. (2021). Live single-cell transcriptional dynamics via RNA labelling during the phosphate response in plants. *Nat Plants* 7(8), 1050–1064.

Harrison, M.J., and van Buuren, M.L. (1995). A phosphate transporter from the mycorrhizal fungus Glomus versiforme. *Nature* 378(6557), 626–629.

He, X.J., Augusto, L., Goll, D.S., Ringeval, B., Wang, Y.P., Helfenstein, J., Huang, Y.Y., Yu, K.L., Wang, Z.Q., Yang, Y.C., and Hou, E.Q. (2021). Global patterns and drivers of soil total phosphorus concentration. *Earth Syst Sci Data* 13(12), 5831–5846.

Hsieh, Y.J., and Wanner, B.L. (2010). Global regulation by the seven-component Pi signaling system. *Curr Opin Microbiol* 13(2), 198–203.

Jackson, R.J., Binet, M.R.B., Lee, L.J., Ma, R., Graham, A.I., McLeod, C.W., and Poole, R.K. (2008). Expression of the PitA phosphate/metal transporter of Escherichia coli is responsive to zinc and inorganic phosphate levels. *FEMS Microbiol Lett* 289(2), 219–224.

Jiang, M.J., Wei, H.H., Chen, H.H., Zhang, J.Y., Zhang, W., and Sui, Z.H. (2019). Expression analysis of three phosphate transporter genes in the fast-growing mutants of Gracilariopsis lemaneiformis (Rhodophyta) under low phosphorus condition. *J Appl Phycol* 31(3), 1907–1919.

Johnston, D.T., Wolfe-Simon, F., Pearson, A., and Knoll, A.H. (2009). Anoxygenic photosynthesis modulated Proterozoic oxygen and sustained Earth's middle age. *Proceedings of the National Academy of Sciences of the United States of America* 106(40), 16925–16929.

Jumper, J., Evans, R., Pritzel, A., Green, T., Figurnov, M., Ronneberger, O., Tunyasuvunakool, K., Bates, R., Zidek, A., Potapenko, A., Bridgland, A., Meyer, C., Kohl, S.A.A., Ballard, A.J., Cowie, A., Romera-Paredes, B., Nikolov, S., Jain, R., Adler, J., Back, T., Petersen, S., Reiman, D., Clancy, E., Zielinski, M., Steinegger, M., Pacholska, M., Berghammer, T., Bodenstein, S., Silver, D., Vinyals, O., Senior, A.W., Kavukcuoglu, K., Kohli, P., and Hassabis, D. (2021). Highly accurate protein structure prediction with AlphaFold. *Nature* 596(7873), 583–589.

Kanno, S., Yamawaki, M., Ishibashi, H., Kobayashi, N.I., Hirose, A., Tanoi, K., Nussaume, L., and Nakanishi, T.M. (2012). Development of real-time radioisotope imaging systems for plant nutrient uptake studies. *Philos Trans R Soc Lond B* 367(1595), 1501–1508.

Kanno, S., Cuyas, L., Javot, H., Bligny, R., Gout, E., Dartevelle, T., Hanchi, M., Nakanishi, T.M., Thibaud, M.C., and Nussaume, L. (2016a). Performance and limitations of phosphate quantification: Guidelines for plant biologists. *Plant Cell Physiol* 57(4), 690–706.

Kanno, S., Arrighi, J.F., Chiarenza, S., Bayle, V., Berthome, R., Peret, B., Javot, H., Delannoy, E., Marin, E., Nakanishi, T.M., Thibaud, M.C., and Nussaume, L. (2016b). A novel role for the root cap in phosphate uptake and homeostasis. *Elife* 5, e14577.

Karl, D.M. (2014). Microbially mediated transformations of phosphorus in the sea: New views of an old cycle. *Ann Rev Mar Sci* 6, 279–337.

Karthikeyan, A.S., Varadarajan, D.K., Mukatira, U.T., D'Urzo, M.P., Damsz, B., and Raghothama, K.G. (2002). Regulated expression of Arabidopsis phosphate transporters. *Plant Physiol* 130(1), 221–233.

Kobayashi, I., Fujiwara, S., Shimogawara, K., Kaise, T., Usuda, H., and Tsuzuki, M. (2003). Insertional mutagenesis in a homologue of a Pi transporter gene confers arsenate resistance on chlamydomonas. *Plant Cell Physiol* 44(6), 597–606.

Konhauser, K.O., Lalonde, S.V., Amskold, L., and Holland, H.D. (2007). Was there really an Archean phosphate crisis? *Science* 315(5816), 1234–1234.

Krueger, R.D., Harper, S.H., Campbell, J.W., and Fahrney, D.E. (1986). Kinetics of phosphate uptake, growth, and accumulation of cyclic diphosphoglycerate in a phosphate-limited continuous culture of Methanobacterium thermoautotrophicum. *J Bacteriol* 167(1), 49–56.

Lambers, H. (2022). Phosphorus acquisition and utilization in plants. *Annu Rev Plant Biol* 73, 17–42.

Lambers, H., Ahmedi, I., Berkowitz, O., Dunne, C., Finnegan, P.M., Hardy, G.E., Jost, R., Laliberte, E., Pearse, S.J., and Teste, F.P. (2013). Phosphorus nutrition of phosphorus-sensitive Australian native plants: Threats to plant communities in a global biodiversity hotspot. *Conserv Physiol* 1(1), cot010.

Lebens, M., Lundquist, P., Soderlund, L., Todorovic, M., and Carlin, N.I. (2002). The nptA gene of Vibrio cholerae encodes a functional sodium-dependent phosphate cotransporter homologous to the type II cotransporters of eukaryotes. *J Bacteriol* 184(16), 4466–4474.

Levitus, S., Conkright, M.E., Reid, J.L., Najjar, R.G., and Mantyla, A. (1993). Distribution of nitrate, phosphate and silicate in the world oceans. *Prog Oceanogr* 31(3), 245–273.

Li, Q., Gao, X., Sun, Y., Zhang, Q., Song, R., and Xu, Z. (2006). Isolation and characterization of a sodium-dependent phosphate transporter gene in Dunaliella viridis. *Biochem Biophys Res Commun* 340(1), 95–104.

Lin, W.Y., Huang, T.K., and Chiou, T.J. (2013). Nitrogen Limitation Adaptation, a target of MicroRNA827, mediates degradation of plasma membrane-localized phosphate transporters to maintain phosphate homeostasis in Arabidopsis. *Plant Cell* 25(10), 4061–4074.

Liu, J.Y., Lou, Y., Yokota, H., Adams, P.D., Kim, R., and Kim, S.H. (2005). Crystal structure of a PhoU protein homologue - A new class of metalloprotein containing multinuclear iron clusters. *J Biol Chem* 280(16), 15960–15966.

Lomas, M.W., Bonachela, J.A., Levin, S.A., and Martiny, A.C. (2014). Impact of ocean phytoplankton diversity on phosphate uptake. *Proc Natl Acad Sci USA* 111(49), 17540–17545.

Lundh, F., Mouillon, J.M., Samyn, D., Stadler, K., Popova, Y., Lagerstedt, J.O., Thevelein, J.M., and Persson, B.L. (2009). Molecular mechanisms controlling phosphate-induced downregulation of the yeast Pho84 phosphate transporter. *Biochemistry* 48(21), 4497–4505.

Lv, Q., Zhong, Y., Wang, Y., Wang, Z., Zhang, L., Shi, J., Wu, Z., Liu, Y., Mao, C., Yi, K., and Wu, P. (2014). SPX4 negatively regulates phosphate signaling and homeostasis through its interaction with PHR2 in rice. *Plant Cell* 26(4), 1586–1597.

Medveczky, N., and Rosenberg, H. (1971). Phosphate transport in Escherichia coli. *Biochim Biophys Acta* 241(2), 494–506.

Michigami, T., Kawai, M., Yamazaki, M., and Ozono, K. (2018). Phosphate as a signaling molecule and its sensing mechanism. *Physiol Rev* 98(4), 2317–2348.

Mimura, T., Reid, R.J., Ohsumi, Y., and Smith, F.A. (2002). Induction of the Na+/Pi cotransport system in the plasma membrane of Chara corallina requires external Na+ and low levels of Pi. *Plant Cell Environ* 25(11), 1475–1481.

Misson, J., Thibaud, M.C., Bechtold, N., Raghothama, K., and Nussaume, L. (2004). Transcriptional regulation and functional properties of Arabidopsis Pht1;4, a high affinity transporter contributing greatly to phosphate uptake in phosphate deprived plants. *Plant Mol Biol* 55(5), 727–741.

Misson, J., Raghothama, K.G., Jain, A., Jouhet, J., Block, M.A., Bligny, R., Ortet, P., Creff, A., Somerville, S., Rolland, N., Doumas, P., Nacry, P., Herrerra-Estrella, L., Nussaume, L., and Thibaud, M.C. (2005). A genome-wide transcriptional analysis using Arabidopsis thaliana Affymetrix gene chips determined plant responses to phosphate deprivation. *Proc Natl Acad Sci U S A* 102(33), 11934–11939.

Morcuende, R., Bari, R., Gibon, Y., Zheng, W., Pant, B.D., Blasing, O., Usadel, B., Czechowski, T., Udvardi, M.K., Stitt, M., and Scheible, W.R. (2007). Genome-wide reprogramming of metabolism and regulatory networks of Arabidopsis in response to phosphorus. *Plant Cell Environ* 30(1), 85–112.

Moseley, J.L., Chang, C.W., and Grossman, A.R. (2006). Genome-based approaches to understanding phosphorus deprivation responses and PSR1 control in Chlamydomonas reinhardtii. *Eukaryot Cell* 5(1), 26–44.

Mudge, S.R., Rae, A.L., Diatloff, E., and Smith, F.W. (2002). Expression analysis suggests novel roles for members of the Pht1 family of phosphate transporters in Arabidopsis. *Plant J* 31(3), 341–353.

Novak, R., Cauwels, A., Charpentier, E., and Tuomanen, E. (1999). Identification of a Streptococcus pneumoniae gene locus encoding proteins of an ABC phosphate transporter and a two-component regulatory system. *J Bacteriol* 181(4), 1126–1133.

Nussaume, L., Kanno, S., Javot, H., Marin, E., Pochon, N., Ayadi, A., Nakanishi, T.M., and Thibaud, M.C. (2011). Phosphate import in plants: Focus on the PHT1 transporters. *Front Plant Sci* 2, 83.

Oganesyan, V., Oganesyan, N., Adams, P.D., Jancarik, J., Yokota, H.A., Kim, R., and Kim, S.H. (2005). Crystal structure of the "PhoU-Like" phosphate uptake regulator from Aquifex aeolicus. *J Bacteriol* 187(12), 4238–4244.

Park, B.S., Seo, J.S., and Chua, N.H. (2014). Nitrogen limitation adaptation recruits phosphate2 to target the phosphate transporter PT2 for degradation during the regulation of Arabidopsis phosphate homeostasis. *Plant Cell* 26(1), 454–464.

Pattison-Granberg, J., and Persson, B.L. (2000). Regulation of cation-coupled high-affinity phosphate uptake in the yeast Saccharomyces cerevisiae. *J Bacteriol* 182(17), 5017–5019.

Pedersen, B.P., Kumar, H., Waight, A.B., Risenmay, A.J., Roe-Zurz, Z., Chau, B.H., Schlessinger, A., Bonomi, M., Harries, W., Sali, A., Johri, A.K., and Stroud, R.M. (2013). Crystal structure of a eukaryotic phosphate transporter. *Nature* 496(7446), 533–536.

Perry, M.J., and Eppley, R.W. (1981). Phosphate-uptake by phytoplankton in the Central North Pacific-Ocean. *Deep Sea Res* 28(1), 39–49.

Pierre, W.H., and Parker, F.W. (1927). Soil phosphorus studies: II. The concentration of organic and inorganic phosphorus in the soil solution and soil extracts and the availibility of the organic phosphorus to plants. *Soil Sci* 24(2), 119–128.

Popova, Y., Thayumanavan, P., Lonati, E., Agrochao, M., and Thevelein, J.M. (2010). Transport and signaling through the phosphate-binding site of the yeast Pho84 phosphate transceptor. *Proc Natl Acad Sci USA* 107(7), 2890–2895.

Puga, M.I., Mateos, I., Charukesi, R., Wang, Z., Franco-Zorrilla, J.M., de Lorenzo, L., Irigoyen, M.L., Masiero, S., Bustos, R., Rodriguez, J., Leyva, A., Rubio, V., Sommer, H., and Paz-Ares, J. (2014). SPX1 is a phosphate-dependent inhibitor of phosphate starvation response 1 in Arabidopsis. *Proc Natl Acad Sci USA* 111(41), 14947–14952.

Pumplin, N., Zhang, X., Noar, R.D., and Harrison, M.J. (2012). Polar localization of a symbiosis-specific phosphate transporter is mediated by a transient reorientation of secretion. *Proc Natl Acad Sci USA* 109(11), E665–672.

Qi, Y., Kobayashi, Y., and Hulett, F.M. (1997). The pst operon of Bacillus subtilis has a phosphate-regulated promoter and is involved in phosphate transport but not in regulation of the Pho regulon. *J Bacteriol* 179(8), 2534–2539.

Rao, N.N., and Torriani, A. (1990). Molecular aspects of phosphate-transport in Escherichia-coli. *Mol Microbiol* 4(7), 1083–1090.

Reid, R.J., Mimura, T., Ohsumi, Y., Walker, N.A., and Smith, F.A. (2000). Phosphate uptake in Chara: Membrane transport via Na+/Pi cotransport. *Plant Cell Environ* 23(2), 223–228.

Reinhard, C.T., Planavsky, N.J., Gill, B.C., Ozaki, K., Robbins, L.J., Lyons, T.W., Fischer, W.W., Wang, C.J., Cole, D.B., and Konhauser, K.O. (2017). Evolution of the global phosphorus cycle. *Nature* 541(7637), 386–+.

Ried, M.K., Wild, R., Zhu, J., Pipercevic, J., Sturm, K., Broger, L., Harmel, R.K., Abriata, L.A., Hothorn, L.A., Fiedler, D., Hiller, S., and Hothorn, M. (2021). Inositol pyrophosphates promote the interaction of SPX domains with the coiled-coil motif of PHR transcription factors to regulate plant phosphate homeostasis. *Nat Commun* 12(1), 384.

Ritchie, R.J., Trautman, D.A., and Larkum, A.W.D. (1997). Phosphate uptake in the cyanobacterium Synechococcus R-2 PCC 7942. *Plant Cell Physiol* 38(11), 1232–1241.

Roomans, G.M., and Borst-Pauwels, G.W. (1979). Interaction of cations with phosphate uptake by Saccharomyces cerevisiae. Effects of surface potential. *Biochem J* 178(3), 521–527.

Rosenberg, H., Russell, L.M., Jacomb, P.A., and Chegwidden, K. (1982). Phosphate exchange in the pit transport system in Escherichia coli. *J Bacteriol* 149(1), 123–130.

Rubio, L., Linares-Rueda, A., Garcia-Sanchez, M.J., and Fernandez, J.A. (2005). Physiological evidence for a sodium-dependent high-affinity phosphate and nitrate transport at the plasma membrane of leaf and root cells of Zostera marina L. *J Exp Bot* 56(412), 613–622.

Rubio, V., Linhares, F., Solano, R., Martin, A.C., Iglesias, J., Leyva, A., and Paz-Ares, J. (2001). A conserved MYB transcription factor involved in phosphate starvation signaling both in vascular plants and in unicellular algae. *Genes Dev* 15(16), 2122–2133.

Saint-Marcoux, D., Proust, H., Dolan, L., and Langdale, J.A. (2015). Identification of reference genes for real-time quantitative PCR experiments in the liverwort Marchantia polymorpha. *PLOS One* 10(3), e0118678.

Sakano, K. (1990). Proton/phosphate stoichiometry in uptake of inorganic phosphate by cultured cells of Catharanthus roseus (L.) G. Don. *Plant Physiol* 93(2), 479–483.

Samyn, D.R., and Persson, B.L. (2016). Inorganic phosphate and sulfate transport in S. cerevisiae. *Adv Exp Med Biol* 892, 253–269.

Schachtman, D.P., Reid, R.J., and Ayling, S.M. (1998). Phosphorus uptake by plants: From soil to cell. *Plant Physiol* 116(2), 447–453.

Sengottaiyan, P., Ruiz-Pavon, L., and Persson, B.L. (2013). Functional expression, purification and reconstitution of the recombinant phosphate transporter Pho89 of Saccharomyces cerevisiae. *FEBS Journal* 280(3), 965–975.

Shane, M.W., Szota, C., and Lambers, H. (2004). A root trait accounting for the extreme phosphorus sensitivity of Hakea prostrata (Proteaceae). *Plant Cell Environ* 27(8), 991–1004.

Shin, H., Shin, H.S., Dewbre, G.R., and Harrison, M.J. (2004). Phosphate transport in Arabidopsis: Pht1;1 and Pht1;4 play a major role in phosphate acquisition from both low- and high-phosphate environments. *Plant J* 39(4), 629–642.

Skolnick, J., Gao, M., Zhou, H., and Singh, S. (2021). AlphaFold 2: Why it works and its implications for understanding the relationships of protein sequence, structure, and function. *J Chem Inf Model* 61(10), 4827–4831.

Thibaud, M.C., Arrighi, J.F., Bayle, V., Chiarenza, S., Creff, A., Bustos, R., Paz-Ares, J., Poirier, Y., and Nussaume, L. (2010). Dissection of local and systemic transcriptional responses to phosphate starvation in Arabidopsis. *Plant J* 64(5), 775–789.

Tisserant, E., Kohler, A., Dozolme-Seddas, P., Balestrini, R., Benabdellah, K., Colard, A., Croll, D., Da Silva, C., Gomez, S.K., Koul, R., Ferrol, N., Fiorilli, V., Formey, D., Franken, P., Helber, N., Hijri, M., Lanfranco, L., Lindquist, E., Liu, Y., Malbreil, M., Morin, E., Poulain, J., Shapiro, H., van Tuinen, D., Waschke, A., Azcon-Aguilar, C., Becard, G., Bonfante, P., Harrison, M.J., Kuster, H., Lammers, P., Paszkowski, U., Requena, N., Rensing, S.A., Roux, C., Sanders, I.R., Shachar-Hill, Y., Tuskan, G., Young, J.P.W., Gianinazzi-Pearson, V., and Martin, F. (2012). The transcriptome of the arbuscular mycorrhizal fungus Glomus intraradices (DAOM 197198) reveals functional tradeoffs in an obligate symbiont. *New Phytol* 193(3), 755–769.

Ullrich, C.I., and Novacky, A.J. (1990). Extra- and intracellular pH and membrane potential changes induced by K, Cl, H(2)PO(4), and NO(3) uptake and fusicoccin in root hairs of Limnobium stoloniferum. *Plant Physiol* 94(4), 1561–1567.

Ullrich-Eberius, C.I., Novacky, A., and van Bel, A.J.E. (1984). Phosphate uptake in Lemna gibba G1: Energetics and kinetics. *Planta* 161(1), 46–52.

Ullrich-Eberius, C.I., Novacky, A., Fischer, E., and Luttge, U. (1981). Relationship between energy-dependent phosphate uptake and the electrical membrane potential in Lemna gibba G1. *Plant Physiol* 67(4), 797–801.

van Kempen, M., Kim, S., Tumescheit, C., Mirdita, M., Gilchrist, C.L.M., Söding, J., and Steinegger, M. (2011). Foldseek: Fast and Accurate Protein Structure Search. *Nat Biotechnol*. doi:10.1038/s41587-023-01773-0.

Vanveen, H.W., Abee, T., Kortstee, G.J.J., Konings, W.N., and Zehnder, A.J.B. (1994). Translocation of metal phosphate via the phosphate inorganic transport-system of Escherichia-coli. *Biochemistry* 33(7), 1766–1770.

Versaw, W.K., and Harrison, M.J. (2002). A chloroplast phosphate transporter, PHT2;1, influences allocation of phosphate within the plant and phosphate-starvation responses. *Plant Cell* 14(8), 1751–1766.

Wang, L., Xiao, L., Yang, H.Y., Chen, G.L., Zeng, H.Q., Zhao, H.Y., and Zhu, Y.Y. (2020). Genome-wide identification, expression profiling, and evolution of phosphate transporter gene family in green algae. *Front Genet* 11, 590947.

Wang, Y., Secco, D., and Poirier, Y. (2008). Characterization of the PHO1 gene family and the responses to phosphate deficiency of Physcomitrella patens. *Plant Physiol* 57, 895–904.

Wang, Z., Ruan, W., Shi, J., Zhang, L., Xiang, D., Yang, C., Li, C., Wu, Z., Liu, Y., Yu, Y., Shou, H., Mo, X., Mao, C., and Wu, P. (2014). Rice SPX1 and SPX2 inhibit phosphate starvation responses through interacting with PHR2 in a phosphate-dependent manner. *Proc Natl Acad Sci USA* 111(41), 14953–14958.

Weiss, M., Haimovich, G., and Pick, U. (2001). Phosphate and sulfate uptake in the halotolerant alga Dunaliella are driven by Na+-symport mechanism. *J Plant Physiol* 158(12), 1519–1525.

Werner, A., and Kinne, R.K. (2001). Evolution of the Na-P(i) cotransport systems. *Am J Physiol Regul Integr Comp Physiol* 280(2), R301–312.

Wild, R., Gerasimaite, R., Jung, J.Y., Truffault, V., Pavlovic, I., Schmidt, A., Saiardi, A., Jessen, H.J., Poirier, Y., Hothorn, M., and Mayer, A. (2016). Control of eukaryotic phosphate homeostasis by inositol polyphosphate sensor domains. *Science* 352(6288), 986–990.

Willis, A., Chuang, A.W., Dyhrman, S., and Burford, M.A. (2019). Differential expression of phosphorus acquisition genes in response to phosphorus stress in two Raphidiopsis raciborskii strains. *Harmful Algae* 82, 19–25.

Willsky, G.R., and Malamy, M.H. (1980). Characterization of two genetically separable inorganic phosphate transport systems in Escherichia coli. *J Bacteriol* 144(1), 356–365.

Wykoff, D.D., and O'Shea, E.K. (2001). Phosphate transport and sensing in Saccharomyces cerevisiae. *Genetics* 159(4), 1491–1499.

Yuan, Z.C., Zaheer, R., and Finan, T.M. (2006). Regulation and properties of PstSCAB, a high-affinity, high-velocity phosphate transport system of Sinorhizobium meliloti. *J Bacteriol* 188(3), 1089–1102.

Zheng, J.J., Sinha, D., Wayne, K.J., and Winkler, M.E. (2016). Physiological roles of the dual phosphate transporter systems in low and high phosphate conditions and in capsule maintenance of Streptococcus pneumoniae D39. *Front Cell Infect Microbiol* 6, 63.

Zhou, J., Hu, Q., Xiao, X., Yao, D., Ge, S., Ye, J., Li, H., Cai, R., Liu, R., Meng, F., Wang, C., Zhu, J.K., Lei, M., and Xing, W. (2021). Mechanism of phosphate sensing and signaling revealed by rice SPX1-PHR2 complex structure. *Nat Commun* 12(1), 7040.

Zhu, J., Lau, K., Puschmann, R., Harmel, R.K., Zhang, Y., Pries, V., Gaugler, P., Broger, L., Dutta, A.K., Jessen, H.J., Schaaf, G., Fernie, A.R., Hothorn, L.A., Fiedler, D., and Hothorn, M. (2019). Two bifunctional inositol pyrophosphate kinases/phosphatases control plant phosphate homeostasis. *Elife* 8, e43582. doi: 10.7554/eLife.43582.

Zvyagilskaya, R.A., Lundh, F., Samyn, D., Pattison-Granberg, J., Mouillon, J.M., Popova, Y., Thevelein, J.M., and Persson, B.L. (2008). Characterization of the Pho89 phosphate transporter by functional hyperexpression in Saccharomyces cerevisiae. *Fems Yeast Res* 8(5), 685–696.

10 A General Perspective of Phosphorus Research in Plants

Hatem Rouached

In plants, research focusing on phosphorus (P) is entering a novel phase involving the routine utilisation of modelling/artificial intelligence, high-throughput phenotyping, and system genomics approaches. Powered by these approaches, the generation/integration of massive data is expected to deepen our comprehension of P biology among plants over the next few decades in close proximity to generating plants that benefit from the much-enhanced P use efficiency (PUE) to ensure sustainable production while optimising soil health (Cho et al., 2021). This chapter underscores the salient features outlined in the chapters that form this book.

While rock phosphate deposits are known to be limited, the need has been recognised to make preparations for an impending global P crisis (Abelson, 1999). The viable solution of new mine discoveries necessitates the emergence of an affordable means of removing heavy metal that ends up contaminating the extracted phosphate. To support phosphate quality/price while ensuring that the production cost stays reasonable, this is a key step. Another plausible option may be to improve the mineralisation of organic P along with its solubilisation by means of plant-secreted enzymes or microorganisms (Tian and Liao, 2018). Indeed, it is possible to enhance P bio-availability through multiple approaches. As a case in point, one mechanism of doing that might be by using a non-metabolisable form of P, phosphite, in conjunction with microbe-secreted phosphite-metabolising enzymes (Fan et al., 2022). However, a major part of soil P is available in an insoluble form that needs to be solubilised prior to root uptake. Accordingly, enzymes and organic acids secreted by the roots can be of assistance. Several genes that encode these enzymes are in the process of being characterised to potentially bioengineer plants via a transgenic approach. Nevertheless, the creation and expulsion of acid phosphatases/organic acids consume significant energy and can escalate plant metabolic costs in P-stressed environments. For this reason, the introduction of P-solubilising microbes to the soil appears to be a more prudent strategy. Knowledge of the intricacy of plant–fungal–bacterial interactions that can be used to enhance the efficiency of plant phosphate becomes critical (Etesami et al., 2021), as cocktails of multiple beneficial microbes impart greater advantages to the plant (functionally) as compared to a single-strain inoculant. While land plants rely on their microbes for nutrient availability and growth promotion under P stresses, deciphering the crosstalk between plant responses to phosphate deficiency and the activation of the defence system against pathogens for appropriately positioning the importance of P in plants' growth-defence trade-off is also significant.

Numerous factors are known to impact how plants respond to low P conditions, including plant growth capacity (above the ground), and root system architecture changes (underground), differing between genotypes and growth environments along with their respective interactions (Poirier and Bucher, 2002). Technologies equipped with high-throughput phenotyping capabilities can help ascertain expressed or desirable genetic traits under P deficiency. It is possible to genetically target these phenotypes via genes as well as environmental conditions to foster suitable factors for growth under limited P conditions. Meanwhile, quantitative imaging and genomics are fast-growing areas for technological innovations, with breakthroughs emerging with great speed and regularity. The development of new methods and instruments has resulted in the onset of high-content quantitative microscopy, allowing for the quantification of various phenotypic readouts within single cells;

DOI: 10.1201/9781003440079-10

their ability to generate large amounts of high-resolution quantitative data characterising biological systems at scales ranging from molecular to plant level is an overlapping attribute of all these techniques. Recently, an even greater surge in sensitivity has served to make transformational changes in genomics with the fast-paced dissemination of measurements at the resolution of single molecules and cells. However, converting this data into predictive quantitative models of biological systems remains a major challenge. Integrating these data will make it possible to derive incisive insights into plants' spatial-temporal responses to P deficiency (Rouached and Rhee, 2017; Bouain, Korte, et al., 2019). For plant biologists, transitioning from functional genomics to systems biology bears significance to exploring their data across other fields and promulgating new concepts to plug the gap between technologies and scales to crops. It is viable to use such approach-based genes in developing transgenic crops with high P-acquisition efficiency through novel technologies of genome editing (Khurshid et al., 2017).

Agronomic/biotechnology growth is expected to make a positive difference in plant P sensing/signalling, transport, and accumulation. Despite the significant progress achieved in uncovering critical molecular constituents in the Pi sensing/signalling pathways that help phosphate homeostasis maintenance in plants (Rouached et al., 2010), the image remains incomplete, as merely changing the identified genes cannot engineer plants that are insensitive to P deficiency. Intensifying research studies on phosphate sensing as well as signalling pathways is likely to help us better decipher how plants detect short-, medium-, and long-term P deficiency. It is also necessary to enhance ways of increasing the capacity of plant uptake. One possible solution could emerge as an observation made in the late 80s, whereby plants were found to over-accumulate phosphate under zinc deficiency (Kisko et al., 2018). These plants were found to lose control of phosphate uptake and the expression of phosphate uptake transporters (Kisko et al., 2018; Bouain et al., 2014; Khan et al., 2014). Since intercellular nutrient coordination is imperative for utilising nutrients, and in wake of the mounting evidence regarding which nutrient uptakes and utilisations are connected with each other, more integrative studies must be conducted in the future to unravel underlying mechanisms where plants coordinate multiple-nutrient stresses at the systemic levels of whole plants (Rouached and Rhee, 2017).

Although P-centred research has long been undertaken using model plants in laboratory-controlled conditions, an increasing number of studies are likely to be directly conducted in the field going forward. While these experiments' reproducibility will be onerous owing to the problems caused by environmental conditions, future research on phosphate would do well to investigate varied situations involving rapid climate changes (increased temperature, CO_2, etc.) (Bouain et al., 2022; Bouain, Krouk, et al., 2019; Nam et al., n.d.). Technologies that can help recreate such scenarios in crops are now available, which include single-cell transcriptome, multi-omics approaches, gene co-expression analysis, gene editing, and efficient nutrient tracing assays (Rouached and Rhee, 2017).

Finally, it is possible that despite the progress on improving PUE, it will be not enough to manage P resources, and accumulated soil P could become lost during runoff events that cause degradation in water quality. To avoid this situation, a step that must be taken is as follows: the soil test P is the point of commencement to examine the P status for agronomic goals. In this context, novel approaches to accurately quantify the P amount that can be added in a safe manner to the soil will be predicated on the crops cultivated within that soil. Specifically, it will be beneficial to opt for crop-specific and locally adapted approaches. Depending on soil types and P bio-availability, co-growing of plant species or alternate cultures could be envisioned (Belgaroui et al., 2016).

REFERENCES

Abelson, P.H. (1999) A potential phosphate crisis. Science, 283(5410), 2015.
Belgaroui, N., Berthomieu, P., Rouached, H. and Hanin, M. (2016) The secretion of the bacterial phytase PHY-US417 by Arabidopsis roots reveals its potential for increasing phosphate acquisition and biomass production during co-growth. Plant Biotechnology Journal, 14(9), 1914–1924.

Bouain, N., Cho, H., Sandhu, J., Tuiwong, P., Prom-U-Thai, C., Zheng, L., Shahzad, Z. and Rouached, H. (2022) Plant growth stimulation by high CO depends on phosphorus homeostasis in chloroplasts. Current Biology, 32(20), 4493–4500.e4.

Bouain, N., Korte, A., Satbhai, S.B., Nam, H.-I., Rhee, S.Y., Busch, W. and Rouached, H. (2019) Systems genomics approaches provide new insights into Arabidopsis thaliana root growth regulation under combinatorial mineral nutrient limitation. PLOS Genetics, 15(11), e1008392.

Bouain, N., Krouk, G., Lacombe, B. and Rouached, H. (2019) Getting to the root of plant mineral nutrition: Combinatorial nutrient stresses reveal emergent properties. Trends in Plant Science, 24(6), 542–552. Available at: http://dx.doi.org/10.1016/j.tplants.2019.03.008.

Bouain, N., Shahzad, Z., Rouached, A., Khan, G.A., Berthomieu, P., Abdelly, C., Poirier, Y. and Rouached, H. (2014) Phosphate and zinc transport and signalling in plants: Toward a better understanding of their homeostasis interaction. Journal of Experimental Botany, 65(20), 5725–5741.

Cho, H., Bouain, N., Zheng, L. and Rouached, H. (2021) Plant resilience to phosphate limitation: Current knowledge and future challenges. Critical Reviews in Biotechnology, 41(1), 63–71. Available at: http://dx.doi.org/10.1080/07388551.2020.1825321.

Etesami, H., Jeong, B.R. and Glick, B.R. (2021) Contribution of arbuscular mycorrhizal fungi, phosphate–solubilizing bacteria, and silicon to P uptake by plant. Frontiers in Plant Science, 12. Available at: http://dx.doi.org/10.3389/fpls.2021.699618.

Fan, Y., Niu, X. and Zhang, D. (2022) Microbial Phosphine Production: The Key to Improve Productivity. Available at: http://dx.doi.org/10.21203/rs.3.rs-2188551/v1.

Khan, G.A., Bouraine, S., Wege, S., Li, Y., Carbonnel, M. de, Berthomieu, P., Poirier, Y. and Rouached, H. (2014) Coordination between zinc and phosphate homeostasis involves the transcription factor PHR1, the phosphate exporter PHO1, and its homologue PHO1; H3 in Arabidopsis. Journal of Experimental Botany, 65(3), 871–884.

Khurshid, H., Jan, S.A. and Shinwari, Z.K. (2017) An Era of CRISPR/ Cas9 Mediated Plant Genome Editing. The CRISPR/Cas System: Emerging Technology and Application. Available at: http://dx.doi.org/10.21775/9781910190630.04.

Kisko, M., Bouain, N., Safi, A., et al. (2018) LPCAT1 controls phosphate homeostasis in a zinc-dependent manner. Elife, 7. Available at: http://dx.doi.org/10.7554/eLife.32077.

Nam, H.-I., Shahzad, Z., Dorone, Y., Clowez, S., Zhao, K., Bouain, N., Cho, H., Rhee, S.Y. and Rouached, H. Interdependent iron and phosphorus availability controls photosynthesis through retrograde signaling. Available at: http://dx.doi.org/10.1101/2021.02.11.430802.

Poirier, Y. and Bucher, M. (2002) Phosphate transport and homeostasis in Arabidopsis. The Arabidopsis Book, 1, e0024. Available at: http://dx.doi.org/10.1199/tab.0024.

Rouached, H., Bulak Arpat, A. and Poirier, Y. (2010) Regulation of phosphate starvation responses in plants: Signaling players and cross-talks. Molecular Plant, 3(2), 288–299. Available at: http://dx.doi.org/10.1093/mp/ssp120.

Rouached, H. and Rhee, S.Y. (2017) System-level understanding of plant mineral nutrition in the big data era. Current Opinion in Systems Biology, 4, 71–77.

Tian, J. and Liao, H. (2018) The role of intracellular and secreted purple acid phosphatases in plant phosphorus scavenging and recycling. Annual Plant Reviews Online, 265–287. Available at: http://dx.doi.org/10.1002/9781119312994.apr0525.

Index

A

ABA-glucosyl ester (ABA-GE) deconjugation gene, 68
ABI5 activity, 61, 67, 71
Abscisic acid (ABA), 61, 67–68, 103
Abscisic Acid Insensitive5 (ABI5)-dependent manner, 60, 61, 67, 68, 71
ACC oxidase (ACO) enzyme, 64
ACC synthase enzyme (ACS), 64
Accumulated soil P, 13
Acid phosphatases, 83, 116
Adenosine triphosphate (ATP), 98, 110, 117
Adventitious roots, 31
Agronomic/biotechnology growth, 152
Agronomic soil test extractant, 5
Al-activated malate transporters (ALMT), 116
ALG2-interacting protein X (ALIX), 22, 138
Alphafold benefits for Pi transporters
    bryophytes, 136–137
    fungi
        mycorrhizal fungi, 134–135
        yeast, 133–134
    green algae, 135–136
    higher plants, 137–138
    phosphate origin and distribution
        phosphate through ages, 129
        present distribution of phosphate on earth, 130
    prokaryotes
        bacteria, 131–133
        cyanobacteria, 133
Aluminum-activated Malate Transporter1 (ALMT1), 19, 53
1-Aminocyclopropane-1-carboxylic acid (ACC), 64
Ammonium transporters (AMTs), 51
Arabidopsis, 23, 60–62, 64, 66, 68, 69, 101, 119, 131, 138
Arabidopsis gene *Phosphate Deficiency Response2* (*AtPDR2*), 35
Arabidopsis Phosphate Starvation Response 1 (*AtPHR1*), 35
Arbuscular mycorrhizal fungi (AMF), 23, 83, 98
Ascorbic acid (AsA), 53
AtNLA, 21
AtPHT1 Pi transporter, 21
AtWRKY42, 113
Auxin biosynthesis, 60
Auxin Response Factor 16 (ARF16), 60, 71, 72
Auxin Response Factor19 (ARF19), 65
Auxins, 61

B

Bacteria, 131
    high-affinity Pi transporters, 131–132
    low-affinity Pi transporters, 132–133
Basic helix-loop-helix (bHLH), 35
Benzoic Acid Hypersensitive 1-dominant (*BAH1-D*), 102
Best management practices (BMPs), 6, 11
    to reduce P loadings to water bodies, 11–12
Biotechnological approaches
    novel P-assimilating strategies, 120
    targeting P uptake and distribution in plants
        organic acids role, P uptake and assimilation, 115–116
        phosphate transporters, 111–114
        root system architecture, 115
        secreted acid phosphatases, 116–117
    targeting reallocation and re-utilisation of P in plants, 117–118
        cytosolic PAPs in P recycling, 119–120
        membrane lipid remodelling, 118–119
Botrytis-induced Kinase 1 (*BIK1*), 102, 105, 138
Brahma (BRM), 20
Brassinazole Resistant1 (*BZR1*), 70
Brassinisteroid Upregulated1 (BU1), 70
Brassinosteroid Insensitive1 (BRI1), 70
Brassinosteroid Insensitive1-ethylmethane Sulfonate-Suppressor 1 (*BES1*), 70
Brassinosteroids, 70
BRI1 Associated Receptor Kinase 1 (*BAK1*), 70, 105
BRI1-LIKE1 and 3 (BRL1/3), 70
BR-regulated U-box40 (PUB40), 70
Bryophytes, 136–137
BU1-LIKE1 (BUL1) complex1 (BC1), 70
Burkholderia multivorans WS-FJ9, 116
bZIP58, 53, 54

C

Calcium-dependent protein kinases (CPK), 48
Casein kinase 2 (CK2), 114
CBL-Interacting Protein Kinase 23 (CIPK23), 51
Cellular membrane lipids, 118
Change point, 7
Chinese Spring (CS), 37
Chrysanthemum, 113
Citrate synthase (*CS*), 115
CLAVATA3 (CLV3)/endosperm Surrounding Region14 (CLE14), 19
Clustered Regularly Interspaced Short Palindromic Repeats-CRISPR-associated protein 9 (CRISPR-Cas9), 72, 114, 119
Cluster root (CR), 64
CLV2/PEP1 Receptor2 (PEPR2) receptors, 19
Colletotrichum *tofieldiae*, 23
Common bean (Phaseolus vulgaris), 42
Constitutive Triple Response (CTR1), 64
Convex hull area (CHA), 36
Coronatine-Insensitive1 (COI1), 69
Cry-for-help hypothesis, 85
Cry-for-help model, 86
Cyanobacteria, 129, 133
Cytokinin (CK), 19, 68

Index

Cytokinin Response 1 (CRE1), 60, 68
Cytokinin Response 1/Arabidopsis Histidine Kinase 4 (*CRE1/AHK4*), 103
Cytokinin (6-BA) treatment, 72
Cytosolic PAPs, 116, 119–120

D

Degree of P saturation (DPS), 8–9
DELLA proteins, 69
Desorption, 2
Diacylglycerol (DAG), 118
Diacylglycerol kinase (DGK), 119
Dicot root systems, 31
Differentiation zone (DZ), 66
Digalactosyldiacylglycerol (DGDG), 118
Dissolved organic phosphorus (DOP), 130
Dissolved reactive P (DRP), 1, 8, 11
Dual PAPs, 119
DWARF 14 (d14), 86
DWARF 17 (d17), 86

E

Effector-triggered immunity (ETI), 98, 104
EF-TU Receptor (*EFR*), 105
EIN3-LIKE1 (EIL1), 64
Elongation/differentiation zone (EDZ), 60
Endophytic bacteria, 87
Endoplasmic reticulum (ER), 19, 23, 114, 138
Endosomal complex required for transport (ESCRT)-III, 22–23
Environmental phosphorus loss assessment, 6
    soil test indicators
        degree of phosphorus saturation, 8–9
        phosphorus saturation ratio (PSR), 9, 10
        soil P storage capacity (SPSC), 9, 11
        soil test phosphorus, 7–8
Epigenetic mechanisms, 20
Erosion, 2
ESCRT-III-Associated Protein, 138
Ethylene (ET), 19
Ethylene-Insensitive 2 (EIN2), 64
Ethylene-Insensitive3 (EIN3), 19
Ethylene-Overproduction Protein 1 (*GmETO1*), 41
Ethylene Response Factors (ERFs), 64
Eutrophication process, 1

F

Fe starvation response (FSR), 51
Flagellin-Sensitive 2 (*FLS2*), 105
Fungi, 83, 85

G

Gallant, 36
Genetically modified (GM) technology, 111
Genomewide association (GWAS) mapping, 39
Gibberellic acid (GA), 20, 69–70
Glucose dehydrogenases (GDH), 99
Glycerophosphodiester phosphodiesterase (GPX-PDE) enzyme, 119
GmEXPB2, 41

GmPAP7a/7b, 117, 119
GmPAP14, 117
GmPT7, 113
Green algae, 55, 135–136

H

H2A.Z, 20
Halobacterium salinarum, 132
Hemerythrin motif-containing Really Interesting New Gene- and Zinc-finger proteins (HRZs), 51, 52
High Arsenic Content 1 (HAC1), 54
Higher plants, 137–138
Hormonal crosstalk, Pi uptake and balance, 71–72

I

Immobilisation, 2
Indole-acetic acid (IAA), 86
Inorganic P compounds, 1, 2
Inorganic phosphate (Pi), 18, 48
    transport under varying Pi levels, 22–23
Inositol Phosphate Kinase (IPK1), 65
Inositol pyrophosphate (InsP$_8$), 20
Inositol pyrophosphates (InsPs), 21, 50
Inositol pyrophosphates (PP-InsPs), 20–22, 84
InsP7-dependent, 21
InsP$_8$-SPX1 complex, 21
Intricate signalling networks, 59
Isopentenyl transferases (IPTs), 68

J

Jasmonate-zim-domain Protein 10 (*JAZ10*) expression, 69, 101
Jasmonic acid (JA), 20, 68–69, 100
JAZ1/2/6, 69

K

Kenong199 (KN199), 37

L

Langmuir sorption maximum, 9
Lateral roots (LRs), 31, 39, 64, 111
    formation, 61, 65, 66
Leaching, 2
α-Linolenic acid (α-LeA), 69
Lipid remodelling pathways, 120
Local phosphate sensing
    hormonal control, 19–20
    molecular mechanisms governing RSA changes, 18–19
    pigenetic control of RSA changes under Pi starvation, 20
    root system architectural (RSA) changes and adaptation, 18
Lotus japonicus, 84
Low Phosphate Root 1 (LPR1), 18, 19, 53, 62, 70
Low Phosphate Root1 AND 2 (LPR1/2), 18, 20, 62
Low phytic acid (LPA), 54
Lysophosphatidylcholine (LPC), 54
Lysophosphatidylcholine acyltransferase (LPCAT1), 54

Index

M

Maize (Zea mays), 37–40
Malate dehydrogenase (*MDH*) gene, 115
MAPK kinases (MAPKKs), 105
Medicago *pt4* plants, 84
Microbe-associated molecular patterns (MAMPs), 104
Mineralisation, 2, 151
Mitogen-activated protein kinases (MAPKs), 103–105
Monogalactosyldiacylglycerol (MGDG), 118
MultAlin software, 130
Multicopper oxidase (MCO), 62
Multidrug and toxic compound extrusion (MATE), 116
Multivesicular body (MVB)-mediated vacuolar proteolysis, 22
MYB30, 64, 65
MYB62 transcription factor, 20, 70
Mycorrhizal fungi, 134–135

N

Never-ripe (Nr), 64
NIN-like proteins (NLPs), 48
Nitrate-Inducible GARP-type Transcriptional Repressor1 (NIGT1), 48, 49, 51
Nitrate reductase (NIR), 48
Nitrate-responsive genes, 48, 49
Nitrate Transporter1.1 (NRT1.1), 48
Nitrogen Limitation Adaptation (NLA), 22, 49, 102
Nitrogen Limitation Adaptation 1 (NLA1), 100
Nod-26-like aquaporin, 54
Nonexpresser of Pathogenesis-Related gene 1 (NPR1), 103
N signalling, 48, 50
N starvation response (NSR), 48
Nucleotide-binding leucine-rich repeat proteins (NLRs), 104
Nutrition Response and Root Growth (NRR), 36

O

Olsen-P level, 38
Organic P compounds, 1, 133
OsCKX2-KO plants, 68
OsGF14b, 35
OsJAZ11 gene, 20, 69
OsPAP10a, 117
OsPDR2 gene, 35
OsPSTOL1, 37, 41
OsPT8 C-terminal end (PT8-CT), 114
OsPT8 dephosphorylation, 23, 114
OsVPE1/2, 118
OsWRKY21 function, 113
OsWRKY108 function, 114
Oxalate extractant, 9
12-Oxophytodienoate reductase 3 (OPR3) function, 69
12-Oxophytodienoic acid (12-OPDA), 69

P

P-acquisition efficiency (PAE), 42, 43, 111, 117
Pathogens, 23, 85, 89, 98, 103
Pattern-recognition receptors (PRRs), 105
Pattern-triggered immunity (PTI), 98, 104
PBS1-LIKE KINASE 1 (*PBL1*), 102, 138
P-deficient soil, 40, 110, 120
P-efficient RAC875, 36
P-Fe crosstalk, 51
  P-Fe influence on photosynthesis, 53–54
  P-Fe influence on plant root growth, 53
  P-Fe-signalling interplay mediated by PHR-HRZ module, 51–52
Pho84, 133–134
Pho89, 134, 136
Phosphatases, 87
Phosphate (Pi), 59
  deficiency impacts, plant growth and development, 61
    abscisic acid (ABA)-mediated regulation of plant growth and Pi homeostasis, 67–68
    auxins role and low-Pi-derived signals, 65–66
    brassinosteroids role under low Pi conditions, 70–71
    cytokinin biosynthesis impact, 68
    ethylene signalling effect, 63–65
    gibberellic acid (GA) biosynthesis role, 69–70
    jasmonic acid (JA) and, 68–69
    low Pi sensing and hormonal control, 62–63
    response on plant growth, 63–65
    strigolactones (SLs) in regulating plant architecture, 66–67
  transporters role, defence responses, 101–102
Phosphate1 (PHO1), 21, 22, 69, 100–102
Phosphate2 (PHO2), 21, 22, 49, 63, 100
Phosphate Deficiency Response 2 (PDR2), 18, 19, 53, 62
Phosphate homeostasis and root system architecture of crops, 31–33
  root development of major crops, 32
    common bean (Phaseolus vulgaris), 42
    maize (Zea mays), 37–40
    rice (Oryza sativa), 32, 34–36
    sorghum (Sorghum bicolor), 41–42
    soybean (Glycine max), 40–41
    wheat (Triticum aestivum), 36–37
Phosphate response (PHR) proteins, 87
Phosphate sensing, 84, 152
Phosphate-solubilising bacteria (PSB), 86, 87, 89, 90, 99
Phosphate-specific transporter (Pst) system, 131, 139
Phosphate starvation-induced (PSI) genes, 49, 63, 68, 87, 100
Phosphate Starvation Response1 (PHR1), 19, 22, 23, 49, 70, 84, 88, 89, 100, 101
Phosphate starvation responses (PSRs), 18, 48, 49, 63, 69, 84, 98, 113
  in plant defence, 100–101
  role in plant–microbe interactions, 23
Phosphate transport (Pit), 132–134
Phosphate Transporter 1 (PHT1), 22, 100, 102, 137, 138
Phosphate Transporter A (PTA), 135, 136
Phosphate Transporter B (PTB), 135, 136
Phosphate transporters (PHTs), 59, 68, 112, 114
Phosphate Transporter Traffic Facilitator 1 (PHF1), 22, 138
Phosphate-utiliser bacteria, 87
Phosphatidic acid (PA), 119
Phosphatidylcholine (PC), 54
Phosphite, 120, 151
Phosphobacteria (PSB), 86, 87, 90
Phospholipids, 111, 118
Phosphorus (P), 1, 30, 110

accumulation in soil, 2–4
fertilisers, 5, 6, 30
management, 4–5
and other elements interaction, 54
and plant immunity, 98
  phosphorylation and MAPK cascades in defence signal transduction, 103–105
  phytohormone cross-talk with phosphorus in defence responses, 102–103
  Pi transporters role in defence responses, 101–102
  PSRs system in plant defence, 100–101
  rhizospheric microbes and phosphorus availability and influence on plant fitness and defence, 98–99
Phosphorus concentrations, 1
Phosphorus deficiency, 1
Phosphorus saturation ratio (PSR), 9
Phosphorus-solubilising microorganisms (PSMs), 98, 99
PHOSPHORUS STARVATION TOLERANCE1 (PSTOL1), 34, 35, 37, 115
Phosphorylation, 23, 103–105, 138
Photo-Fenton reaction, 19
PHR1 Interactor F-box1 (PHIF1), 54
PHT1 phosphate transporters, 84
PHT5-type, 21
Pi acquisition efficiency (PAE), 111, 114, 117
PiBP/PstS, 131
Pi homeostasis, 18, 21, 23, 35, 43, 49, 60, 114
P-inefficient Wyalkatchem, 36
PIN-FORMED2 (PIN2) degradation, 66
PIN-FORMED (PIN)–auxin, 19
Piriformospora indica fungi, 134, 137, 138
Pi-solubilising bacteria (PSBs), 111
Pi-solubilising fungi (PSF), 111
Pi-solubilising microbes (PSMs), 99, 111, 120
Pi starvation, 20–23, 35, 41, 51, 53, 114, 118, 133, 136
Plant disease, 98
Plant growth-promoting rhizobacteria (PGPR), 86
Plant growth regulators (PGRs), 72, 73
Plant PAPs, 117
Plant roots, 30, 62, 116
Plasma membrane (PM), 114
P-N crosstalk, 48
  NIGT1s-dependent regulatory cascades, 49
  regulating Pi homeostasis in plants in N-dependent manner, 49
  regulating root growth, 50–51
  SPXs as pivotal nodes in integrating P-N-signalling crosstalk, 49–50
Point sources, 1
Polyphosphate granules, 83
Poultry litter (PL), 4
Precipitation, 2
Primary nitrate response (PNR), 48, 50
Primary root length (PRL), 62
Programmed cell death (PCD), 105
Protein Kinase A (PKA), 134
Protein phosphorylation, 105
Proteobacteria, 85
PRU1, 22
Pseudomonas syringae strains, 102
PSR impairment, 89
Pup1-containing line, 35
PURPLE ACID PHOSPHATASE 10C (OsPAP10c), 36, 117

Purple acid phosphatases (PAPs), 116, 118
P-use efficiency (PUE), 36, 111, 118, 151, 152
PvPAP3, 117
Pyrabactin resistance1/Pyr1-like receptor components (PYR/PYL), 67
Pyrroloquinoline quinone (PQQ), 99

Q

qpe9-1, 35
Quantitative trait loci (QTL), 35, 37, 39, 41
Quiescent centre (QC) activity, 62

R

Reactive oxygen species (ROS) production, 19, 53, 105
Receptor-like kinases (RLKs), 105
Recombinant inbred line (RIL), 37
Regulator of Leaf Inclination 1 (RLI1), 48, 50, 70
Rhizodeposition, 85, 89
Rhizophagus clarus, 83
Rhizosphere, 59, 66, 85, 86, 116
Rhizosphere acidification-triggered STOP1 accumulation, 51
Rice (Oryza sativa), 32, 34–36
Rice *Pi Starvation-induced Transcription Factor 1 (OsPTF1)*, 35
Root components, 31
Root cortical aerenchyma (RCA), 120
Root Elongation under Phosphorus Deficiency (REP), 35
Root growth angle (RGA), 69, 111
Root Hair Defective 6-Like 2, 60
Root hairs, 20, 31
Root surface area (RSA), 31, 32, 35, 42
Root system architecture (RSA), 18, 31, 60, 102
Root-to-shoot growth ratio, 32

S

S-adenosylmethionine (AdoMet) synthetase, 63
SbPSTOL1 genes, 41, 42
SCARECROW (SCR), 19, 62
Secondary roots, 31
Seed bank, 85
Sensitive To Proton Rhizotoxicity (STOP1), 19, 51
SgPAP23, 117
Shaping beneficial plant-microbe interactions
  cry for phosphate, 86
  integrating prokaryotes-based solution, nutritional challenges, 85–86
  long-lasting interaction, tricks and tips of, 83–84
  Pi-mediated bacterial accommodation inside host, 87–89
  rhizobacteria at work, 86–87
  SYNCOM development, increasing Pi mineralisation and uptake by plant, 89–90
SHORT-ROOT (SHR), 19, 22
SHR–SCARECROW pathways, 19
Signalling translators, 50
Single nucleotide polymorphisms (SNPs), 39
SMARTBLAST, 139
Snf1-related kinases 2 (SnRK2s), 67
Soil biodiversity, 85
Soil pH, 2, 59

Index

Soil phosphorus, 1–2
   assessment for agronomic use, 5–6
   cycle, 2, 3
   testing methods, 5
Soil PSR, 9
Soil P storage capacity (SPSC), 9, 11
Soil test P (STP), 5, 7–8, 13
Somatic Embryogenesis Receptor Kinase1 (*TaSERK1*), 37
Sorghum (Sorghum bicolor), 41–42
Sorghum RSA, 41
Sorption, 2
Soybean (Glycine max), 40–41
Split-line model, 8
SPX domain, 20, 21, 23
SPX-InsP-PHR complex, 48
Stem cell niche (SCN), 19
STP threshold approach, 7, 8
Strigolactones (SLs), 20, 66, 103
Sturdier immune response, 104
Sulfolipid sulfoquinovosyldiacylglycerol (SQDG), 118
Sulfolobus acidocaldarius, 140
SULTR-like P Distribution Transporter (SPDT), 114
SUMO E3 ligase (SIZ1), 65
Surface runoff, 2, 5
Synthetic microbial communities (SynCom), 89
Systemic phosphate signalling
   inositol pyrophosphates (PP-InsPs), 21–22
   intracellular P sensors, 20–21
Systemic signalling responses, 62

T

TaEXPB23, 37
Ta-PHR1-A1, 37
Tap root system, 31
Threshold PSR values, 9, 10
Topsoil foraging, 111, 115
   in crops, 31
Transgenic lines, 35
Transport Inhibitor Response 1 (TIR1), 60
2,3,5-Triiodobenzoic acid (TIBA), 66
Triticum aestivum Aluminium activated malate transporters 1 (*TaALMT1*), 116
Tryptophan Aminotransferase 1 (TAA1), 65
Two-component regulatory system (TCS), 131

V

Vacuolar ion transporter (VIT) genes, 35
Vacuolar PAPs, 119
Verticillium dahlia, 103
VIH1/2, 20, 21
Vitamin C 4 (VTC4), 53

W

Water-soluble P (WSP), 9, 11
Weathering, 2
Wheat (Triticum aestivum), 36–37
White lupin (*Lupinus albus*), 64, 111, 117
WRKY transcription factor (WRKY6), 22, 54, 113

Y

Yangmai 16/Zhongmai 895, 37
Yeast, 133
   high-affinity Pi transporters, 133–134
   low-affinity Pi transporters, 134

Z

ZmPSTOL genes, 39

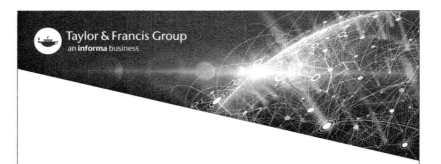